Women in Biotechnology

Francesca Molfino • Flavia Zucco
Editors

Women in Biotechnology

Creating Interfaces

Springer

Editors
Francesca Molfino
Women and Science Association (Italy)
Rome
Italy

Flavia Zucco
Institute for Neurobiology and Molecular Medicine (INMM)
The National Research Council (CNR)
Rome
Italy

ISBN 978-1-4020-8610-6 e-ISBN 978-1-4020-8611-3

Library of Congress Control Number: 2008930761

Cover image: *Esercizi di equilibrio (Exercises in equilibrium)* (1999, biro e pastello su carta) remains with the artist Paola Gandolfi

Printed on acid-free paper

9 8 7 6 5 4 3 2 1

springer.com

Foreword

Johannes Klumpers

Biotechnologies, such as genetic engineering, cloning and biodiversity, raise many legal and ethical concerns, so it is important that people understand these issues and feel able to express their opinions. This is why the European Commission has been, for a number of years, supporting actions to improve communication among scientists in these diverse areas.

The project 'Women in Biotechnology' (WONBIT), financed under the 6th Framework programme of the European Commission, is an excellent example of what can be done to target opinion-formers such as scientists, economists and lawyers in bottom-up activities, and to encourage a debate on gender issues triggered by developments in the life sciences.

WONBIT gave rise to a successful international conference highlighting the importance of adopting good practices and ethical considerations in parallel with the rapid pace of progress in biotechnology – from a woman's point of view. In particular, the conference addressed women in decision-making positions in biotechnology with specific reference to scientific excellence, social competencies and management qualities as well as issues relating to environment, society and the younger generation.

But it did not stop there: a key part of the conference was dedicated to stimulating public debate among non-specialists, which has led to a number of recommendations to policy-makers on better communication in biotechnology, on taking better account of the gender aspects of research, and on involving more women in the decision-making process that surrounds developments in biotechnology.

I am sure that this publication on the outcome of the WONBIT conference will contribute to enhancing the significance of women's role and presence in biotechnology, as well as changing outdated attitudes that view biotechnology as a simple production tool to a view that recognises its use and development to be both environmentally sustainable and socially acceptable.

Head of Unit 'Scientific Culture and Gender', European Commission, DG Research

Foreword

Annamaria Simonazzi

Biotechnology is a 'hot topic'. Many of the great problems facing humanity – from how to cope with lethal infections or diseases to economic development issues – are susceptible to biological intervention. Biological technology opens up great opportunities and raises formidable challenges. Simple genetic engineering is now routine: with the filing of the first patent application for an artificial living organism, the idea that someone might own the rights to it raises legal and ethical concerns.

Despite countless conferences held throughout the world, an authoritative woman's perspective on these issues is still lacking. The international conference on 'Women in Biotechnology', jointly organized in Rome, Italy (June 2007) by the Fondazione Brodolini and the Women and Science Association (Associazione Donne e Scienza), Italy and under the auspices of the European Commission, was intended to remedy this shortcoming.

The conference's multidisciplinary approach made it possible to profit from interaction among different specializations and bodies of expertise in addressing a range of relevant issues: What is women's role in directing and shaping the main lines of research in universities and laboratories? To what extent are women participating in the economic development and industrial exploitation of research results? What have women to say concerning the application of these results to such a wide range of problems? Finally, how do biotechnologies affect, for better or worse, women's lives and opportunities?

The interaction among women active in various fields – from the frontier of scientific research to investigation of the consequences on human bodies, economies and societies – produced lively debate. The papers collected in this book evidence the richness of this debate. Rather than putting forward a single solution, they develop a shared methodology for the analysis of problems, seeking to furnish a framework to guide the decisions of policy makers and institutions. The promotion of women's role in society has been the focus of the social policy research activities of the Fondazione Brodolini and I am sure that the construction of this shared methodology will greatly contribute to the increased visibility and influence of women in the policy debate on biotechnologies.

Head of the Scientific Board of the Fondazione Brodolini Rome, Italy

Acknowledgements

We thank the European Commission, Research Directorate-General for their financing of the WONBIT project, under Contract No. 036675 (SAS6). We want to thank Johannes Klumpers and particularly Ekaterina Charvalos, Project Officer in charge of the project, for assisting us at each stage in preparation both of the Conference and the book. We are deeply thankful also to the Brodolini Foundation, an organization that has been active for many years in promoting women's endeavors in social sciences and that has promoted and coordinated the Wonbit project; especially we thank Annamaria Simonazzi, Diego Teloni and Barbara Leda Kenny who have been throughout very helpful, friendly and collaborative.

Last, but not least, we are grateful to the friends of the Women and Science Association (Associazione Donne e Scienza), Italy – of which both editors are members – for discussions over the years on post-academic science and biotechnologies, making it possible to lay the groundwork for this project.

Thanks also to Stefano Bolelli Gallevi, who has efficiently assisted us in preparing the book for the publisher.

Contents

Contributors

Judit Acsády, Ph.D. in Sociology. She has worked for the Sociology Research Institute of the Hungarian Academy of Sciences since 1995, specializing in gender issues and history of feminism. After her studies at ELTE University, Budapest, she did postgraduate work at the University of Amsterdam and EHESS, Paris. She is the mother of a ten year old son.

Elvita Alvarez, Ph.D. candidate in political economy. She is a sociologist and lecturer in Gender Studies on the Faculté des Sciences Economiques et Sociales of the University of Geneva, Switzerland.

Katalin Bartók, Ph.D. in Biology, Associate Professor at the Department of Taxonomy and Ecology – Biology and Geology Faculty, Babes-Bolyai University, Romania. She studied the role of humus in mineral nutrition of plants, the taxonomy, and the ecology of lichens and their significance in the monitoring of pollution. Author of five books and numerous (104) scientific publications.

Amalia Bosìa, Full Professor of Biochemistry at the Medical School of the University of Torino, Italy. From 1996 Coordinator and member of the board of professors of the Ph.D. program in Biochemistry and Cellular Biotechnology (Department of Genetics, Biology and Biochemistry, University of Torino). Principal investigator of many research projects. Research stage periods: Physiologish Chemisches Institut (University of Marburg, FRG); Max-Plank-Institut fur Biophysik (Frankfurt/Main, FRG); Research Center of Vascular Biology and Medicine (University of Jena, FRG); Department of Biological Chemistry, Institute of Life Sciences, The Hebrew University of Jerusalem, Israel. Major recent research topics: regulation of NO pathway in normal and transformed murine and human vascular cells; molecular mechanisms of insulin action in vascular smooth muscle cells; mechanisms of induction and reversion of 'multidrug resistance-MDR' and of 'apoptosis MDR' in transformed cells.

Fabienne Crettaz von Roten, Senior scientist awarded a Ph.D. in statistics. She is now a senior scientist, head of a research unit 'Relations science-société' in the Observatoire Science, Politique et Société of the University of Lausanne, Switzerland.

Giovanna Di Chiro, Research Associate and Visiting Professor in the Environmental Studies Department at Mount Holyoke College, USA. She has published widely on the intersections of gender, race, scientific expertise, and environmental justice. She is co-editor of the volume, *Appropriating Technology: Vernacular Science and Social Power* (University of Minnesota Press) and is completing a book titled *Embodied Ecologies: Science, Politics, and Environmental Justice*. She is the co-founder of the Pioneer Valley Community Environmental Health Coalition and collaborates with environmental justice organizations to conduct community-based action research on environmental health disparities in the low-income communities of Western Massachusetts.

Heidi Diggelmann, Graduated in medicine at the University of Berne, Switzerland. She rapidly developed a passion for research. Her interest focuses on viruses capable of inducing cancer in animals and humans. After several years of postdoctoral research in Chicago and Zürich, from 1971 to 1991 she was group leader and Department Head at the Swiss Institute for Experimental Cancer Research in Lausanne, Switzerland. In 1991 she became Director of the Institute of Microbiology of the University of Lausanne. From 1997 to 2004 she was President of the National Research Council of the Swiss National Science Foundation.

Barbara Duden, Professor Dr., is an historian who teaches in the Institute for Sociology and Social Psychology at the University of Hannover, Germany. For more than a decade her writing and research have focused on the history of experienced somatics and the demise of common sense perception through experts' intrusion into personal experiences and deliberations. See: *Disembodying Women. Perspectives on Pregnancy and the Unborn* (Harvard University Press, 1993) and A historian's biology: On the traces of the body in a technogenic world. In: L. Passerini, & A. Geppert (Eds) *European Ego-Histoires: Historiography and the Self, 1970–2000. Historein*, 3 2001.

Zoltán Ferencz, MA in Economics and Political Science is a research fellow in the Institute of Sociology at the Hungarian Academy of Sciences. He organised numerous empirical projects in the field of environmental, water and risk issues, focussing especially on the analysis of databases. In the last three years he has been involved in a number of international projects (NATO CCMS; FP6 (SAFEFOODS, COWAM2, TLM-NET)).

Sarah Franklin, Professor of Social Studies of Biomedicine and Associate Director of the BIOS Centre at the London School of Economics, UK. Her work focusses on the social and cultural aspects of new reproductive and genetic technologies, as well as cloning and stem cells. Her most recent book is entitled *Dolly Mixtures: the remaking of Genealogy* (Duke University Press, 2007).

Elena Gagliasso Luoni, Associate Professor of Philosophy of Science at the University of Rome 'La Sapienza', Italy. Her research area is located at the borders of contemporary epistemology of biology, specifically on relations between body-objectivation theories and evolutionary developmental theory. She is a member of and teaches at the 'Summer Specialization School of History and Philosophy of Biology'.

She is a founding member of 'Interuniversitary Centrum of Research in History and Epistemology of Biosciences' (Res Viva), of the board of CERMS (Interdepartmental research on the methodology of scientific research). She is president (p.t) of the 'Methaphor Club'. She has served on the Scientific Committees of the following journals*: Ecologia politica, Rivista di Storia della Scienza, Sofia.*

Wendy Harcourt, Social scientist and editor of the quarterly journal *Development* at the Rome-based international nongovernmental development organisation, the Society for International Development, where she is a senior advisor. Her field of work is the contemporary policy and practice of social and economic global development from an international relations and gender perspective. She is currently Chair of Women in Development, Europe and visiting fellow at Clare Hall, Cambridge, England. She is the author of numerous journal articles and book chapters on different facets of gender and development as well as editor of four collective books, the latest of which is *Women and the Politics of Place, co-edited* with Arturo Escobar (Kumarian Press, 2005).

Katherine Harrison, BA in English and French from University of Wales, Swansea, and MA in English from King's College, University of London, UK. Her MA dissertation was concerned with representations of new technologies in the work of Donna Haraway and Jeanette Winterson. She is currently undertaking Ph.D. research at Birkbeck, University of London, and also holds a Marie Curie Fellowship for Early Stage Training in Gender and Women's Studies at Linköping University, Sweden, where she is researching intersections of language, gender and new technologies.

Nancy Hawkins, Assistant Professor at Simon Fraser University in the Department of Molecular Biology and Biochemistry in Canada. She obtained her Ph.D. in Molecular Biology from Princeton University in 1996 and undertook postdoctoral training in the Department of Molecular and Cell Biology at the University of California, Berkeley. She currently holds a University Faculty Award (UFA) from the Natural Sciences and Engineering Research Council of Canada (NSERC). The goal of the UFA program is to enhance the recruitment, retention and early career progression of women and Aboriginal people in tenure-track faculty positions. She also has an ongoing interest in promoting and encouraging young girls to pursue math and science. She spent four years on the Board of Directors for the Society of Canadian Women in Science and Technology (SCWIST), a non-profit organization established to promote, encourage and empower women working in science and technology.

Georgina Kosturkova, Graduate of the University of Sofia, Bulgaria and was awarded a Ph.D. from the Academy of Sciences in Moscow; she is a senior scientist/researcher and a group leader in the Department of Plant Biotechnology, Institute of Genetics, Bulgarian Academy of Sciences. Expert in all spheres of *in vitro* cultures with main interest in genetic engineering, stress modeling. Author of more than 75 publications, leader and partner of 20 national and international projects. FAO consultant on plant biotechnology and biosafety. M.Sc. and Ph.D. supervisor.

Floriana Marin, Ph.D. in Environmental Economics at the University of Trento, Italy, is a junior researcher at the Cell and Molecular Biology unit of the Fondazione E. Mach - IASMA Research Centre (Italy), in the framework of the Projects *OSSERVA*3 and *EcoGenEtic.Com*. Her fields of research concern the study of the public perception of GM food and the application of economic methodologies for the valuation of non-market goods, such as public health or environmental benefits.

Lucia Martinelli, Ph.D. at the Wageningen Agricultural University Netherlands, is the head of the Cell and Molecular Biology unit of the Fondazione E. Mach – IASMA Research Centre (Italy). She is expert in genetically modified crops, from their production to traceability in food and feed on both laboratorial and 'cultural' aspects, among them security, risk assessment and communication. For her research on grape biotechnology she was awarded with the 1994 First Prize of the Rudolf Hermanns Foundation (Germany). Involved in gender studies, she is the president of the Commission for Equal Opportunities of the Trento Autonomous Province.

Francesca Molfino, Psychoanalyst in private practice and feminist. Member of the Italian Psychological Association. Cofounder of the Women and Science Association (Associazione Donne e Scienza), Italy. For over thirty years her interests have focused mainly on issues concerning women: the psychological aspects of abortion, incest, violence, transsexuality, the image of the female body and the alterations it undergoes with food disorders. She has published several articles on these topics and on the application of psychoanalysis to art and literature. She is the author of a book on women politicians: *Women, Politics and Stereotypes*, co-author of a book on art collectors: *The possession of beauty,* and edited with Claudia Zanardi *Symptoms, Body and Femininity*.

Manuela Perrotta, Ph.D. in Information Systems and Organization at the Department of Sociology and Social research of the University of Trento, Italy. She is a member of the Research Unit on Communication, Organizational Learning and Aesthetics (RUCOLA) at the University of Trento. Her doctoral research examined the relationships among technologies, practices, and organizations in the field of Assisted Reproductive Technologies (ART). Her current research investigates the practical knowledge of women in small business.

Alexandra Plows, Research Associate, Cardiff University, UK, Centre for Economic and Social Aspects of Genomics (Cesagen), part of the Economic & Social Research Centre funded Genomics Network. Her research currently focuses on identifying and theorising public engagement with genetics, genomics and related areas of bioscience. Her research relates issues emergent in bioscience to multiple themes and fields including sustainable development, political economy, citizenship, identity, feminism, and social movements. Her previous research and Ph.D. work was focused on UK environmental direct action networks. Research methods relating to qualitative research and ethnography in particular have been a long standing interest. She has published a number of single and coauthored papers and book relating to her research.

Qamar Rahman, Dean of Research & Development at the Integral University, Lucknow(India), Emeritus scientist at the Industrial Toxicology Research Centre, Lucknow. Adjunct Professor at the Hamdard University, Delhi. Visiting scientist in several institutions in the USA and in Europe. Fellow of the National Academy of Sciences, India. Actively engaged in the research of risk assessment of fibers, particles and nanoparticles. Conducted in depth studies in Asbestos-Based Units in India. Author of numerous article (128) in journals of international reputation.

Elettra Ronchi, Ph.D. from the Rockefeller University (New York, US) in neuroendocrinology and genetics and has held research and teaching position in the US, Italy and France. Her other academic qualifications are in the fields of health care systems management and risk management and she has more than fifteen years experience in science and biotechnology policy. She began her policy career in 1992 as a consultant on biomedical research technology and transfer to the United Nations. In 1995 she joined the OECD to lead a programme of work on new biotechnologies related to human health. Most recently, in 2007, she was assigned to the Health Policy Division of the OECD, where she is responsible for a programme of work on health information technologies. Elettra Ronchi sits as expert and as OECD representative on a number of bioethics committees and advisory boards. She is the co-founder of a non-profit association of Italian women in international careers based in Paris.

Suman Sahai, Ph.D. in Genetics from the Indian Agricultural Research Institute in New Delhi. From 1981 to 1989, she was a faculty member at the University of Alberta in Canada, University of Chicago in the U.S., and at the University of Heidelberg in Germany. She returned to India in 1989 and organized Gene Campaign, a non-governmental organization that plays a key role in formulating Farmers' Rights and has been at the forefront of generating awareness on issues relating to trade, intellectual property rights, and genetic resources conservation and sustainable use. She has published extensively in science and policy issues related to food security, She is a member of the National Biodiversity Board and serves on the Research Advisory Committees of national scientific institutions, the high-powered National Commission on International Trade and the Ethics Committee of the Indian Council of Medical Research. In 2001 she was appointed Knight of the Golden Ark (Netherlands) for establishing Gene Campaign and in 2004 she was honored with the Borlaug Award for her outstanding contribution to agriculture and the environment.

Silja Samerski, Biologist, awarded the Ph.D. in Social Sciences with a thesis on 'The Mathematisation of Hope. On Autonomous Decision-Making Through Genetic Counseling'. She is Assistant Professor at the Institute for Sociology and Social Psychology at the University of Hannover (Germany), where she just finished a project in collaboration with Barbara Duden on the social and cultural effects of the 'release of genetic terms' into everyday language. Her research focuses on the latent and symbolic functions of genetic literacy and professional counseling. See: The 'decision trap': How genetic counseling transforms pregnant women into managers of fetal risk profiles. In: P. O'Malley and K. Hannah-Moffat (Eds.) *Gendered Risks* (London:

Routledge Cavendish, 2007) and The paradoxes of taught self-determination: How genetic counseling enables pregnant women to make an impossible decision. *Signs* (in print)

Marsha J. Tyson Darling, Ph.D. is Professor of History and Interdisciplinary Studies, and Director of the Center for African American & Ethnic Studies Program at Adelphi University, New York. She is a recognized scholar focused on the areas of race, gender, and justice in relation to genetics and biotechnology, among other areas. She has published and presented extensively on these topics. In addition, Darling has chaired a UN expert group on gender and racial discrimination. She is also involved in several organizations, including the Association for Women's Rights in Development and the Global Network for Women's Reproductive Rights.

Susanne Uusitalo, Has two Master's Degrees from the University of Turku, Finland; in English Philology in 2004 and in Philosophy in 2006. She is working on her doctoral dissertation in practical philosophy, focussing on the philosophy of addiction. She is also doing research as a member of the Department of Philosophy, University of Turku, Finland. Her research interests are in bioethics, ethics in general, feminist philosophy, and those fields of philosophy that touch upon the issues of addiction. She also has a sound background in Women's Studies at the University of Turku, Finland.

Gillian Youngs, Ph.D. is currently Senior Lecturer in the Department of Media and Communication, University of Leicester, UK. Her research interests include the social implications of technological developments. She has been involved in associated consultancy and report writing for several sections of UNESCO and national and international NGOs. She has written numerous books, chapters and articles for scholarly journals: her latest monograph is *Global Political Economy in the Information Age: Power and Inequality* (Routledge, 2007). She was a founding co-editor of *International Feminist Journal of Politics* in 1999–2005 and continues to serve as a board member. Her other journal board memberships include Political Geography and Journal of Global Ethics.

Flavia Zucco, Head of Research at the Institute of Neurobiology and Molecular Medicine (CNR) in Rome, Italy. She graduated in biology in 1969 at the University of Naples. Since 1971 she is a member of the permanent staff of the National Council of Research (CNR). Her research interests cover in vitro toxicology, in vitro cellular differentiation, and more recently, bioethics. She is the author of a number of scientific publications and congress communications and has served as the editor of several congress proceedings and collective books. She has organised courses and congresses at national and international levels, and has given invited presentations in Belgium, Greece, Germany, Ireland, Italy, and India. She has been active in the field of women and science since 1988. She is President of the Associazione Donne e Scienza (Women and Science Association), Italy. In that area too, she has published several papers and organized a number of meetings and conferences.

Abbreviations

ABI	AgroBioInstitute
ACU	Assisted Conception Units
ADA	Anti-Drug Antibodies
ADHD	Attention Deficit Hyperactivity Disorder
AHRA	Assisted Human Reproduction Act
ANT	Altered Nuclear Transfer
ARPANET	Advanced Research Projects Agency Network
ART	Assisted Reproductive Technologies
ATP	Adenosine Triphosphate
AWID	Association for Women's Rights in Development
BAS	Bulgarian Academy of Sciences
BBB	Blood Brain Barrier
BBSRC	Biotechnology and Biological Sciences Research Council
BSE	Bovine Spongiform Encephalopathy
Bt	Bacillus thuringiensis
CAT	Chloramphenicol Acetyltransferase
CAUT	Canadian Association of University Teachers
CDWRI	Cotton & Durum Wheat Research Institute
Cesagen	Centre for the Economic and Social Aspects of Genomics
CEU	Central European University
CF	Cystic Fibrosis
CGD	Chronic Granulomatous Disease
cGMP	Current Good Manufacturing Practice
CMS	Cytoplasmic Male Sterility
CNR	Cell Nuclear Replacement
CNS	Central Nervous System
CNT	Carbon Nanotube
CNT	Cell Nuclear Transfer
CORE	Comment on Reproductive Ethics
CPCB	Central Pollution Control Board
CRC	Canadian Research Chairs

DAI	Dobrudja Agricultural Institute
DC	Direct Current
DC	Discrete Choice
DH	Department of Health
DNA	Deoxyribonucleic Acid
EB	Eurobarometer
EGP	Environmental Genome Project Protection Agency
EJ	Environmental Justice
EJM	Environmental Justice Movement
ELAD	Extracorporeal Liver-Assist Device
ELSA	Ethical Legal Social Aspects
ELSI	Ethical, Legal and Social implications
ENWISE	ENlarge Women In Science to East
EPA	Environmental Protection Agency, USA
EPWS	European Platform of Women Scientists
ES	Embryonic Stem
ESF	European Social Forum
ESRC	Economic and Social Research Council
ET	Embryo Transfer
ETC Group	Action Group on Erosion, Technology and Concentration
FAO	Food and Agriculture Organization
FAQ	Frequently Asked Questions
FDA	Food and Drug Administration
FGRI	Fruit Growing Research Institute
GE	Genetically Engineered
GFP	Green Fluorescent Protein
GI	Gastrointestinal
GKT	London School of Medicine at Guy's, King's College and St Thomas' Hospitals
GM	Genetically Modified
GMO	Genetically Modified Organism
GMP	Good Manufacturing Practice
GSCE	General Support Center Europe
GUS	β-glucoronidase
hES	Human Embryonic Stem Cell
HESCCO	Human Embryonic Stem Cell Coordinators
HFCS	High Fructose Corn Syrup
HFE act	Human Fertilisation Embryology Act
HFEA	Human Fertilisation Embryology Authority
HGDP	Human Genome Diversity Project
HGP	Human Genome Project
HOOO	Hands Off Our Ovaries

HRT	Hormonal Replacement Therapy
HT	Herbicide Tolerance
HTA	Human Tissue Act
ICPD	International Conference on Population and Development
ICSI	Intracytoplasmic Sperm Injection
IDRC	Institute of Development Research Canada
IFIs	International Financial Institutions
IGen	IGeneration
INN	International Nonproprietary Names
IPCB	Indigenous People's Council on Biocolonialism
IPGR	Institute for Plant Genetic Resources, Bulgaria
IRB	Institutional Review Board
ISO	International Organization for Standardization
IVF	In Vitro Fertilisation
Karnobat AI	Agricultural Institute of Karnobat
Kyustendil AI	Agricultural Institute of Kyustendil
LGBTQ	Lesbian, Gay, Bisexual, Transgender, and Queer Persons
LMOs	Living Modified Organisms
MAGIC	Myoblast Autologous Grafting in Ischemic Cardiomyopathy
mESC	Mouse Embryonic Stem Cells
MHRA	Medicines and Healthcare Products Regulatory Agency
MIUR	Ministry of University and Research (Ministero dell'Istruzione, dell'Università e della Ricerca)
MON 810	Genetically Modified Corn Line by Monsanto
MP	Members of Parliament
MRC	Medical Research Council
MRI	Maize Research Institute
ms infinity	Math and Science Infinity
MS	Mass Spectrometry
MT	Metric Ton, A Measurement of Mass Equal to One Thousand Kilograms
MWNT	Multi-Walled Carbon Nanotube
NBS	National Blood Service
NCAS	National Center for Agricultural Sciences
NCT	National Center for Toxicogenomics
NESCI	North East England Stem Cell Institute
NGO	Non-governmental Organization
NGS	National Geographic Society
NHGRI	National Human Genome Research Institute
NHMRC	National Health and Medical Research Council
NIEHS	National Institute of Environmental Health Sciences

NIH	National Institutes of Health
NMR	Nuclear Magnetic Resonance
NNI	National Nanotechnology Initiative
NPs	Nanoparticles
NPT	Neomicin Phosphotransferase
NRT	New Reproductive Technology
NSERC	Natural Sciences and Engineering Research Council of Canada
OECD	Organisation for Economic Cooperation and Development
OSHA	Occupational, Safety and Health Administration
OTC	Ornithine Transcarbamylase
PEG	Polyethylene Glycol
PGD	Preimplantation Genetic Diagnosis
PUS	Public Understanding of Science
QD	Quantum Dots
R&D	Research and Development
RCTs	Random-Controlled Trials
RNA	Ribonucleic Acid
ROS	Reactive Oxygen Species
S&T	Science and Technology
SCID	Severe Combined Immunodeficiency
SCNT	Somatic Cell Nuclear Transfer
SCWIST	Society for Canadian Women in Science and Technology
Septemvri RCES	Regional Center for Extension Services in Septemvri
Shoumen AI	Agricultural Institute of Shoumen
SNP	Single Nucleoti Polymorphism
STS	Science and Technology Studies
SWNT	Single-Walled Carbon Nanotube
UFA	University Faculty Award
UKSCB	UK Stem Cell Bank
UNESCO	United Nations Educational, Scientific and Cultural Organization
US PTO	United States Patent & Trademark Office
V. vinifera	Vitis vinifera
VCRI	Vegetable Crops Research Institute
VEI	Viticulture & Enology Institute
WHO	World Health Organization
WISEST	Women in Scholarship, Engineering, Science and Technology
WSF	World Social Forum
X-EDS	Energy-Dispersive X-ray Spectroscopy

Introduction

Francesca Molfino and Flavia Zucco(✉)

1 Post-academic Science

In the area of natural and physical science, biology, as a life science, seeks to account both for the conservation of individuals and the invariance of certain laws of the discipline, and for their transformations. It seeks to construct laws and at the same time account for processes that often seem to be incompatible with their absolute, generalised nature. On the inclusion of man in contemporary science we see a form that we might define as 'phagotization of the human dimension in its biological sense, and a rootedness of our humanness in the material bases common to all living things (e.g. DNA, physical-metabolic functions, cognitive neural connections, knowledge of the outside world dependent on belonging to the species)' (Gagliasso 2001). The aim is to account for the entire human person (psychic characteristics, forms of societal and economic organisation) through the *apparatus of the laws* of biological evolution, recently integrated with the laws of molecular genetics and population genetics.

However, once the human species found a place as an object of the natural sciences, it should have brought with it the relevant set of humanities but, since they were deemed insufficient for the production of 'scientific' truths, and thus unreliable, no intermediate space was created for exchange between the various ways of thinking about humankind. The rift between the various forms of knowledge that had begun to open in the last few centuries grew wider and deeper. This separateness was imposed in part because it was believed that science could draw upon the 'absolute truth' (credibility turned to faith), thereby substituting the certainties belonging to religion, philosophy and psychology.

Francesca Molfino
Via di Monte Giordano 5, 00186 Rome, Italy
f.molfino@mclink.it

Flavia Zucco
via Lorenzo il Magnifico 61, 00162 Rome, Italy
flavia.zucco@inmm.cnr.it

F. Molfino, F. Zucco (eds.) *Women in Biotechnology,*

A great impact on the view of science from the inside has come from theories in the history of science and philosophy of science. The critical work of influential thinkers about science (Popper, Lakatos, Kuhn, Latour) has weakened the positivistic picture of it and done away with the idea of 'Pure Science' as representation of the 'Eternal Truth'.

This representation of science was mainly due to Robert Merton (1973), the first and best-known science sociologist, who depicted it as communitarian, universal, original, sceptical and neutral. Many critics of Merton's model contest that science had never strictly corresponded to that description. However what is sure is that nowadays it is not even remotely applicable to contemporary science. According to John Ziman (1996, 1998), contemporary science belongs to scientists and their institutions and it is exclusive to the highly developed countries, bound to main research financed trends, assertive and commissioned.

Science is no longer a relatively uncontested practice within society (Levitt 1999). The change is sometimes revealed in the idea that the modern contract between science and society has broken down and needs updating (Nowotny et al. 2002; Ziman 2000).

The internal and external causes of change in the practice of science should be distinguished. The internal causes have to do with the increasing theoretical, technological and practical complexity of science as a practice. The external causes have to do with the financial, institutional and societal conditions of science.

Science and technology appeared as knowledge and applications that solved problems, found solutions, and extended the scope of human skills and actions. And yet it is precisely because of the apparently limitless growth of technological applications and the difficulty of forecasting the consequences that the latter are no longer determined by precise needs but take the form of highly complex negotiations between those who devise them, control them (the scientific community) and the various social parts concerned. This complex negotiation will determine the winning theories and technology, as well as the theories to be discarded, although not necessarily less true than the 'winners'. Not only are the ends of technologies complicated and compromised by chance as well as social influences, but the origins of the technologies are also thus conditioned. Kevin Kelly (1998) suggests that in contemporary science it is *the pursuit of novelty* that prevails, rather than of knowledge, that *the production of tools* has taken the place of that of theories. The difference there used to be between science and technology has disappeared: what matters is not to increase knowledge but to produce new opportunities.

In addition, scientific practice has become vastly more complex. The financial, social and institutional conditions of doing science have changed. For instance, many scientists are public employees (with all the related implications); there is much more investment from corporate business, much more science policy steering on the political level and much more attention from the media and the public. Thus the societal climate with which science is confronted is more pressing and influential, both at the institutional level and in respect of relations with the public.

Paradoxically, the growth and expansion of science have led to criticism of the very assumptions it rests on, above all through the endless proliferation of diverse methodologies and results.

As Ulrich Beck (1992) remarks: 'The access to reality and truth once ascribed to science gives way to decisions, rules and conventions that could have had different outcomes'. Science 'becomes *increasingly necessary*, but at the same *decreasingly adequate* for socially binding definition of the truth'. This loss of function 'derives from the triumph of differentiation in claims of scientific validity'. The differentiation of science 'rises the uncontrollable tide of particular results, conditional, uncertain and unrelated. This *hypercomplexity* of hypothetical learning can no longer be mastered with methodical rules of verification'.

The hypertrophic development of technologies leads towards a paradoxical outcome: the initial purpose of applying technology to create an environment suited to human beings is turning into a serious threat to the environment itself. The practice of science together with specific scientific results and activities has become far more controversial. A great many meta-studies have brought the achievements of science into perspective and made society much more aware of the drawbacks and side effects of scientific and technological progress.

These changes contribute to a situation in which science is more interconnected with a democratic society. Because science meets with criticism from outside, it is bound to be more self-critical and to deliver more value for the money, support and other facilitating conditions that flow from society to the practice of science. The challenge is to accommodate the demand for moral justification, public accountability and transparency without jeopardizing scientific freedom, creativity and progress. For it is undeniable that the processes of communication between science and society and evaluation of prospects and results are faced with many pitfalls and obstacles. At the very time that the threats and risks appear increasingly serious and evident, scientific applications are growing ever less accessible to attempts to determine proofs, attributions of responsibility and indemnifications (Beck 1992).

Various political movements have taken a critical stance on the consequences of biotechnologies, but in the course of time the need has arisen – given the impossibility of taking a plausible position with scientific necessities and the positive results of many technologies – for the scientists themselves to take the initiative in rediscovering their self-reflecting function. Change in the negative consequences of biotechnologies can only come about by starting out within technological knowledge, since they can neither be rejected outright nor passively endured.

2 The Encounter Between Feminism and Science: Some Brief Notes on Fox Keller, Harding and Haraway

Women began to take a stance on science over thirty years ago, producing a vast number of publications over the years (for detailed examination of the literature see the exhaustive bibliography in Rosser 2000); here we will simply offer some thoughts and pointers on certain contributions that seem to us representative of the feminist approach to science.

The attitude taken by feminists to science is coherent with the position women have adopted towards culture, taking the form of both rejection and critical endeavour to change it. Feminism has had a hermeneutic action, leading to a completely new re-interpretation of the tradition; it becomes a form of deconstructionist philosophy, calling into question the subject, the assumptions at the epistemological, ethical and political levels: scientific knowledge does not form in a vacuum, and does not simply and solely extend the theoretical context of previous researches.

Over and above the various critical points of view, the great merit of the feminist positions lies in having addressed an array of assumptions in our culture of which we had been unaware – for example, the recognition of science and culture as based on the male view of the difference between sexes as domination and subordination.

In the early 1960s various feminist scholars dedicated biographies to women scientists to demonstrate that they had made original contributions, and that these contributions could to some extent be seen as exemplary of femininity (Barbara McClintock[1]), while certain other studies have shown that female research could be cancelled by male-managed science (consider the well-known episode of scientist Rosalind Franklin recounted by Sayre 1975).

The first step towards science was thus a matter of accrediting women as rightful members of the scientific community, revealing just how many fundamental – and often unacknowledged – contributions they had made to science. Studies have since continued from these examples on the hurdles, barriers and gender stereotypes that scientific organisations and academies put in the way of women scientists. Behind this remains the idea of science as a field of learning that is in itself objective, devoid of values, open to men and women, and that the inequality of access, results and career can be mainly a matter of cultural prejudice and policy.

Subsequently, at the end of the 1980s, some feminist historians, science philosophers (Harding 1986; Haraway 1989) and scientists (Hubbard 1990; Fausto-Sterling 1992; Birke 1986) called into question the alleged lack of values in science; for example, the generalised validity of certain experimental results concerning heart diseases, where women had for thirty years been excluded in 82% of studies (Rosser 2000).

Feminist criticism went on from the ‘woman question’ issue in science to the question as to whether science may in some way be influenced by the values of women.

---

[1] Barbara McClintock dealt with the genetic development of maize, and was awarded the Nobel Prize. Fox Keller (1983) argues that Barbara McClintock conceived a model differing from the standard models of her time, less hierarchic, more diversified, with a system of genetic transmission in a field centred on a guide-molecule. Fox Keller does not hold that McClintock took a different approach to genetics because she is a woman, but that being a woman she has a sensibility or finds a certain affinity with certain models in scientific research, departing from the dominant models of the time, thanks to which she won the Nobel Prize for the discoveries related to them. This is a particularly subtle and complex argument, holding that there can be a way of addressing issues, a perspective, a sensibility vis-à-vis certain models, certain metaphors, an ability to devise a type of plausible explanations that can be correlated with gender differences.

Evelyn Fox Keller (1985) sees promise in crossing the traditional disciplinary confines, which, risky as it may be, could open up fresh opportunities. It offered all the participants – male and female – different, new tools for interpretation, deconstruction and reconstruction of that tangle of influences between cultural norms, technical-scientific development and their metaphorical expression. Without this disciplinary crossover, Fox Keller doubted that a greater number of women would be able to modify this type of science since, while one aspect of the cultural revolution of the last two decades is the striking increase in the number of women scientists, above all in the life sciences, nevertheless the presence of women does not in itself guarantee real change.

In *The century of the gene* Fox Keller (2000) analyses the 'Genome Project' from an anti-reductionist standpoint. The idea was passed down by biologists from the decades following the discovery that the molecular foundations of genetic information would disclose 'the secret of life'; that it was only necessary to decode the message in the sequence of nucleotides to understand 'the programme' that makes an organism what it is.

Things were soon found to be rather more complex: a gene can be involved in the synthesis of many proteins (tens at times, or even hundreds), while a protein can have to do with various genes, and a certain fragment of DNA can be reorganised and transcribed in all sorts of ways. But this is not the end of the story, for genes do not determine the destiny of an organism: in fact, their activity depends crucially on the environment, and the priority of the gene as the fundamental concept to account for biological function and structure is now part of the last century, and increasing importance is being attributed to epigenetic control of the genome.

Genetic reductionism has proved an immensely useful strategy for scientific research, and often still is, and it is true that the molecular biologists are not so green as to use the gene concept without due consistency: they know that there is no longer any univocal definition, but use partial definitions that work in operation and offer the possibility to identify (and patent) DNA sequences associated with certain proteins, to formulate hypotheses and explain experimental observations.

However, Fox Keller holds that this line on genes has reached the limits of its efficacy, and has not only generated confusion (especially among non-specialist readers and users) (see Chapter 7), but has also limited the imagination of biology researchers.

Fox Keller points out that the greatest significance of feminist criticism of science is political, but also, as a scientist, criticizes that science, with its reductionism, presented itself as the source of absolute certainties. For this reason, in the last few years she has moved away from investigation into the relations between women and science to concentrate on a more sustained critical scrutiny of biology.

In a recent article Fox Keller (2005), ever attentive to the metaphors of science, notes that in biology there has been a movement away from reductionism to 'systems biology', or in other words 'treating biological entities as complex living systems rather than an amalgam of individual molecules'; thus 'new methods are required perhaps to be borrowed from other disciplines'. 'The new paradigm grows out of rapid advances in instrumentation for the biosciences, the vast improvements in computing speeds and modelling capabilities, the growing interest from physical and information scientists in biological problems, and the recognition that new approaches are needed for biology to achieve its full promise of improving human well-being'. Nevertheless Fox Keller does not for now see any common 'appropriate theoretical framework' shared by engineers, computer scientists, physicists and mathematicians. Biology has become a driving science for technological investments and attracts physicists and, above all, mathematicians, but the meaning of certain terms in the other disciplines needs to be changed, for example: essential and fundamental. No longer 'is the essence of a process to be sought in abstract or simple laws, but in the messy specificity of particular adaptations that have come into existence by the haphazard processes of evolution'. 'Biology throws a serious monkey wrench into all our traditional assumptions about what ought to count as deep or fundamental, about what counts as explanation, or even about what we will count as progress' (Keller 2005).

On the future of biology Fox Keller wonders, for example, whether the importance has been evaluated 'of the temporal dynamics of our systems', if 'we paid enough attention to the time keeping of our regulatory systems'. A fundamental focus to address the various issues raised by scientific complexity has become, for Fox Keller, reformulation 'of the linguistic habits that have underlined our existing theoretical traditions'.[2]

As we have mentioned, feminists have a different point of view on science; a radical tendency has developed to reject science as a means for man to dominate

[2] 'What I am suggesting is that, prior to the need to construct an appropriate theoretical framework may well be the need to construct a more appropriate linguistic framework, one that takes us beyond the paradigm of building the whole out of the parts, and begins to accommodate the historical co-construction of parts and wholes that is so central a theme of evolutionary biology. Indeed, one of the greatest benefits of the remarkable technical developments we have seen in recent years is that it has begun to be possible to explore the dynamic interactions that not only bind parts into wholes, but equally, that reveal the ways in which those interactions constitute the parts themselves. The beginnings of a new lexicon is already evident as geneticists seek to forge new ways to think about biological function, looking for the clues to that function not in particular genes, nor in the structure of DNA and its protein products, but rather in the communication networks of which the DNA and the proteins are part'.

'Communication has become the new buzz word in biology, and it captures the discovery by traditionally reductionist life scientists of the powers of sociality. This is a definite good, but communication is just one term. The more we learn about how the parts work not only in interaction and versatile systems of gene regulation, about the signals mediating all the different levels of organization, and about the variety of epigenetic mechanisms of inheritance at play and the evolutionary feedback between the different mechanisms, the more compelling the need for an entire new lexicon, one that has the capacity for representing the dynamic interactivity of living systems, and for describing the kinds of inherently relational entities that can emerge from those dynamics. To repeat, time is crucial here: it is the medium in which interactions occur.

For too long we have tried to build a biology out of nouns, a science constructed around entities. Perhaps it is time for a biology built out of verbs, a science constructed around processes. Perhaps even genes can be revived for the twenty-first century by reconceptualizing them as verbs. I envision, in short, a conceptual framework that rests on a dynamic and relational epistemology' (Keller 2005).

nature – as exploitation of the earth's resources and learning functioning reciprocally with the industrial and capitalist revolution (Merchant 1979). More recently this approach has been emphatically espoused by women working and living in the third world, underlining the interconnections between biology and social, political and economic plans, and actively seeking to do something about the exploitation of the countries of the South, whose resources are sapped and learning misappropriated (Shiva 2001). Insofar as it is mainly produced and managed by male western culture, science invades and destroys the indigenous cultures of countries like India and Africa in the name of a learning that is supposed to be thoroughly universal and objective but is, rather, partial, situated and local.

Sandra Harding (1986, 1998a) argued that, just as in the past the development of science was inextricably interwoven with the expansion of Europe, so one might hypothesise a different, female science deriving from the women's movements in which, starting from real, everyday life the gap between the dominant conceptual schemes and bodily experience might be closed. However, according to Harding (1998b) it is not a matter of exposing the 'hard' conception of science, proposing, as in the case of many critiques, a version somehow 'weakened' or softened. While Harding sees the illusions of realism and universality, the dreams of neutrality, intrinsic to the image of science, as memories of an epistemological innocence, she is equally opposed to any conclusions that smack of relativism. Harding thus advances her proposal of *strong objectivity*, a fuller objectivity for more extensive and comprehensive scientific learning, able to take up and absorb the challenge of what had hitherto been seen as a threat, and held defensively at a distance from its limits: the world that produces it, the subjects, bodies, values, cultures, society, relationships and differences.

Women and postcolonial movements have given birth to the demand for 'equal respect and equal material resources, equal recognition and distributive justice'. Harding calls for a new form of subjectivity, 'leadership by the idea of the Other', based on the communication that has now reached every component of the global economy.

A new objectivity can make use of the prospect deriving from the experience and voices of those who are on the dominated, subjected side of the dominant science. Thus the power disadvantage becomes epistemic advantage; the subjected, marginalised individual would be recognised as a vantage point inaccessible to those in the dominant position.

Harding keeps as the central core of her thought the feminist starting point of consciousness raising, of the nomination and re-signification of female experience.

This new epistemology is open to a strong self-reflexivity which Harding holds necessary to construct a strong scientific objectivity that does not presume to overcome its partiality by denying it.

The feminist theoreticians Harding (1986) and Fox Keller look to a science that could offer better and richer explanations, that allows us to live better in the world, and give critical, reflective relations with both our and other people's domination practices. So the problem of women in science may seem more ethical than political or epistemological. Donna Haraway plunges into technological discourse, criticising the pursuit of purity and non-contamination characteristic of certain ecofeminist positions as too close to colonialist, racist argumentation.[3]

Haraway introduced the figure of the 'Cyborg'[4] at the end of the 1970s as an image of a new femininity, taking a distance from the feminism that considered the body and its procreative function as the basis of the 'difference', in that it provided the structure of femininity with characteristics associated with care and naturalness. This should not mean adopting the cyborg model, but rather represent a stimulus to imagine and propose new ways of perceiving the body consequent to the technological invasion.

For feminist philosophy altering the living body–machine duality produces changes in the programme of categories of thought. Modifying perception of that which is living, biotechnology tends to transform human relations and culture, above all when manipulations of the genetic code or artificial reproduction are at issue. On the other hand, Haraway continues with the feminist critique of science, seeing it as learning bound up with the social structure in that it represents forms of life and of practices shared by a community, and is based on and makes use of set narrative patterns. Thus she interprets science as domination over nature, identified with the maternal female figure – as male learning asserting itself in breaking away from nature, in neutrality and objectivity.[5]

Scientific thought, human sciences and culture are also based on certain other dualities such as: mind/body, objectivity/subjectivity, public/private, man/animal, natural/artificial, etc.

While criticising the scientific assumptions advanced as vehicles of absolute truths, Haraway refuses to fall back on a relativist position. What she proposes

---

[3] As Judith Wajcman points out in her book *Technofeminism* (2004), Haraway's optimism is 'a refreshing antidote to the technophobia that characterizes much radical feminist and ecological thought. Indeed, in stressing the liberatory potential of science and technology, she is rephrasing an old modernist theme linking science with progress. While critical of many aspects of the way this happens, such as extending private property to include life forms (patenting), she warns against a purist rejection of the 'unnatural', hybrid entities produced by biotechnology. Sharing her 'frank pleasure' at the introduction into tomatoes of a gene from flounders, which live in cold seas, which enables tomatoes to produce a protein that slows freezing, she revels in the very difficulty of predicting what technology's effects will be'. 'Haraway's ground-breaking work has transformed feminist scholarship on technoscience' (Wajcman 2004).

[4] The term 'cyborg' or bionic organism indicates a being, also human, of humanoid form consisting of a set of artificial and biological organs. The term comes from contraction of cybernetic organism (an organism that is a self-regulating integration of artificial and natural systems). It was popularised by Manfred E. Clynes (who was the chief research scientist in the Dynamic Simulation Laboratory at Rockland State Hospital in New York) and Nathan S. Kline in 1960, conceiving of a human being enhanced to survive in inhospitable extraterrestrial environments. They held that close relations between human and machine were the key to cross the new frontiers of space exploration in the near future.

[5] The occurrence of certain forms of rhetoric peculiar to scientific texts is addressed, albeit somewhat differently, in an interesting chapter (*Literature*) of the book *Science in Action* (Latour 1987).

is to consider science as 'situated knowledge',[6] whose contingency and partiality are to be recognised, while stressing that contingency and partiality do not necessarily mean falsity.

Haraway proposes a strong female subject incarnating the new forms of learning – a hybrid bodily image of an organism at once human and mechanical. Given that in reality membership of a social group has become changeable, without limits and irrelevant to individual biographies, the 'cyborg' metaphor reformulates a theory of the subject uniting particular-body with universal-machine, and marks a breakaway from belonging to a class, sex or ethnicity. Thus it is possible to represent the individual in her most personal aspect: the body, which, however, through union with technology, becomes a universal subject, and bodiliness compenetrated by the technological factor becomes second nature. Transforming himself with these technological elements shared by the community, the human being leaves all metaphysical leanings behind, abolishing those dualities on which culture had been based up to our own times.

If technology has given humankind a means to evolve from the very beginning of its existence on earth, then the human being is not 'given', as it were, but evolved over the millennia; shifting technology within the human being is thus a consequent operation, by no means perverse. Technological interventions occur at such an elaborate and highly developed level that they become co-production with the evolutionary potential of the species. With the advent of technologies women can transcend the biological body as the basis of gender difference and redefine themselves 'outside the historical category of woman, other, object'.

With a background in biology, but also with parallel study of philosophy and literature, Haraway declares that it was impossible for her to be a biologist 'without a kind of impossible consciousness of the radical historicity of these objects of knowledge'. Since we are going through a phase of radical reconfiguration of species categories, which are essential tools in biology, Haraway's interest is now to explore the ways we create and participate in categories and find the limits and connections of species categories as 'ongoing kin-kind work that has very important kinds of instrumentalization these days'.

Instead of falling back on science fiction, Haraway transforms and transposes into philosophical considerations and examples designed to provoke the

[6] Although it is now a well-known concept, we will quote the definition given in the Wikipedia: Situated knowledge is knowledge specific to a particular situation. Imagine two very similar breeds of mushroom, which grow on either side of a mountain, one nutritious, one poisonous. Relying on knowledge from one side of an ecological boundary, after crossing to the other, may lead to starving rather than eating perfectly healthy food near at hand, or to poisoning oneself by mistake.

Some methods of generating knowledge, such as trial and error, or learning from experience, tend to create highly situational knowledge. One of the main benefits of the scientific method is that the theories it generates are much less situational than knowledge gained by other methods. Situational knowledge is often embedded in language, culture, or traditions. Retrieved September 22, 2007, from http://en.wikipedia.org/wiki/Knowledge#Situated_knowledge.

consequences of contemporary biology, manipulations and experiments on the cells.

In her latest manifesto on 'companion species', with her characteristic paradoxical style Haraway deals with the relations between species (Haraway 2003), with how the borderlines between what is human or animal have become contingent, so open to change as to be continually crossed. We're part of a crowd, with other species and environments; nature, the other living species, are not something separate from the person, but are to be seen rather as part of us.

Of the metaphors applied to evoke the person, the 'cyborg' seems to have proved the most lasting and widespread over the past years.

This is possibly because describing the human as a highly sophisticated *biochemical and informational* machine which will be capable of *re-engineering* mind and body has transposed perception of the manipulation of life in images of exceptional bodies, reduction of human beings to chemical materiality, to a re-programmable information system. The plethora and diffusion of science-fiction images in films, novels, videogames and comics suggests there is not only a problem of public ignorance of science, but also of how science knowledge can be received and processed by the public (Selinger and Crease 2006). In the 1970s in biology, too, cell activity was described solely with the same metaphor of a machine that combines a mechanical model with a more recent model of machines inherited from cybernetics: a mechanism without intentionality under the control of a program.[7]

The movement going under the name of 'cyberfeminism' has taken up Haraway's themes, but shows greater interest in processing gender images through science

[7] 'Nowadays each entity active in the cell is described as a machine: ribosomes are assembly lines, ATP synthases are motors, polymerases are copy machines, proteases and proteosomes are bulldozers, membranes are electric fences, and so on'.

'Subsequently, in the nanotechnologies, other metaphors have marked a possible change in these images; phenomena are described according to a dynamic model of nature in which the biomaterial components are 'multifunctional composite structures'. 'Whereas engineered materials are usually processed for a single property, biomaterials are multifunctional composite structures. The interest of material scientists, especially chemists working on high performance composites, is to learn something about the art of associating heterogeneous structures from nature itself. In their effort to design composite structures at the molecular level, they either turned their attention to such familiar materials as wood, bone, or mucus, or to mollusk shells, insect cuticles, spider-silk, etc. These composite structures – associating hard and soft, combining inorganic and organic components, and capable of high performance – appeared to be ideal models for human technology for various reasons. They are models of functional diversity, being adapted for a variety of tasks including growth, repair, and recycling'.

'Consequently, the focus is less on the ultimate components of matter than on the relations between them. Interfaces and surfaces are crucial because they determine the properties of the components of composite materials and how they work together. Nanochemistry distinguishes itself from the culture of purity and high vacuum chambers by advancing an impure process of composition and hybridization that mimics natural materials'.

'The top of their 'art' consists in making heterogeneous components converge in the right location and assemble into larger aggregates without any external intervention spontaneously' (Bensaude-Vincent 2004).

fiction and looks mainly to artistic forms of expression, working on possible imaginary models of the female figure and body.[8] Other feminists have criticised Haraway, seeing the ambiguously androgynous 'cyborg' body images as hybrid. Utopian machine-human beings dreaming up physical omnipotence and stressing the stereotypes on women, 'militarism' and 'patriarchal capitalism' (Balsamo 2000), instead of freeing femininity from the status of 'other' than male. Moreover, attributing technologies with a transforming, liberating power could also be put down to the fact that Haraway is a white woman of Anglo-Saxon culture; other women in different positions and with different cultures may not be motivated to take up this approach, and might indeed feel discriminated in turn if the position were to become hegemonic. For example, African-American women maintain that race is the primary oppression and view the scientific enterprise as a function of White Eurocentric interests.

Not surprisingly, scientific research, biotechnologies, and reproductive technologies reflect the varying complex aspects of the interrelationship between developed and developing countries in general and between the particular cultures of colonized and colonizing countries. General themes include the underdevelopment of the Southern continents by Europe and the other Northern continents (Harding 1993); ignoring, obscuring, or misappropriating earlier scientific achievements and history of countries in Southern continents; fascination with so-called 'indigenous science' (Harding 1998b); the idea that the culture, science, and technology of the colonizer or former colonizing country remain superior to those of the colony or postcolonial country; and the insistence that developing countries must restructure their local economies to become scientifically and technologically literate so they can join and compete in a global economy (Mohanty 1997). In Northern, formerly colonizing, countries the concurrent restructuring effects of multinational corporations and other forces of globalization are evidenced in downsizing, privatization, and widening economic gaps between the poor and the very wealthy. The particular forms and ways that these general themes take shape and play out vary, depending on the history, culture, geography, and length of colonization for both the colonized and colonizing countries (Rosser 2000).

In parallel with development in other fields of learning, the last ten years have seen increase and diversification in the positions and critiques of the feminists. In the case of science we once again find contrasting positions, albeit with differences within

[8] In the WONBIT conference some artists evoked with videos relations between human and machine (Eleonora Oreggia), laboratory and home (Catherine Fargher), contemporary biomedical engineering and representations of corporeality (Trish Adams), skin/time time system and sex/gender system (Nicole Pruckermayr) and between various popular cultural settings and biotechnology (Ruby Sircar).

Thanks to the low cost of digital production and distribution, young artists are able to show on the Internet their creations of new female images. Cyber-feminism, too, seems to repeat or imitate the stages and strategies of the movements of the 1960s–1970s. There exist women-only lists, self-help groups, women networks and inevitably discussion and adoption of positions regarding the new technologies and the various points of view of the other women. See: www.cyberfeminism.net.

them, too. Some feminists suggest that women, having been marginalized by a dominant male discourse, may be in a privileged position, that of outsider to the discourse, to find the holes in what appears solid, sure, and unified. Otherwise, the dominant discourse threatens to rigidify all thought in society along previously established lines.

As Rosser (2000) noted: ‘Post-modern feminism may explain why many feminists oppose In Vitro Fertilisation and other reproductive technologies, while many women eagerly seek them out in their desire to overcome infertility and produce “perfect” children’.

The methods of study, and questions asked by feminists adds layers of complexity to previously developed theories on science, but there is no achieving that ‘universalism’ that women had hoped to found either through biology or through their position as ‘object’ of patriarchal culture. At times ‘gender’ itself is not taken as a central element that could mean unifying foundation – other factors are placed on the same plane as sex/gender issue: class, race, ethnos.

However: the lenses of feminist theories provide a variety of perspectives through which research and its implementation may be viewed to ensure broader inclusion in both research design and application. These lenses also provide glimpses of the power and political struggles surrounding these technologies, which lead to potential dangers for women. Feminist perspectives may signal early warnings for scientists and physicians, who may unwittingly participate in developments that harm people, especially men of colour and women.

This diversity of positions does not mean that women have no common languages or principles, but that the different objects of knowledge and the different positions of the knowers generate contrast on certain developments in biotechnological researches. Indeed, as we will see in relation to biopolitics, today’s complexity of technological knowledge and openness to manifold choices are characteristic of every form of knowledge.

In a variety of positions it is difficult to maintain a central thread so as not to stray in the wide range of feminist critical thought, but the concept of a ‘sexuated corporeal subject’ is a starting point, albeit ‘mythical’, and yet sufficient to find a common, sharable principle of knowledge and practice that can orientate future developments in biotechnology.

3 In Search of New Values and Subjects

Environmental catastrophes, terrorism, worry about risks, a generalized feeling of uncertainty about the future may all be symptoms of a process of transition concerning social and cultural values (Nowotny 2005; Bauman 2007).

There is also a growing awareness that policies should be open to ways of solving problems other than by investing in science and technology.

A way to solve conflicts between science and society, is suggested, i.e., bridging the gap between the two cultures (humanistic and scientific). Actually, this debate dates back to 1950, and particularly to the book *Two cultures and the scientific revolution* (Snow 1959). The essence of this debate is that science can no longer be considered as the only engine of progress: it has to be accompanied by humanistic approaches to readdress the question of progress in civilization. Technological progress (application of science) needs to be combined with responsible ethical analysis, grounded in humanistic approaches (sociology, history, psychology, philosophy).

Scientists are called upon to take on greater responsibility with regard to the outcomes of their studies: the statement that science is neutral, while only its application can be judged in the light of social values, is no longer valid, due to the above-mentioned aspect that science and technology are now overlapping. In this respect it is worthwhile to mention that even a 'Scientist's oath' has been advocated, by the Nobel prize-winner Rotblat (1999) in order to cope with this problem.

The problem is not, however, purely ethical but deals with the conceptualisation of the world in which we are living and the fact that science offers no solution to this problem, since it has developed specialist representation/definitions, that cannot be translated into everyday language. Unity of knowledge (Damasio et al. 2001) is still a basic need of the reason. To achieve this end it is necessary to create a new community combining scientific thought and a subject representative of common sense or new forms of solidarity (Gadamer 1983).

For scientists: the cultivation of an *extended self-consciousness*, is necessary to cope with new problems arising in their disciplines and deriving from the rapid and continuous increase of knowledge gained.

It is ever more widely recognized that analysis of theoretical concepts (and thus assumption of the framework of references) needs adequate words to be represented, and this is indispensable to the progress in specific fields of science. Several disciplines are affected by this problem, which contribute not only to the difficulties in communication among scientists but even more with society by increasing the lack of understanding and thus of transparency (Appleby 1999; Bensaude-Vincent 2001; Fraser 1999; Kiessling 2001; Porter 1992; Sloviter 2002).

In addition to these, another important aspect should be taken into account, namely the accelerated process of producing applicable knowledge leaves no time for cultural adaptation and ethical evaluation. We learned from Eric Hobsbawm (1994) that from an historical point of view, we should consider the twentieth century a short one, lasting from 1920 to 1989. For biology, the same century was even shorter – only twenty years as from the discovery of DNA and the related biotechnological applications.

In his opening address to the Ethical Committee of the French Government in 1984, French President (1981–1995) François Mitterand underlined the fact that, faced with progress of science, we need time to think, and this becomes essentially time for ethics. Thus philosophical and ethical processing should gain not only space but also time in scientific research, in order to cope better with the challenges of the contemporary world. These conditions will lead to a more constructive dialogue among the different stakeholders instead of the unproductive yes/no positions.

This also means accepting the possibility of thinking in utopian terms, producing neither immediate goals nor concrete solutions in good practices, but building a real form of solidarity, at the level of that invoked by Hans Georg Gadamer (1983) over thirty years ago. Faced with the chaotic development of our technological civilization we fail to notice those stable, immutable elements present in our social life. 'This is why I truly believe that *new forms of solidarity* are now being rediscovered and that *solidarity could indeed become the cohesive element for humankind in the future society: I see certain characteristic features in the Latin world, or surprising resistance to the logic of industrial profit* ' (italics added). Gadamer also added:

> I can only wonder whether beneath that technological European and American patina there may not survive in remote, ancient civilizations like those of China, Japan and India, above all, some trace of that religious and social tradition that characterised their culture over the millennia, perhaps hidden behind a room furnished in the European style or in a job organised in the American way – elements that might perhaps when the occasion arises once again inform the conscience with new forms of solidarity, common and binding, and let practical reason reign once again.

4 Women in Scientific Research

Contemporary science poses a paradox for women who engage in research:

- First, women have at last started to have free access to science. For the first time, in the last decade they represent a large percentage in so-called hard-science, as students, Ph.Ds., researchers. The situation is worse at higher levels and on boards, where the cooptation criterion is still operative: however several institutions and NGOs, at national and international levels, are working to enhance the presence of women in science, also at decisional levels.
- Second, the presence of women in science is acknowledged for the values they bring to it, and women scientists have become more self aware of their potentialities (Gilles de Gennes 1991 Nobel Prize winner for physics in a 2005 interview listed the quality of women scientists as being very good in keeping a research team cooperative, in showing a high degree of responsibility in their research, and in controlling their egos better than men!).
- Third, they want to express their creativity and curiosity in exploring unknown areas of science, they want to elaborate theories and feel free to comply with their wishes of self-fulfilment.

But … it is precisely now that science does not allow such a freedom. As we mentioned above: if you want money you should fit in the mainstream of research, you should commit yourself to producing something useful for society (read: for the market). You should enter a competition that often leaves dead and injured on the ground (and this is not just a metaphor: cases of suicide due to stress among Ph.Ds. have been reported). Moreover, in terms of family care society proves no more friendly towards women than in the past.

In addition, society at large is criticizing science with respect to several features: the possible risks inherent in new technologies; their problematic impact on values deeply rooted in the people's culture and in society's fundamental institutions; the lack of transparency in science planning and communication. Moreover science is also taken as the main reference point by the political authorities, not only for practical (economical, technological) decisions but also in matters concerning the governance (ethical, cultural and social issues). As a consequence, science is representing *the power*, as never before and peoples do not accept this wider role, particularly concerned by the protection of their private life. It is often objected that individually women will not usher in a different science, but this attitude fails to take into account the enormous change in mentality that freed women from their domestic confines in the last century. Moreover, women scientists offer certain characteristics that make them particularly useful to overcome the more insidious aspects of post-academic science. The secondary aspect is that they, as new entries, are less contaminated by the cultural models dominant in academia. They usually bring a more holistic vision than men of their work, thus naturally positioning themselves across different disciplines (this attitude being one of the most promising for science).

Finally they are truly creative, in the sense that they add to scientific rationality the curiosity and intuition of a genuine researcher.

Science may just gain from such a kind of contribution and the worse mistake women scientists can make is to mimic the behaviour of the majority of their male colleagues: it is not just a critical mass that is needed, but the quality of their contribution is relevant.

Demonstration that women may change science is to be seen in the fact that several disciplines, such as primatology, anthropology, medicine and social sciences in general, have already been modified by research performed by women scientists, thanks to their scientific integrity, their strong commitment to science, their responsible research work and, last but not least, due to the influence of the feminist movement on their cultural background. According to Londa Schiebinger (1989, 1999), feminism helped women scientists in finding a gender bias in the innermost core of science, and led them to challenge it. It is evident that the male subject, conceptualized as a neutral and universal subject, has in fact often seriously compromised the most sacred value of science: its objectivity.

In Italy the encounter between women and science took place in two stages: in the 1970s the feminist critique came into line with the non-neutrality of science critique proposed in general by political movements of the left. It was only in the 1980s that denunciation of the non-neutrality of science began also from the point of view of the gender difference.

Certain catastrophic events (Chernobyl, and Seveso in Italy) led to the birth of an eco-feminist current (Donini 1990; Scienza e potere, coscienza del limite 1986). The eco-feminism movement was then one of the most active and fundamentalist as far as respect for nature is concerned, and this is not surprising if we recall that Rachel Carson's *Silent Spring*, published in 1962, is recognized to be the starting point of environmentalism. Subsequent to a Conference on Women Science Researchers a study group was formed at the Women Documentation Centre (Centro di Documentazione delle Donne), Bologna. The group rapidly became a permanent National coordination, engaged in activities of research and processing on contemporary science and the presence of women in it (Alicchio and Pezzoli 1988; Benigni et al. 1988). At the same time it produced didactic material and carried out workshop activities, organising meetings, conferences, etc.

The peculiarity of the National coordination lay in its composition: in fact, from the outset it was made up of scholars from scientific and humanistic fields, with exchange, comparison of the different points of view and some entirely new thinking in terms of international feminism.

The Women and Science Association (Associazione Donne e Scienza) took up the coordination heritage and thus has a long tradition (since the late 1980s) of collective interdisciplinary confrontation on the burning issues in contemporary science.

Biotechnology has been a major topic, because many women are involved in biomedical research, much more than in any of the other so–called hard sciences. Moreover, biotechnology has to do with bodies, lives, birth, diseases, death, topics that have always been very close to women's daily life experience and charged with symbolic significant, derived from the cultural heritage.

The WONBIT conference in Rome (21–23 June 2007) was set up with the expertise accumulated over the years by the Women and Science Association (Associazione Donne e Scienza) in discussing the specificity of biotech: the basic aspects characterizing the contemporary post-academic science, the changes in societal assets and the increased new presence of women in the hard sciences and technologies, within the framework of feminist thinking.

The basic choice in organizing the conference was to have a wider involvement of women experts from the most various disciplinary and geographical areas, in order, at least, to touch on the various facets of biotechnology in different contexts. Then we aimed at having an open debate, allowing all the speakers of the lectures by invitation and of the selected papers to express themselves freely on specific aspects of Biotech according, not only to their personal expertise, but also to their personal feelings.

When we picture a different scientific community, a major priority is obviously to attribute value, and thus power and authoritativeness, to the work of women scientists.

Part I Women Scientists in Biotechnological Research contains essays that verify roles and status of women scientists. With the aim of seeing what positions women hold in science, Wendy Harcourt (see Chapter 1), Elettra Ronchi and Nancy Hawkins (see Chapter 2) have carried out exhaustive research. The percentages of women in academic positions in the various countries range between 15% and 20%. On the other hand, some encouragement comes from countries like Canada, where the wage gap between men and women is narrowing (see Chapter 2), and where a particular programme was 'designed to help universities recruit and retain talented researchers and thereby strengthen research excellence' in science, starting from the principle that to close the gender gap in science, education must begin at the earliest levels of schooling.

Achieving gender equity is the first essential step for women to be able to play an active role in producing, controlling and orienting biotechnologies. In the European panorama, in the countries of the East we find contrasting positions in the field of biotechnologies, especially in relation to Genetic Modified Organisms (GMOs): while Bulgaria (see Chapter 4) shows a favourable attitude, in Romania the extraordinary biodiversity of the environment has not been taken into consideration, and biotechnologies are seen as dangerous, invasive tools (see Chapter 18), partly because areas that should have been protected have been wrested from the community without consultation or transparency. Different again is the position of Hungary, where the dilemmas and questions about biotechnology, with the exception of genetically manipulated food production, are rarely aired in public debates. And yet the last few months have seen the information void filling up thanks to the civil organisations (see Chapter 3).

Unlike procedure in the usual conference of 'experts', where pro and contra position are present in a sort of competition for the most brilliant advocate, we have debated in an open and friendly manner, anxious to widen our knowledge and share new approaches. An example of the fruitfulness of this way of interacting can be seen in the choice of two speakers, one from Europe, the other from Canada, to combine their contributions in a single paper to be published in this volume. With the WONBIT conference and this volume, instead of communicating biotechnologies to the experts and the directly involved, the aim has been to get the various parties to interact among themselves.

This was the response to the call by Haraway to practice a form of exchange which she defines as 'thinking technology' outside the laboratory – the practice of coming to learn how to focus on each other, and do something that neither of us could do before and cannot do alone, and do it in a rule-bound way by playing a specific game that has arbitrary rules which allow you to play, or to invent something new, something beyond functional communication, something open. In fact, that is exactly what play is: a game given a safe enough space to do something that would be dangerous otherwise (Gane 2006).

5 Woman's Body as Object and Subject of Biotechnology

When the WONBIT conference (2007) was being organised it was clear and indispensable that the 'body' should be the central theme in the interplay of biotechnologies, the humanities and the female subject. It is worth pointing out that some of the papers, mostly collected here in Part II: Bodies, Cultures and Scientific Metaphors (see also Chapters 16 and 17), highlight certain areas such as interviews for genetic tests or informed consent for the use of embryos, where the encounter between women and biotechnologies takes place, where reactions of estrangement, rejection or mute acceptance of such technologies are produced, while what is lacking are means of exchange with the women scientists who deal with such issues.

Technology linked to the body, whether associated with the biology, chemistry or mechanics, always comes about as repair to some damage or disease, to something missing or non-functioning. However, over time it has developed to the point of acquiring value in itself, in the frenzied pursuit of novelty, to the degree of extending applications to every body, and devising mutations in the species.

These technologies have great implications for the female body, not only for reproduction, but because in the family it is the women who manage health and nutrition, hereby becoming primary targets for the industries producing biotechnologies, from food to cosmetics. It is the women, as main food managers for the family, who have to know whether this or that is healthy, safe or harmful.

Precisely on account of their bodies and management of the bodies of others, women are at the same time, and in a very particular way, object and subject of biotechnologies. They occupy a particular position in relation to scientific knowledge, different from that of the men who have constructed it on the principle of distance from the object and the need to possess and control it.

Nevertheless, making the aim the understanding of phenomena while eschewing domination or omnipotence is indeed a complicated business, calling for both curiosity and sustained reflection – theoretical reflection, but articulated over the reciprocal influences between lab life, everyday life and social life (which is very different from the preferential exchange between biotechnologies and industry).

On the side of humanities, from the philosophical point of view, the body in general, and the female body in particular, became the 'new subject' where manifold social, symbolic, discursive and scientific threads are at play together. It is an entity that Michel Foucault analysed with great lucidity as a node of power and knowledge relations.

The body maintains a precise identity and at the same time expresses significations that can be shared by a more or less extensive group of people. From having been a sign of precariousness in religion, in the last twenty years the body has been seen as the authenticity of self, as bearer of memory. The use of tattoos and piercing has served to inscribe on a surface the fleeting data of experience, and at the same time to recognise membership of a group.

In the 1970s feminists founded a new line of research into subjectivity, basing it on the sexed body. If the aim was to trace out a female identity, absent as active subject in history and always defined in antithesis to the male, the idea was to seek it through the sexual difference inscribed in the body, but always with the understanding of the body as biological, symbolic and sociological superimposition.

Feminists have had to come to terms in their thinking with the biological sex that has both relegated and limited women in their lives, and enshrined the positive female differences when associated with maternity.

In virtue of its singularity and at the same time collective representativeness the body is chosen as foundation of the critical theories of our 'postcolonial' society (feminism, homosexual, lesbian and queer theories), where gender and racial difference underlie all cultural claims.

When we speak of sexual and gender difference, it immediately occurs to us that favouring one term or another brings us back to the difficulty of explaining the exchange between psychic and biological, and defining what belongs to one field or the other. Re-biologising gender has often entailed a traditional kind of view of the difference between the sexes, while dealing with gender as a social construct indicated an attempt to make the dividing lines and characteristics of the two sexes flexible and modifiable. There have been continual shifts in the meanings of gender and *sexual difference*, where the former has to do with social constructs on sex and the latter with a fundamental difference, which is in fact biological. The difficulty still remains of setting limits, finding connections, exchanges between the biological, the psychical, the social-cultural and language. As Judith Butler (2004) says, even though we do not find a final answer, sexual difference represents the area in which the question of the relationship between biology and culture is formulated and reformulated and can be configured better as the necessary background for the possibility of thought, language and being a body in the world. It represents something that cannot be wholly enunciated, and that disturbs the grammar of enunciation, remaining, more or less permanently, to be questioned.

Technology and biotechnology have changed thoughts and perceptions of the body: the representation that seems most evident today is that of a body as an organised system, a functioning organism, and indeed a set of functioning organisms down to the tiniest cells that interact together, and that in turn exchange information with the environment.

Biotechnologies produce contrasting effects on representation of the body: they dissolve the visibility into invisible components, then make them concrete again, but fragmented through visualization tools and techniques. However, when that which can be detected with the senses (pain, wellbeing, sickness, health) is not an experience of real human beings but a matter of cells and populations of cells conditioning perception of the self, a modified, technical-mechanical perception is produced.

In the clash between the 'body subject' of feminist theories and the 'body object' of biotechnology, feminist criticisms still 'function only in a derivative relation to science', and women are not at present subjects 'which explore new ways to engage with biotechnologies in a challenging not derivative way' (Fantone 2003). However, in the articles by Alexandra Plows (see Chapter 10) and Marsha Tyson Darling (see Chapter 16) we sense the pressure that the manifold divergent positions of the feminists exert in the direction of biotec. The two authors note, for example, how the right and wish to have a child through artificial reproduction, upheld by many women, can at the same time work in the direction of technologies that tend increasingly to manipulate and exploit the female body. In fact, collecting eggs for experimentation has raised the issue of respect for the autonomy of the individual together with the problem of social justice.

Thus Marsha Tyson Darling (see Chapter 16) shifts the focus to the importance of fair, equitable treatment for all women, above all according to an American logic, in which the state has issued no coordinated national guidelines and which turns individual freedom (wanting a child) into exploitation of women from the deprived classes and Afro-American women.

According to Franklin (see Chapter 5) writing about the widespread practice of In Vitro Fertilisation (IVF) over the last thirty years, in the entire world about three million babies so far have been born with artificial fertilisation. From this large-scale experimentation and the accumulation of embryonic cells derive the technologies of cloning somatic cell nuclear replacement and the use of human embryonic stem cells.

An occasion for direct encounter between women and biotechnology lies in the demand for eggs for experimentation. The question is, what typeof informed consent can be expected for the donation of egg and embryo that is not merely formal, that entails real exchange between doctor and patient? The embryos produced, as Franklin says, 'are complex embodiments of reproductive hope, reproductive labor, and often reproductive loss', and all this mainly concerns a couple. Precisely when having a child is so overwhelmingly felt as a goal as to admit to tiring, frustrating procedures, something that belongs to them will have a different fate, and may be used for any purpose, from research into new forms of therapy to procreation of another child.

In these cases informed consent has to do with a very wide range of consequences, with no unique purpose, since it is involved with other scientific researches that often risk not being carried out. Equally difficult is the feedback to patients on possible forecasts of future illnesses since, for example, 'genetic predisposition to diseases susceptibility, may only come to light through research many years after donation'.

Also the system of 'egg sharing for research', which cuts the costs of IVF in exchange for egg donations means potential exploitation, like any sale of body tissues or organs.

The question one cannot help asking is, what are the two talking about?

Franklin arrives at what we see as the significant conclusion that time and attention must be dedicated to negotiating procedures, to improvement of the informed consent procedures since 'patient information are not only practical', but 'are as likely to determine the eventual success or failure of the human Embryonic Stem Cell fields (in which far more public than private funds currently invested) than the scientific knowledge or technology on which it is based'.

While the trend in feminist thought runs in the direction of using biology (as Haraway had attempted to do) to build up a strong female subject, biotechnologies have so far been working in the opposite direction, towards fragmentation of the body.

The body dealt with by biotechnologies is the body in the laboratories, and thus a body not only as organism, but also as organisation (see Chapter 8). In laboratories for artificial reproduction, the female body – the person, that is – is fragmented and reassembled: attention is on the 'microreproductive process, in which eggs, sperms and embryos absorb all the care'. The goal to achieve becomes the product of conception, the embryo, the foetus, and not the 'baby in arms'.

The female/maternal body assimilated to technological procedures in cultural collective imaginings takes other meanings. The body was represented as a combination of flesh and machine (cyborg), but then the combination gradually turned into monstrous hybrid figures, in which enhancement of the human body saw the addition of organs from other animal species. Besides expressing the fear of unknowable, uncontrollable transformations, the images recall the way of representing the world in fairy tales: the bad deformed and monstrous, bizarre creatures seen as alarming hybrids of animal and human beings. A description of this contiguity between science and surrealism is offered in the article by Harrison (see Chapter 9), who refers to a novel by Margaret Atwood to describe the transformations that experiments in the biotechnological laboratories can undergo in collective imaginings.

Value is attached to body parts in that they can be detached and transferred from their context. According to the reports of people who have undergone transplants, at the level of images a transplanted organ can be sensed as a malign creature within one, an invader, or as a child to be fed and cared for. In terms of reality it is extremely distressing to have prostheses: 'the wounds and privations suffered by the bodily subject are such as to demolish any fantasy of machine substitution' (Braidotti 2002).

The body has become material reality in fragments, in increasingly infinitesimal systems that in turn function in complex ways. Moreover, biotechnologies depend on visualization techniques and computers for the retrieval and display of complex, layered data. Biotechnology moves on from the concrete, the cell or the infinitely small, to the level of extreme abstraction relating to digital data systems and probabilistic causality.

In between these two systems are the body and mind of the man or woman, whose lives seem now to occupy a non-place, in a space and time uncontemplated in this knowledge.

If no subjectivity – no possibility of self-awareness and reflection – has been constructed between laboratory experimentation and the abstraction of models and mathematical laws, then the scientific results are interpreted at the level of magic, at the same time inducing the recipient of the technologies to regress towards expectations of total resolution or salvation. Human beings fall back on the idea of magic to order a certain sequence of events according to a given regularity leading to a positive outcome; thereby the potentially devastating effects of the negative can be contained, relativising the uncertainty of the future.

When women and technology collide, as Barbara Duden and Silja Samerski tell us (see Chapter 7) in relation to genetic counselling, the technological prerequisites are not set on account of their capacity to provide new understanding, or to repair the damage suffered by the body, or for medical improvement of life span, but for their financial profit logic or to enhance the expansion of the biotechnologies themselves, and so to enhance fantasies of omnipotence. On the one hand, having no perceivable or visible result in the short term scientific technology follows the will for power that proves strongest at the moment, contemplating no limits whatsoever, be they ethical, social or psychological.

On the other hand the recipient, faced with means and learning she is excluded from, finds herself at the mercy of a technology and attributes to, indeed asks of this technology the power to declare damnation or salvation.

Amalia Bosìa (see Chapter 17) takes a concrete, but also lucid and critical look at the consequences and contradictions of gene therapies, and shows how hard it can be for a scientist to distinguish them from 'genetic enhancement', or in other words playing God. Often the therapy overlaps with a genetic enhancement that exceeds the boundaries of normalcy and health. Many medical interventions like vaccination or 'cholesterol lowering drugs, cardio-pulmonary resuscitation, and hormone replacement therapy, are designed to prevent, forestall, or counteract the normal aging process', and so fall into line with genetic enhancement. The difficulty of distinguishing or marking out the borderlines between medical interventions lie both in the mixing up and overlapping between the concept of 'enhancement' and what constitutes a 'normal' healthy human being, and in the ignorance and difficulty of knowing what constitutes a 'normal' healthy human being. In other cases the quality of life is reduced to medical parameters, confusing the risk of a future illness with the certainty of its eventuality and proposing medical intervention as the solution for women.

'The more radical the novelties are from the technical-scientific point of view, the greater must be the social learning, so that society can adjust culturally and work on them so that they take on sense and meaning' (Nowotny 2005).

Thus the need is for a different way of thinking, constructing a subject starting from new discoveries about living things. For example, as Elena Gagliasso writes: 'the forms of collective thinking seem to be unaffected by the ontological fact that some beings – women – can be *singular individuals* for a long part of their life, may become *dual* (containing and interacting with the other) for limited and possibly repeated periods of time, and go back (with altered body and psyche) to an individual state by a separation/loss of duality' (see Chapter 6).

6 Biopolitics

Having discussed the main points concerning contemporary science, and having admitted that this is one of the most authoritative institutions of the twenty-first century, the political arena becomes the obvious ground to develop analysis over biotechnology control in order to implement environmental and gender justice. Science, as we have said, in parallel to its advancement should give room and time to philosophical and ethical analysis. We have also described the complex relationships between women and science and how women may act as moral pioneers on various occasions.

To approach the political problem and the relationship between science and society, further aspects are taken into account in Part III (Environmental Effects of Biotechnology) and Part IV (Facing Impact: Society and Biotechnology): interactions between biotechnologies, environment, public health, scientific communications, and wider definition of experts, and the role of groups, communities, etc.

Communication here is not only between scientists and the public or the politicians, but also the use the media make of scientific discoveries, often producing distorted information and false hopes. Another example of attention to the way scientific research is presented to the public is to point up false advertising for its effects on important choices (and possibly negative effects) concerning health and wellbeing, and of course for skewed, manipulated criteria in the use of available resources. Equally important in the process of exchange and communication between the world of science and the public is lifelong learning in our societies: lifelong learning as crucial to adult women's responsibilities and practices.

Precisely because in contemporary society we are engaged in and capable of continuous learning, due weight must be given to the possibility of cooperation/interplay between providers and users of knowledge and information in areas of crucial relevance.

Science has often used metaphors to represent scientific issues to the public, by translating into commonly shared representations more specialized ones. This procedure has been useful and productive, but on the other hand in some cases has gone too far, letting people assume aspects which were not included in the original proposition (think of the metaphor of the code used for the genome). Today's science seems so specialized that the use of this powerful tool seems no longer possible, to the extent that some experts in communication claim that the *black hole* in physics was the last metaphor (Wilczek 2001).

Thus it was not surprising that the need for new words to convey new concepts was felt and shared by many of the speakers, in the WONBIT conference, practising interdisciplinary communication. A great deal of work needs to be done in this respect if dialogue is to be taken further, to settle (at least) some issues. The public pressed for distinction between the actual and symbolic meaning of words, calling more for an enrichment of the languages with concepts of metaphorical relevance than asking for a better understanding of the scientific/technical words. More, in fact, than for scientific explanations, the demand emerged for a common path of reflection from the laboratory to everyday life.

This aspect cannot be underrated or dismissed but should be approached as central to social relations, with the aim of making communication honest, clear, and of coping not only with scientific but also rational, and emotional aspects (see Chapters 13–15).

In toxicology, for instance, in order to avoid odd expectations or inappropriate roles, there are scientists who insist on distinguishing the science of toxicology from the art of toxicology, risk assessment belonging to the latter (Doull 1984, 2003; Malmfors and Rosing 2002). It is clear, from this example, that this conceptual distinction calls for different wording to be used in the two areas. The practical outcome would be a diverse kind of communication to the public, which, in return would produce more balanced expectations. Moreover, this example provides evidence of the need for a wider cultural background.

In many scientific issues we should be aware that we are facing the philosophical problem of the *soros*. This word, in ancient Greek, means heap; it stands also for a philosophical problem raised by Eubulides of Miletus (fourth century BC): his argument is that, since we are not able to define when a certain number of grains becomes a heap, it is a concept that does not exist. In other words we face a continuum where it is impossible to establish a clear-cut limit – in which it becomes difficult to define and decide in what direction and according to what parameters scientific research should be moving.

Now, if we think of embryo development and the need of setting the change from the condition of cells to that of an individual, we are indeed dealing with the problem of the *soros*. The same can be said about setting a safe threshold in risk management. Outside science we can apply the same concept to the setting of the voting age.

The aim of this example is to show that in science as well as in other areas of decision making, we may encounter information that is not clear-cut (that doesn't provide certainty) so that we are obliged to reach agreements among the different stakeholders, who must take on the responsibility to decide in the most sensible way (which means to take into account scientific data, but much more than that, as we said above). In other words the issue is no longer scientific but political.

If the concept of *soros* became a commonly shared concept, a lot of endless and time wasting discussion would be avoided, and the problem under discussion would come into better focus. Evaluation of biotechnology risk (see Chapter 17) or of nanoparticle (see Chapter 12) belong to this area and call for data on their possible adverse effects, but also for elaborating models to predict those effects, making the uncertainty related to them explicit.

Giovanna Di Chiro (see Chapter 11) appropriately addresses the question of who defines which research questions are most important for improving people's health. Describing the role of women in the US Environmental Justice Movement (EJM) she underlines that those women coming from the poorest (and thus more polluted) areas of the towns are indeed increasingly collaborating with the 'expert scientists' to readdress the issue of science in real public interest. Women activists have become 'proactive' technoscientist actors by increasing their scientific literacy and insisting on actively participating in forefront decision-making about the scientific 'revolution' that affects their lives. In fact, the example of the Environmental Genome Project raises concern about the effort made in studying the genetic basis of certain diseases, rather than reducing the exposure to dangerous compounds and improving the poor life conditions of those peoples. Moreover, this project entails the risk of spreading a reductionist view of the complex mechanisms of several diseases, which may open the way to a new science-based racial classification. To cope with those problems and to answer to the pressure of the interested people, the NIH has set up a program (ELSI) that encourages dialogue with and engagement of the community of colour. Marsha Tyson Darling (see Chapter 16) too, invites us to explore 'how the mystique of genetic manipulation leads to genetic reductionism and the geneticization of health disorders' and invites women (feminists and researchers) to be engaged in research on the 'precautionary principle'.

That women are more cautious in evaluating new technologies has been demonstrated by the study by Crettaz and Alvarez (see Chapter 14), which shows that in Switzerland 'women find Genetically Modified (GM) food less useful than men, equally risky, less morally acceptable, and less to be encouraged', moreover 'women feel more concerned by the negative effects of biotechnology and GM food on the environment'.

The moral issue concerning GM food is also approached by Susanne Uusitalo (see Chapter 13) when she states that 'genetically modified food is not automatically morally relevant, but depending on its constitution and the situation it can be'.

People's integrity should be respected, in the sense that it should be possible for them to follow their authentic belief and desires. The labelling of GM products, for instance, may be done correctly, thus coping with this need.

Also women working in biotechnology are concerned about the work they do: Georgina Kosturkova (see Chapter 4) describes the reasons why almost exclusively women are doing research on plant biotech in her institute: their 'skilled hands, quick fingers, gentle action, patience and creativity' are essential to their experimental work. Moreover they share the view that new technologies could not be misused in women's hands. But, in Romania, a country close to hers (Bulgaria), Katalin Bartók (see Chapter 18) describes a serious misuse of GMOs: biodiversity has undergone great stress with the spread of GMOs, in the absence of any democratic and transparent decision process: species and land fertility have been dramatically reduced even in the Danube Delta, a protected area. Agriculture has been seriously compromised and, in this area, the precautionary principle is advocated in order to reject GM techniques and maintain a sustainable and ecologically friendly production system. In this respect Lucia Martinelli and Floriana Marin (see Chapter 15) presented a good example of creating an environmental friendly science in the area of GMOs.

Suman Sahai (see Chapter 19) expands further on the problem, underlining that also the socio-economic impacts, related to new technologies, should be included in risk evaluation procedures. In fact it is very important to evaluate the impact of the introduction of new technologies on 'the ability of the concerned communities to make use of the biological diversity upon which their survival and traditional livelihood depends'. Again the point is made that this evaluation should be approached according to the different context and with the essential contribution of the people involved. The case of herbicide tolerance is used to illustrate the disadvantages of innovation in particular in small and isolated communities, in which agriculture is mainly conducted by women.

In all these writings the difficulties of choice and definition of risks in biopolitics appear evident, with the consequent impossibility of arriving at solutions to the problems raised by biotechnologies in a reasonably short time. This brings us back to the complexity of science and the ambivalent position of the stakeholders towards the technologies. Complexity and ambivalence are words that occur with great frequency in writings on biotechnologies and scientific researches, implying that no easy solution can be found to problems that involve people experiencing the same problem with contrasting attitudes and values.

Complexity is a property of certain types of systems; it distinguishes them from the simple or complicated systems.

Today we see the biological and physical sciences characterised by a crisis of simple explanation, and consequently what seemed to be the non-scientific residues of the human sciences – uncertainty, disorder, contradiction, plurality, complication, etc. – enter into the basic problems of scientific knowledge.

Complexity appears with renunciation of the intent to reduce the particular to the general; like the irreversibility of time and so the introduction of history and events; like the impossibility of isolating a single element not linked to a system; the impossibility of explaining in simple terms and delineating, for certain systems, the problems of self-organisation; foregoing linear causality in favour of a complex causality that involves retroactions, deviations, synergic aspects, internal and external causalities; distinguishing a system from its environment but without separating or disjoining it; reintroducing the observing agent in relation with the object observed.

We come up against limitations to the explicative powers of logic, quantification and formalisation, which does not mean not recognising their explicative capacities, but rather not considering them as the only methodologies. Together with this we have the possibility of accepting contradiction as expression of complementary concepts when arrived at through a procedure of coherent, rational research.[9]

When we speak of ambivalence[10] we are referring to different or contradictory positions that do not find synthesis (in the dialectical sense), nor issue in elimination of one of the two parts (as held by the principle of non-contradiction).

If ambivalence arises in the psychological field, Robert Merton (Merton and Barber 1963) identified it in the different roles of the social actor and in the twofold identity of the individual as 'existence for the self and social existence'.

Verification of 'sociological ambivalence' was in fact made on the Apollo mission scientists in the early 1970s (Mitroff 1974). From interviews with physicists it emerged that they moved within a field of tensions which saw in play at the same

[9] Let us take the classical example of the human genome. Identification has been achieved of 30,000 genes, or basic sequence sections that codify proteins, but combining all the sections we have just 1% of the total length of the genome; the – non-codifying – 99% perform functions that we are just beginning to clarify. When correlations occur between different sections (for example, a gene formation that depends on the proteins codified by other genes) the problem becomes complex, in that we have not exhausted its significance with simple listing of the genes.

[10] There are various concepts of ambiguity and ambivalence:

Ambiguity is the property of words, notations and concepts (within a particular context) as being undefined, indefinable, or without an obvious definition and thus having an unclear meaning.

Ambiguity is in contrast with definition, and typically refers to an unclear choice between standard definitions, as given by a dictionary, or else understood as common knowledge. Ambiguity is the use of words that allow alternative interpretations. In factual, explanatory prose, ambiguity is considered an error in reasoning or diction; in literary prose or poetry, it often functions to increase the richness and subtlety of language and to imbue it with a complexity that expands the literal meaning of the original statement. Ambiguity is distinct from vagueness, which arises when the boundaries of meaning are indistinct.

Ambivalence means simultaneous and contradictory attitudes or feelings (as attraction and repulsion) toward an object, person, or action, a continual fluctuation (as between one thing and its opposite), uncertainty as to which approach to follow.

Ambivalence means a state of having emotions of both positive and negative valence or of having thoughts or actions in contradiction with each other, when they are related to the same object, idea or person (for example, feeling both love and hatred for someone or something). The term is also commonly used to refer to situations where 'mixed feelings' of a more general sort are experienced or where a person experiences uncertainty or indecisiveness concerning something.

time norms, values and beliefs of opposite sign, equal in intensity and interdependent among themselves (Calabrò 1997). Ambivalence corresponds to situations in which choice between two opposite positions is rejected, which means that they can be seen as the ends of a continuum in which it is impossible to eliminate either the effect of one position or the other and to settle the conflict.

Introducing ambivalence in the way of thinking means hanging in alternation between two poles, in a swinging movement that produces continuous redefinition of the point of equilibrium between the two.

It is a new way of thinking about realities removed from radical choices and the dualities that always led to an opposition with consequent elimination of one of the two terms in a false dialectical synthesis.

Ambivalence determines that opposite needs be interdependent, like concern for the individual and for the community, like recognising diversity and the urge for integration. The characteristic of ambivalent judgement is having to choose between positions that are opposite but entail – both of them – negative and positive elements. The illusoriness of the concept of progress is set aside and the actors themselves are brought into play, and their ability to construct their identity amid social changes.

In contemporary society the various interests in play and the progressive differentiation of the functions of science are so intertwined as to be continually producing situations in which the actors cannot but find themselves in ambivalent positions.

Social realities and science face the individual with the impossibility of making unequivocal choices; the various alternatives are often equivalent, and the consequences only partially perceived. So it happens that, for a variety of reasons relating to both the subjects and objects of the actions, we are faced with profoundly ambivalent situations. In such cases the preconditions for a certain type of rational choice – that is, the possibility to calculate costs-benefits in such a way that only *one* expedient choice emerges – simply evaporate.

Faced with the difficulty of choosing, there are a number of possibilities: the wait-and-see strategy, the escape strategy, or coming to grips with the ambivalence. It is in fact in relation to the latter possibility that women come into the picture as subjects who have had a long experience of dealing with ambivalent situations in which there is no question of choosing one way rather than another, as for example with job or maternity. Women are more subject to diverse normative codes, both within the manifold roles they play and in relation to their responsibilities towards themselves and others. Unlike men, they are more exposed to the demands of others, and so less individualist in their choices, not because less autonomous, but because immersed in relationships.

The manifold nature of female experience is an obligation dictated by the social organisation, which expects women to be present on various fronts; these they arrive at and cross, maintaining at the same time contrasting and apparently irreconcilable styles. Today, for example, they find themselves having to govern something as concrete and individual as their own bodies, and the laws and regulations of artificial reproduction. Thus adult women have the main responsibility in activities related to everyday life – as consumers, caregivers, and also as users of a variety of technological tools. As we have seen, the great quantity of technological applications together with the difficulty of foreseeing the results drives 'science in action' to the need to find criteria for choice and 'technical decision making' that are shared by the public at large and the politicians, as well as the industrialists and investors. Society is ever driven to have to choose and evaluate, and so efforts have lately gone into promoting research and reflection on how to define the experts and 'expertise', and on extending participation in the decision-making processes to the 'lay' people (Selinger and Crease 2006).

The ceaseless repetition of concepts like 'sustainable development' and the 'precautionary principle', have shown that science involves various different planes at the same time – economic, political, cultural and ethical. Through technological interventions on food, the body and the environment, science has entered into the lives of individuals, into their identity and their survival, mostly because the individual has become directly responsible for the decisions of life, outside traditional institutions.

Into the biotechnologies are also added other profoundly individual levels associated with the functions of the female body, such as procreation, which place women again as preferential subjects of the new awareness of human beings.

In this area feminist experts should be involved in building a new 'situated knowledge' and in empowering women as crucial and informed actors in the 'game' played in terms of market interests. On the social level, science needs a collective subject that can keep together these various planes, or at least manage their reciprocal influences, over and above the creation of experts, often more involved with concrete instances and applications.

In his book *Le monde des femmes* sociologist Alain Touraine (2006) wonders whether the western world possesses any principle that might dominate particular interests and generate great initiatives and involving debate. The answer is that a principle of this type exists and that it is already exerting a mobilising action: it lies in the pursuit of reconstruction and reintegration of those elements that have been separated by the European model of modernisation, and the ecologists no-global movements are working in this direction: 'the category that plays the major role in inventing the new cultural model, profoundly contrasting with the model that has hitherto dominated our experience of modernization, is that of *women*'. It is women 'who formulate the great issues of reconciling body and spirit, past and future, private and public, interest and emotion, order and movement, and above all of women and men'.

What we ambitiously proposed with the WONBIT conference and wish to continue proposing with this book, too, is thus the creation, or, better, consolidation, of a 'scientific community' of women, carrying forward the developments that have been achieved in the last few years either with political action or through a rich feminist literature.

In this Introduction we have sought to describe the characteristics of contemporary biology, listing some of the issues and theoretical references regarding relations between women and biotechnologies, and endeavouring to construct a framework to take in the various topics proposed in the individual contributions to the book and offer some further food for thought.

Our conclusion returns to practical issues, proposing a more concrete perspective on the realities with an Afterword suggesting to policymakers what immediate developments and practical applications were needed to enhance the standing and authoritativeness of women in the area of biotechnologies.

References

Alicchio, R., & Pezzoli, C. (1988). *Donne di scienza: esperienze e riflessioni*. Torino: Rosenberg & Sellier.

Appleby, M. C. (1999). Tower of Babel: Variation in ethical approaches, concepts of welfare and attitudes to genetic manipulation. *Animal Welfare, 8*, 381–390.

Balsamo, A. (2000). Reading cyborgs writing feminism. In G. Kirkup, L. Janes, K. Wooward, & F. Hovenden (Eds.), *The gendered cyborg: a reader* (pp. 148–158). London/New York: Routledge.

Bauman, Z. (2007). *Modus Vivendi. Inferno ed utopia del mondo liquido*. Roma, Bari: Laterza.

Beck, U. (1992). *Risk society: Towards a new modernity* (p. 221). London: Sage.

Benigni, L., Menniti, A., & Palomba, R. (Eds.) (1988). Pubblici scienziati: la carriera imperfetta. *Sistema ricerca* (Suppl. 9).

Bensaude-Vincent, B. (2001). Chemical analysis: Language reform played an integral role in the development of a discipline. *Nature, 410*, 415.

Bensaude-Vincent, B. (2004). Two cultures of nanotechnology? *International Journal for Philosophy of Chemistry, 10*, 65–82.

Birke, L. (1986). *Women, feminism, and biology: The feminist challenge*. New York: Methuen.

Braidotti, R. (2002). *Metamorphoses. Towards a Materialist Theory of Becoming*. Cambridge: Polity Press.

Butler, J. (2004). *Undoing gender*. New York/London: Routledge.

Calabrò, A. R. (1997). *L'ambivalenza come risorsa*. Bari: Laterza.

Damasio, A. R., Harrington, A., Kagan, J. et al. (Eds.) (2001). Unity of knowledge: The convergence of natural and human science. *Annals of the New York Academy of Sciences, 935*.

Donini, E. (1990). *La nube e il limite*. Torino: Rosenberg & Sellier.

Doull, J. (1984). The past, present, and future of toxicology. *Pharmacological Reviews, 36*, 15S–18S.

Doull, J. (2003). The "Red Book" and other risk assessment milestones. *Human and Ecological Risk Assessment, 9*(5), 1229–1238.

Fantone, L. (2003). *From dissection to digital genetic maps: Representations of the (female) body and the political implications of genetic research*. Paper presented at Gender and Power in the New Europe, the 5th European Feminist Conference, Lund University, Sweden, August 20–24, 2003.

Fausto-Sterling, A. (1992). *Myths of gender*. New York: Basic Books.

Fraser, D. (1999). Animal ethics and animal welfare science: Bridging the two cultures. *Applied Animal Behaviour Science, 65, 171–189*.

Gadamer, H. G. (1983). *Reason in the age of science*. Cambridge, MA: MIT Press.

Gagliasso, E. (2001). *Verso un'epistemologia del mondo vivente* (pp. 216–217). Milano: Guerini Studio.

Gane, N. (2006). When we have never been human, what is to be done? Interview with Donna Haraway. *Theory Culture Society,* 23, 135–158. DOI: 10.1177/0263276406069228

Haraway, D. (1989). *Primate vision: gender, race, and nature in the world of modern science*. New York: Routledge.

Haraway, D. (2003). *The companion species manifesto*. Chicago, IL: Prickly Paradigm Press.

Harding, S. (1986). *The science question in feminism*. Ithaca, NY: Cornell University Press.

Harding, S. (Ed.) (1993). *The "racial" economy of science: Toward a democratic future*. Bloomington, IN: Indiana University Press.

Harding, S. (1998a). *Is science multicultural?*. Bloomington, IN: Indiana University Press.

Harding, S. (1998b). Women, science and society. *Science, 281, 1599–1600.*

Hobsbawm, E. J. (1994). *Age of extremes: The short twentieth century 1914–1991.* New York: Pantheon Books/Random House.

Hubbard, R. (1990). *The politics of women's biology.* New Brunswick, NJ: Rutgers University Press.

Keller, E. F. (1983). *A feeling for the organism.* San Francisco, CA: Freeman.

Keller, E. F. (1985). *Reflections on gender and science.* New Haven, CT/London: Yale University Press.

Keller, E. F. (2000). *The century of the gene.* Cambridge, MA: Harvard University Press.

Keller, E. F. (2005).The century beyond the gene. *Journal of Biosciences, 30*, 3–10.

Kelly, K. (1998). The third culture. *Science, 279*(13), 992–993.

Kiessling, A. A. (2001). In the stem-cell debate, new concepts need new words. *Nature, 413*, 453.

Latour, B. (1987). *Science in action.* Cambridge, MA: Harvard University Press.

Levitt, N. (1999). *Prometheus bedeviled: Science and the contradictions of contemporary culture.* New Brunswick, NJ: Rutgers University Press.

Malmfors, T., & Rosing, H. (2002). Introduction–risk from philosophy of science point of view. *Toxicology, 181/182*, 109–113.

Merchant, C. (1979). *The death of nature.* New York: Harper & Row.

Merton, R. (1973). *The sociology of science.* Chicago, IL: Chicago University Press.

Merton, R. K., & Barber, E. (1963). *Sociological ambivalence.* New York: Columbia University Press.

Mitroff, J. I. (1974). Norms and counter–norms in select group of the Apollo moon scientists: A case study of the ambivalence of scientists. *American Sociological Review, 39*, 579–595.

Mohanty, C. T. (1997). Women workers and capitalist scripts: Ideologies of domination, common interests, and the politics of solidarity. In M. J. Alexander & C. T. Mohanty (Eds.), *Feminist genealogies, colonial legacies, democratic futures.* New York: Routledge.

Nowotny, H. (2005). *Unersättliche Neugier: Innovation in einer fragilen Zukunft.* Berlin: Kulturverlag Kadmos.

Nowotny, H., Scott, P., & Gibbons, M. (2002). *Re-thinking science: Knowledge and the public in an age of uncertainty.* Cambridge, MA: Blackwell.

Porter, D. G. (1992). Ethical scores for animal experiments. *Nature, 356*, 101–102.

Rosser, S. V. (2000). *Women, science and society* (pp. 46–47). New York/London: Teachers College Press.

Rotblat, J. (1999). A hippocratic oath for scientists. *Science, 286*, 1475.

Sayre, A. (1975). *Rosalind Franklin and DNA.* New York: W.W. Norton.

Schiebinger, L. (1989). *The mind has no sex? Women in the origin of modern science.* Cambridge, MA/London: Harvard University Press.

Schiebinger, L. (1999). *Has feminism changed science?* Cambridge, MA: Harvard University Press.

Scienza e potere, coscienza del limite (1986). *Quaderni di Donne e politica* (Suppl. 5).

Selinger, E., & Crease R. P. (2006). *The philosophy of expertise.* New York: Columbia University Press.

Shiva, V. (2001). *Protect or plumber? Understanding intellectual property rights.* London: Zed Books.

Sloviter, R. S. (2002). Apoptosis: A guide for the perplexed. *Trends in Pharmacological Science, 33, 19–24.*

Snow, C. P. (1959). *Two cultures and the scientific revolution.* Cambridge/London: Cambridge University Press.

Touraine, A. (2006). *Le monde des femmes* (p. 190). Paris: Fayard.

Wajcman, J. (2004). *Technofeminism* (p. 81). Cambridge: Polity Press.

Wilczek, F. (2001). When words fail: Scientists have to struggle with words that don't fit reality. *Nature, 410*, 149.

Ziman, J. (1996). Is science losing its objectivity? *Nature, 382*, 751–754.
Ziman, J. (1998). Why must scientists become more ethically sensitive than they used to be? *Science, 282*, 1813–1814.
Ziman, J. (2000). *Real science: What it is and what it means*. Cambridge: Cambridge University Press.

Part I
Women Scientists in Biotechnological Research

Chapter 1
Heading Blithely Down the Garden Path? Some Entry Points into Current Debates on Women and Biotechnologies

Wendy Harcourt(✉)

Abstract: This chapter aims first to identify some of our reasons for wanting to speak about 'women and biotechnologies' in the current debates on biotechnology. Secondly it reports on issues related to women's status in biotechnology today in Europe in research, academia and industry. And thirdly it reflects on the link between women and biotechnology in Europe and the global policy debate on biotechnologies.

Keywords: Feminist, gender, biotech researchers, environment, development, poverty

'We need a careful analysis of the benefits and risks in biotechnology, particularly in the relationship between science and business. The pace of scientific advance always threatens to outstrip regulation. We need to stand back and analyse the legal and ethical problems raised before we proceed blithely down the garden path.' Donna Dickenson, Emeritus Professor of Medical Ethics, University of London.[1]

1.1 Introduction

Women and biotechnology is a vastly complex subject ethically and politically as well as scientifically, requiring a multidisciplinary approach that takes into account different viewpoints, if not new knowledge frameworks. The Women in Biotechnology (WONBIT) conference held in Rome 21–23 June 2007 boldly brought together the cutting edge of scientific knowledge and practice, feminist studies and European policy-making on women and biotechnology. Biotechnology is a frontier area in

Wendy Harcourt
Society for International Development, via Panisperna 207, 00184 Rome, Italy
wendyh@sidint.org

[1] Personal communication by email March 6, 2007.

scientific development and its importance spans ethical, environmental and economic issues as it re-designs the possibilities of transforming life.

Our aim was to open up discussions about this frontier in relation to both the findings of scientific research and the institutional activities that inform research policies and decisions. It responds to the sense of public uncertainty about biotechnology as a science that is promoting radical change to the biological cultural heritage of humanity. Most of all it tries to promote a dialogue among the people doing science and the people who must deal with the cultural, socio-political and economic impacts of science. Scientific language is often incomprehensible to non- scientists and the critique of science engages a rhetoric that scientists can find alienating.

The stated aim of the WONBIT conference was to promote an interdisciplinary and broad debate among feminists and scientists on biotechnology. The ambition was to map out a common ground of concern on 'women and biotechnologies' in order to 'bridge the communication gap among science and feminism and among different feminist approaches to science and, in particular, to biotechnologies'.

The four areas chosen for discussion indicate the breadth of the debate WONBIT was aiming to cover:

1. Women scientists in academic and industrial biotechnological research
2. Bodies, cultures, scientific metaphors
3. Environmental effects of biotechnologies
4. Facing impact: society and biotechnologies

Within these four areas there are two main threads that one would expect to find in a feminist analysis of any knowledge or discipline which essentially is concerned with gendered power relations. One is the status and power of women as actors in biotechnical research and policy making. The other is how biotechnological research and policy is impacting women's lives. A third, which is specific to biotechnology, is how the manipulation of life by biotechnology creates particular fears or hopes at a very fundamental level where we define the boundaries of life, humanity and nature.

In this overview paper I outline some of the debates informing the discourse on women and biotechnology being held by women policy makers and scientists working in biotechnological research and the feminist movement in Europe and the Global South concerned with the impact of biotechnology on women's bodies, lives and society at large. I am deliberately not addressing the broader bioethical or religious arguments for or against biotechnologies, nor am I able to address the scientific scope of any particular biotechnological study.

The core theme of this paper is the tensions within and among women scientists and feminist debates on biotechnology. It maps out some of the questions being tackled by European women who are both scientists working in biotechnology and feminists who are critiquing biotechnology, as well as by women working in the European scientific policy arena. In doing so, the paper suggests ways in which we can move beyond the rhetoric both of scientific neutrality and feminist political standpoints in order to promote an innovative dialogue that will help us navigate the vastly complex field of women and biotechnologies.

1.2 Why Women and Biotechnology?

Biotechnology is, as a foundational thinker in this field Patricia Spallone[2] described it, 'everywhere these days'. Biotechnologies are conditioning the quality of life in medicine, agriculture, animal husbandry, zootechnical production, the pharmaceutical industry, energy and the environment; in short it permeates all areas of our lives. Biotechnology is found in all scientific disciplines, and many areas of industrial research and production; in addition it has powerful social ramifications with a strong public policy concern. Our topic, 'women and biotechnology', therefore is relevant to science, industry, and social policy and development issues, as well as to gender and women's studies, and is subject to feminist activism and critique. Given this vast field, what are our working assumptions?

We need to look at women and biotechnology from several vantage points. One is the issue of women in professional careers, for example, the numbers of women involved in biotechnology as research scientists and in industries. Here our focus is on the role and status of women biotechnologists, the women professors and students in academia and women technicians as well as the women decision makers working on national and transnational government policy on biotechnology. The main interest here is in gender relations and equity issues related to professional recognition and rewards. A second focus is on the new knowledge being created by feminist social scientists' analyses of biotechnology. Here we raise questions about biotechnology as a science and practice from a social justice, gender and ethical perspective, as seenfrom various disciplinary positions. Thirdly there is a political focus on women and biotechnology within a broader framework which looks at the economic, social, political, environmental and cultural implications of biotechnology for women both in Europe and globally.

All three approaches start from the assumption that we need to find and encourage biotechnologies that promote and improve the quality of life for all with priority given to ensuring that gender difference is acknowledged and understood in biotechnology. We need to understand as thoroughly as possible the social and political consequences of genetic modification of the genetic heritage of living organisms. The focus, I would propose, is therefore on women and their role in the control, design, access and use of biotechnologies as well as on understanding the impact of biotechnologies on women biologically, medically and socially. Another assumption is that if women are well represented in biotechnology as professional scientists, as decision makers in relevant policy arenas and as leaders in the social dialogue with

---

[2] Pat Spallone worked as a biochemist in medical research before entering her current area of interest in social studies of science and technology. She has written widely on the social, ethical and public policy implications of reproductive technologies and genetic engineering, particularly developments of In Vitro Fertilisation, human embryo research, genetic testing, behavioural genetics, and genetic engineering in agriculture. She has investigated these in relation to global development concerns, and women's health and reproductive rights (Spallone 1989).

other ethically concerned social scientists and opinion makers, then we are moving in the right direction towards gender, social and technological justice.

These approaches also assume that we see biotechnologies as non-neutral. It therefore follows that we need to ensure that the conception, development, dissemination and application of biotechnologies are held to ethical democratic standards which ensure gender equality. In order to do this we need not only to be technologically and scientifically able to understand and engage in biotechnology, but also to understand policy implications and to have a big picture vision that gauges the implications of biotechnologies for communities, environments, bodies, food, work and safety. In building this broader context it is important to see the economic interests at play, national interests as well as private interests of medical and pharmaceutical technologies, conscious of possible gender inequalities and gaps in how and where biotechnological research is happening. We therefore need to have a dialogue about the context in which biotechnology is practiced here in Europe and have more information on the effects of biotechnologies on women's bodies and our environment, as well take a broader global responsibility as European scientists and social scientists.

Let me start by sketching three main debates that are framing the women and biotechnology debate: (a) the international policy debate on biotechnologies from a gender perspective, (b) feminist scientists and social scientists engaged with biotechnology, and (c) feminist advocates on women and biotechnology.

1.2.1 International Policy Debate on Biotechnologies from a Gender Perspective

Understanding of the logic of exclusioninclusion of women in the sciences is crucial for development of effective strategies to achieve gender equity, equal partnership and the sustainable development of science. Quoted in UNESCO (1999).

The UN and different international research centres studies such as the Institute of Development Research Canada (IDRC)[3] have been pioneering international policy discussions on women and biotechnology in both industrialized and developing countries. Their aim is to put in place a gendered international policy response to the myriad of practical applications of biotechnology in many areas (health care, pharmaceuticals, agriculture, agro-food system) that have produced new products and adapted existing ones. Their approach is to ensure that women as both creators and users of biotechnologies participate in the debates on the new developments of biotechnology. These policy debates since the late 1990s have aimed to review policy on biotechnology from a gender perspective in order to ensure that there is a responsible policy environment that is able to respond to the complex set of research, transfer and product marketing cycles of biotechnology.

[3] See OECD (2006), IDRC (2005), and UNESCO (1999)

The policy debate places biotechnology within a regulatory and behavioural context, arguing that much more scrutiny is required about how biotechnology is developing, particularly given that decisions currently being made about biotechnology are in the hands of a very male environment operating at the high echelons of scientific research and industry.

The recommendations coming out of these international discussions are on several levels. One is to define the policy and methodology requirements to allow women scientists working in biotechnology to play an active role in identifying and defining ethical issues related to biotechnology from a gender perspective. This includes: defining what research questions need to be studied and what avenues pursued; deciding who defines the priorities and on what basis (scientific curiosity or human good or commercial gain); and agreeing in genetic research if people have the right or duty to know or not know their genetic make-up, whether or not positions should be codified or left up to individual choice (for example the legal ruling that only one or two embryos can be implanted In Vitro Fertilisation). Another is to ensure that women scientists take up important positions in biotechnology fields in order to participate in educating politicians, decision makers and the general public about the limitations of science and its ability to understand the complex interrelationships between political culture and science (for example the human genome project) and the threats that exist today to science from economic interests in research. Thirdly they argue for the need to understand how trade and international agreements and intellectual property rights impact women's ability to manage and benefit from their knowledge of biodiversity and the use of biotechnologies. And fourthly, within industry, policies are required to look at how to develop gender approaches to bioethics and health commercialisation of biotechnologies and the rise of women entrepreneurs in white and green biotechnologies.

The studies promote the need to change the conception of science itself so that it becomes part of an overtly politically process aimed to eliminate what UNESCO calls 'intolerables' in the pursuit of sustainable development. These intolerables include the problems of reproductive health, poverty, environment, degradation and violence along with a commitment to a gendered understanding of these problems. UNESCO argued for a new approach to knowledge and learning processes that will responded to innovation and change in science and technology as well as the diverse realities cultural contexts and gender needs:

> Women cannot be excluded from the cognitive heritage of humankind either in the realm of scientific knowledge or more significantly as agents in the construction of science and its understanding. In the new understanding of science and values, the holistic approach to science engages with the need to incorporate a gender vision to its redefinition and commitment. In the changing understanding of scientific and technological process, women in science contribute crucially to a different understanding of the social image, content and the role of science and technology. The changing paradigms in scientific understanding require incorporation of the diverse expressions of women's voices, concerns and scientific expertise ... the egalitarian model of human development and sustainable scientific research and knowledge requires gender equality and quantity in its commitment (paragraph quoted in UNESCO 1999).

The discussions and papers produced through UNESCO studies underline the importance of higher level education in science and technology to encourage girls and for more women to be engaged with management responsibilities at all levels. In order for this to be done there needs to be a scientific multidisciplinary culture, the end of gender inequalities in education and training and a change in the unbalanced distribution of domestic responsibilities for women. UNESCO argues:

> Authorities, administration and management must stop the gender brain drain [….] And adopt equal opportunity policies based on scientific criteria in order to benefit the community and all humanity' ending the explicit and implicit barriers to women in biotechnology both in research and in policy and decision making fields including economic and financial fields.

UNESCO also underlines that scientists in biotechnology need to take into account the ethical issues related to biotechnology as well as recognize the threats that exist today to excellence in science from economic interests in research and indicates that in Europe there is considerable diversity that impacts a gendered understanding of science and technology.

1.2.2 Feminist Scientists and Social Scientists Engage with Biotechnology

Social and humanities research can meet scientific research on equal terms. One is not 'prior' to the other. Patricia Spallone.[4]

There are new forms of engagement with the biological sciences from women scientists that break down some of its empirical rigidity and acknowledge the non-neutrality of science.[5] There are also examples of how feminists are engaging in productive relationships with science that hopefully will produce new feminist thinking and theorising forward on biotechnology that builds from Donna Haraway (1991) and Rachel Carson (1962), two feminist thinkers who have led the way in productive, particularly Northern feminist, engagement with biological, medical science and technology.

In this debate around biotechnology we find the whole area of biotechnology discussed in a highly innovative and challenging way. One argument, for example among feminists looking at the impact of biotechnology on disability, is that we need to unsettle the concept of progress and improvement within science and medical science that informs biotechnological research and modern society by questioning stereotypes of perfection. In defence of difference they propose a vision based on who we are rather than a vision of what we might be (embedded within a particular concept of what is perfect). They question assumptions about perfection and

---

[4] Personal communication by email 2 February 2007.

[5] For example, Franklin (2007).

women's bodies in relation to genetic and other invasive technology and the expanding market of research into new biotechnologies in the pursuit of a better life through technical fixes that could fundamentally and permanently change humanities and the environment (Rockhold 2006; Hurst 2006).

In this debate major concerns are raised around how to monitor the research, issues of transparency, commercialization ethic review boards and how to nurture a critical consciousness. To quote Boscoe and Tudiver (2000), 'Biotechnology research has transformed the impossible into the possible. What appeared unthinkable several years ago is now considered inevitable: buying and selling genetic materials; patenting life forms; specifically, genetic transfer; selling Iceland's gene pool for research purposes'.

A very strong voice in this debate is the concern at the risky and invasive genomics based biotech interventions where women's bodily tissues, particularly, ova are at the centre of industry's planned development of *in vitro* red (medical) biotechnology – an embryo based genomics industry that promises to provide products – enhancements to genetically inheritable characteristics.

Tyson Darling (2006) points out that it is important to consider the gendered dimensions of 'red biotechnologies'. This includes the use of women's bodies to develop the most recent generation of new human genetic and reproductive technologies which will alter human reproduction and move it into the realm of the medical-industrial complex. She argues that this move 'essentially situates the experimental use of women's bodies (especially their reproductive organs and tissues like ovaries, ova, wombs and embryos) – through the aggressive marketing of assisted reproductive technology and pre-conception genetic testing options – as processes, materials and products of a male-dominated medical-industrial production process'.

On the one hand, women's bodies are essential to provide the raw materials for the 'production' of genomics based processes and products, while on the other women are sought after as consumers of emerging biotechnologies. The use of women's bodies – the biological processes and interactions of the cells, tissues and genes of women's bodies, and the chemical processes of women's body's hormonal and gestational processes – provide the raw materials of red biotechnologies.

To quote in full from Tyson Darling (2006):

> Privatizing human DNA and commercializing genetic processes intrinsic to women's bodily reproductive functions (including cells, tissues and genes) is an important, indeed pivotal step in usurping women's bodily integrity … Cultural and societal mores have sustained pressure on women to engage in biological reproduction, and women's bodies have been increasingly subjected to experimental and risky procedures. A new consumerism heavily laden with eugenics propaganda is being marketed and directed towards women and the social norms and values expectations that affect women regarding motherhood … plans that would privatize human DNA sequences and processes as property, threaten to require reformulation of the status of women's bodies, weakening of bodily integrity and the denigration of the moral status of women's body parts to the level of replicable processes, raw materials, spare parts … We face serious ethical and equity challenges to protect democratic processes relating to reproductive autonomy, social justice, women's human rights and women's bodily integrity.

1.2.3 Feminist Advocacy Response

At the forefront of feminist responses to biotechnology are women activist groups which are concerned about how the rapid changes by the new science and technologies are changing gender equality or women's rights.[6] They ask questions about increasingly proficient genetic technology as well as the underlying racism and eugenics in some of these practices. They question the drive behind the industrial and medical interest in these technologies, querying possible genetic manipulation and the push for 'designer babies', they also point to the 'virtual holocaust of girls' in Asia where access to sonograms and amniocentesis has allowed families to carry male fetuses to term and terminate female ones.

At the feminist advocate level the concern is how biotechnology is changing the way women experience the world, the choices they make and the work they do (AWID 2004a, b, c) including on how bio-nanotechnologies are increasingly assumed to be key components of women's life and livelihoods in relation to health (including reproductive health) agriculture, food security and the environment. And how biotechnological research is incorporated into social and economic production and marketing processes through government policy, trade and aid relationships.

The use of genetically modified food is one of the most contested areas of women in environment movements in both the North and South. For Northern consumers the blurring of lines between medicine and food with use of terms like functional foods and neutraceuticals, enticing food processors agbiotech firms and drug companies to merge complementary interests in food, biotechnology and pharmaceuticals, are of increasing concern (Delahanty 2000).

As for Southern consumers, feminist advocates are raising concerns about genetically transformed food and agricultural systems and the long-term effects on environment, biodiversity and human health as well as the benefits to women subsistence farmers in developing countries. They point out how new high yield crops often displace subsistence farms, putting their livelihoods and their families into jeopardy. With the industrialization of agriculture more women perform hard labour in the fields and are exposed to chemicals and pesticides specific to GM crops. The gendered implications of their livelihoods and health issues are impacting poor women's health, food and security rights. Feminist advocates argue that such genetic engineering is not easing the food insecurity of poor communities but rather helping prosperous larger-scale farmers and harming natural environments and magnifying hunger by shifting to cash crops. They propose for example that land reform would do more than technology to alleviate hunger.

Nano-biotechnology (exploring the merging of living and nonliving at the nano scale) is an area of intense investment and research and development. Feminist advocates raise the concern that the effects of these extremely small synthetic particles on environment and health are still not fully understood. Nor are the ethical implications of the fact that at the nano scale all matter is the same. They hold that

[6] See Jones (2005); Kerr et al. (2004); Vargas et al. (2006).

the implications of scientists being able to construct atom by atom new structures with new characteristics and functions are profound.

A specific problem for feminist advocates entry into biotechnology debates is the perceived level of expertise and specialization that is required or even 'technophobia'. The ability to dialogue with women scientists who are engaged in the knowledge production of biotechnologies is therefore very important.

1.3 Women and Biotechnology in European Policy, Research and Industry

1.3.1 The European Public and Biotechnologies

It is important to situate our debate within the European context. Biotechnology is emerging as a key policy issue for the European public. There are ongoing public debates on stem cell research, the co-existence of GM conventional and organic farming, the use of genetic information, innovations in nanotechnology and pharmacogenetics, reproductive technologies. All of these topics are regularly discussed in the media. The most recent Eurobarometer survey (64.3) 'A Report to the European Commission's Directorate-General for Research, May 2006' (Gaskell et al. 2006) presents an interesting review of public perceptions of a range of biotechnologies covering medical (red), industrial (white) and agricultural (green) biotechnologies in 25 European countries. The report is the sixth in a series undertaken for the Commission by George Gaskell at the London School of Economics. The European scientific team works with survey questionnaires that monitor patterns and trends in public perceptions. The Eurobarometer Surveys aim to bring public voices to policy makers, representatives of industry, journalists, scientists and social scientists, and also to the public.

The research team reports that in general European citizens are optimistic about biotechnology and are relatively well informed about biotechnology innovations and the promise to deliver benefits in health, agriculture and foods in industrial production. On the other hand the survey indicates public concern that what is scientifically possible is not always socially, ethically or environmentally desirable. The strong European public resistance to GM food the study suggests is not 'a manifestation of a wider disenchantment with science and technology in general'. They surmise that overall there is a utilitarian approach to biotechnologies with most Europeans not averse to taking risks and willing to delegate decisions on biotechnologies to experts based on scientific evidence. Though, on the other hand, there is a substantial minority that would like to see greater weight given to moral and ethical considerations.

They conclude that as a whole Europeans are generally optimistic about the contribution of biotechnology to a European way of life. The majority support the development of nanotechnology, pharmacogenetics and gene therapy and the industrial application of biofuels, bioplastics and bioplants for pharmaceuticals. Support for genetic data banks are not as clear, 58% are willing to support genetic banks for research but access to genetic information for governance and commercial purposes are perceived by the majority as unacceptable.

One of the most controversial topics of stem cell research has European support. Providing it is regulated, there is 59% support for embryonic stem cell research. Interestingly in the survey the highest support for stem cell research came from Belgium, Sweden, Denmark, Netherlands and Italy. Low approval is in the Baltic States, Slovenia, Malta, Ireland and Portugal. The report underlines that 'although it might be expected that the differences would be greater in those countries where the public is more familiar with the stem cell issue, or in the countries where the national religion is Roman Catholicism, the findings do not support these expectations'. The majority 53% of Europeans think that stem cell research is valid provided the potential benefits and possible risks of embryonic stem cell research are clarified and regulation and ethical oversight is sufficient (Gaskell et al. 2006).

The survey addresses the dilemma of moral versus utilitarian arguments around stem cell research and concludes that, though there is a divide, Europeans lean towards the contingent view that the possible benefits for health and the alleviation of disease cannot be ignored. Interestingly the report adds that what counts most is scientific knowledge yet, 'Europeans generally do not consider it important to be informed about scientific details, perhaps because they are content to leave these to the experts. They want to know the societal consequences of stem cell research the risks and benefits and whether regulations and ethical oversight is sufficient' (Gaskell et al. 2006).

In relation to regulation and governance what emerges is that 60% of Europeans want scientific criteria to inform decision making. In relation to how decisions should be made in governing biotechnology, there are strong differences here among the 25 European countries with 72% in Hungary supporting science and only 36% in Austria, which favours moral delegation (see the section on Principles of Governance across Europe, in Gaskell et al. 2006).

Along the same line of enquiry it is also interesting to see who the public trusts in making decisions involving biotechnology.

In relation to public confidence in the actors who create and regulate biotechnology research scientists, industrial scientists, industry and national and European regulators, there is a high level of trust in university and industry researchers in the EU, with the exceptions of Sweden and Germany. It is interesting to note that many of the new member states are amongst the most confident in the biotechnology system (see Fig. 1 in Gaskell et al. 2006).

The report suggests it is important for scientific policy to ensure that moral and ethical considerations and public voices inform discussions and decisions. In a comparison among Europeans, Canadians and US publics, the report emphasizes that Europeans are optimistic and interested in biotechnology including nanotechnologies (which in Canada is far less accepted); with the exception of GM food, they are confident in the use of biotechnology as long as regulation is in place. They postulate that there are four sorts of European publics: the active

10% (who have heard about biotechnology and probably searched the internet and attended a meeting about it) the attentive 15% (who have heard about biotechnology and have textbook knowledge about biology and genetics) the spectator 35% (who have read about it and might have talked about it) and the unengaged 40% (who have not read or heard about technology and have little knowledge about biology and genetics).

They go further and divide European countries into different modes of engagement: Nordics and the Dutch are attentive, Germany, Italy, Spain and Austria tend to be more active, Slovakia, Slovenia, France, Belgium and Luxembourg are spectators, Hungary, Greece, Ireland, UK, Lithuania, Poland, Cyprus, Malta, Estonia, Latvia and Portugal are all closer to unengagement (see Fig. 28 in Gaskell et al. 2006).

Unfortunately their questionnaire did not consider the data by sex so that the very interesting responses and observations are not as useful as one would have liked for WONBIT with our specific interest in women and biotechnology. Rather than streaming gender throughout to measure gender differences, there is just one short section at the end of the survey on women and science.

This section measures women's interest in science and technology, concluding that there is a strong difference between men and women's interest in science (16% of women are often interested as opposed to 36% of men). In relation to questions that measure respondents' knowledge on biology and genetics, men score higher than women, correctly answering overall 5.4 as opposed to 5.0 for women. However on the question related to pregnancy women score higher. Women and men approve more or less the same on genetic engineering, nanotechnologies and pharmacogenetics but women are considerably more concerned about GM food. The research points to strong attitudinal differences and to distinct differences in engagement with biotechnology, particularly men in higher education seem more interested than women in biology and genetics. They concluded that typically 'women are less interested in and less supportive of science and technology'.

This survey raises some important concerns in support of our approach. It is important to note a general interest and acceptance of biotechnology across European countries and that the public in general is looking to scientists to guide the debate on both what is risky and what therefore needs to be regulated. The focus though is not on detailed scientific knowledge but more about the societal implications, the risks and benefits for the public. Science is seen as outside public scrutiny (even beyond religious concerns) but decisions made around biotechnologies, while based on scientific knowledge, need to be framed within ethical standards which have to be more open to public scrutiny. The failure of the survey to look at gender as a cross-cutting concern and to limit questions to women and science indicates the importance of a gender-aware design in surveys in general. It is important to note that though the conclusion was that women were less interested in bioscience and technology than men, the point that women were interested in biotechnologies that they perceived to be related to their own body (i.e., pregnancy) suggests that it is important for women to understand better how biotechnologies impact their lives. It also suggests that there is a gender difference in the time available to devote to understanding complex science and technology debates.

1.3.2 Women and Science Policy in Europe

In seeking to collect some relevant facts, such as figures on women working in biotechnology today, I found empirical information on the numbers of women working specifically in biotechnology research and industry very hard to come by. Some of the figures are being collected now by the Organization for Economic Cooperation and Development (OECD), Elettra Ronchi (who presented a key paper report to the WONBIT conference) mentions that, surprisingly enough, the OECD has only recently started to collect biotechnology statistics but that the statistics gathered through OECD governments are following a standard approach, which does not actually include a breakdown by gender. In relation to women and science, OECD findings to date are that women are concentrated more in biology, health, agriculture and pharmaceuticals and at the lower level of the scales only 20% of women are among the senior staff in the EU (OECD 2006).

UNESCO[7] has fostered an important discussion, specifically on Women and Biotechnologies in the context of the Mediterranean and the future of science, engineering and technology in the late 1990s involving all countries in the European region, in order to establish policy that would achieve equality for women scientists and integrate a gender perspective into sciences.

The findings are based on studies of the diverse situations of women in science in Europe. The studies indicate that even if there is an increasing trend of women scientists undertaking training in higher studies, there is a lack of women faculty members and at top management positions in both education and research institutions as well as at the ministerial level. UNESCO found that relative to other parts of Europe (apart from the Nordics) East and Central European countries had far more women scientists integrated into science institutions. However the contextual analysis showed that numbers did not necessarily reflect gender equality as the women were still in the minority in higher management positions. The conclusions underline the need for not only better statistics in each country but also stronger networking among women scientists as well as collaboration among UNESCO, European Union and OECD women and science units. These strategic processes are needed in addition to policy agreements at national and international levels in order to shift the dominance of a patriarchal culture built on the subordinate position of women and androcentric models which determined gender codes and values that deterred women from entering science, and science from responding to women's needs.

Considerable knowledge and statistics have been collected since 1998 by the European Commission on the position of women scientists in Europe in research and industry, which includes women working in life sciences and biosciences and technology. This section relies on the data collected by the European Commission unit for Women and Science. According to women involved in setting up the unit and pushing for the data, the statistical material was gathered after a long campaign

---

[7] See Introductory note to the science agenda: Framework for action: version 15.06.99. Retrieved October 20, 2007, from www.unesco.org/science/wcs/eng/intro_framework.htm

by women social and natural scientists and technologists who managed to secure funding for a good feminist social statistician[8] (Rees 2005).

In Europe since the 1990s there has been considerable professional effort to analyze gender in science as well as to measure and understand the situation of women scientists engaged in science in research and industry (and within this biotechnology). The European wide studies have aimed to gauge the considerable inequalities between women and men in the research and development of biotechnology in academia, policy and industry.

The ENlarge Women In Science to East (ENWISE) expert group set up among women in the European Union and Baltic States has examined women in science, finding that most women are employed in areas where R& D are the lowest.

Gender policy, child care, legal protection and support for working mothers have been reduced so that, given the lack of funding and rigid patterns of promotion as well, the prospects for young scientists are very bleak. This situation is leading to a brain drain of women scientists from the EU and Baltic States to other parts of Europe.

A key policy body in relation to women and science as a policy focus is The Helsinki Group on Women and Science established in November 1999 by the European Commission to bring together national representatives from all the EU Member States and countries associated to the Science and Society in the European Research Framework programme. The aim of the Helsinki group is to ensure a correct gender balance in science by keeping women constantly in the research policy sights. Maintaining such high visibility requires long-term monitoring and expert advice provided by the Helsinki Group on Women and Science. The Group aims to promote participation and equality of women in the sciences on a Europe-wide basis. They meet twice a year and provide an important forum for dialogue about national policies. Recognising the value of networking and mutual support among women scientists, the group also helps explore the ways in which the potential, skills and expertise of women could best be secured and for sharing and comparing experiences.

The Helsinki Group helps the Commission build a clear picture of the situation on the ground at the national level. It has, in particular, appointed national statistical correspondents to help the Commission gather and compile sex-disaggregated statistics and build gender-sensitivity indicators. The national reports on the situation

---

[8] *Women in scientific careers: unleashing the potentials* 2006 and *Women entrepreneurs in SMEs: realizing the benefits of globalization* and *the knowledge-based economy* report of the OECD and *Women in Industrial Research: Good practices in companies across Europe*, *She Figures: Women and Science* 2006 point to an acknowledged and growing need to understand the role of women in science, including bioscience and technology and the type of policy environment required to engage more women. The Women and Science Commission Staff Working Document (EUR 21784) looks at the activities of the EC to promote gender equality in science since 1999. *Women and Science: Excellence and Innovation: Gender Equality in Science* 2005 gives a detailed account of the milestones passed to reach this level of information and consequent policy making to increase the participation of women in the scientific community and to encourage research bodies, funding institutions and the scientific community at large to think in a more systematic way about promoting research environment that overcomes gender bias.

of women scientists are a synthesis reviewing the scientific infrastructure, equality measures and the climate for women wanting to pursue scientific careers and the level of gender balance in decision-making bodies as well as the success of gender mainstreaming in the systematic integration of gender equality in all policies and programmes. The reports underline the importance of sex aggregated statistics, as the studies show that strong sex segregation is a feature of scientific careers. For example women constitute the majority of undergraduates in medical and biological sciences but even in these fields the nearer the top of the academic hierarchy the fewer the women.

The Helsinki Group's examination of gendering of science and scientific excellence has led to an important discussion in European policy on the use of patronage and nepotism in scientific appointments' procedures, the social construction of scientific excellence and the measures used to exclude women by scientific elite bodies as well as the importance of engendering and modernising human resource management processes in science and the need for gender proofing science education along with the need for more gender studies to ensure that happens. The Helsinki Group repeatedly has called for European scientific research to integrate the gender dimension in new and emerging areas of science. (The Helsinki Group on Women and Science 2005)

The Gender Watch System monitors progress of EC women in science framework programmes in order to ensure that the gender gap in research and development is closing, monitoring gender mainstreaming across the European research framework programme and the increasing knowledge base of women in science (European Commission 2005).

Gender studies and gender research are emerging in many countries though there are differences among member states, particularly as resources and infrastructure for gender research are lacking in new EU member states (in European Commission 2005). The glass ceiling is still in evidence but there are important advances among younger women graduates with an increase in the number of women scientists graduating.

Another professional group that acts as a watch dog for gender and science in Europe is the European Platform of Women Scientists (EPWS) formed in 2005 as a group of women scientists and organisations committed to gender equality in scientific research. The group is set up to monitor and promote gender equality by providing a forum in which to discuss issues of scientific excellence, examine the gender dimensions of scientific projects and prepare position papers and host expert groups on women and science. The group also looks at measures to facilitate the balance of women scientists' work/private life. According to members of the group they see that there is progress but it is still patchy. In their 2005 report they acknowledge that the main challenges continue to be to empower women in decision making in research and technology, to relieve the pressure on women to be visible and to progress in their careers, to adjust the inequalities in career advancement and significant differences in dropout rates. In addition they aim to challenge the existing system's way of defining and evaluating scientific excellence to show it is not gender neutral. They also call attention to the differences within Europe.

The EU enlargement and changing roles and life plans for women scientists in Europe has major changes and impacts on scientific research, for example gender analysis and research exists in some countries but not in other countries. As state and market forces shift in newer EU member countries it is hard to convince women to stay in industrial research and higher education in science engineering and technology.

The stated future priorities for women in science in future research in the EC (European Commission 2005). This research agenda includes:

- Improve scientific excellence by promoting gender awareness and fairness
- Boost women in leading positions
- Strengthen gender research and the gender dimension in research in new areas of scientific research such as nanotechnologies
- Ensure Gender budgeting across programmes
- Give Gender research a clear budget item
- Increase women's engagement in innovation, entrepreneurship and patent creation
- Reconcile professional and private life (for a healthier work life balance)
- Ensure maternity and parental leave

1.3.3 Women and Science in Europe, Statistics and Indicators

She Figures 2006, published by the European Commission Directorate General for Research, gives some of the figures on women in science in European research, government and industry. *She Figures* 2006 is the broadest collection of European data on a range of gender sensitive indicators developed to measure and compare success rates of women and men in obtaining senior positions in research and their access to scientific research and development in the tertiary sector. It is designed as a benchmarking tool gathered from 17 European states. The data comes from Eurostat and was commented on by the Helsinki Group of Women and Science led by the Women and Science Unit of the Directorate General for research. The aim of the publication is to highlight the gender imbalance in tertiary level science in Europe. The data shows that women's participation is dramatically low in certain branches of natural sciences and engineering and technology. It is also intended to show systematic gender imbalances where women are under-employed in research generally and have poorer access to research and development resources.

As the introduction states, the figures speak for themselves. Across the EU as a whole only 29% of scientists are women and only 18% of the business and enterprise sector are women. Only 15% of women are A grade level positions in tertiary institutions (in technology just 5.8%), in only 2 of the 17 member states are the proportion of female members of scientific committees over 40%, and over half of the committees have less than 20% women.

In terms of who is setting the scientific agendas the numbers tell an interesting story also. In the nine countries where there is low research and development per capita expenditure there is the highest proportion of women in decision-making positions. In the countries with the highest per capita expenditure in science research and development, Luxembourg and The Netherlands have the lowest numbers of women in management positions (see Fig. 4.2 in European Commission 2006). In the government sector where women constitute 35% of all researchers, the lowest proportion is in engineering and technology, 22.3%, with natural sciences at 31%, social and agricultural sciences 44.3% and near parity in medical sciences 49.7%. In scientific fields women are more likely to choose health and welfare agriculture and veterinary science. It is striking that in life sciences women constitute more than 50% and that there is near parity in medical (see Tables 2.5 and 2.2 in European Commission 2006).

In relation to jobs in life sciences women make up 54.4%. Pharma companies have 43.7% of women employed with seven countries (Turkey, Norway, Bulgaria, Sweden, Spain, Sloveia, Latvia) over 50% of women researchers in the 'business enterprise' area (see Table 2.5 in European Commission 2006). The glass ceiling is evident in research, industry and policy but tellingly so in academia (see Proportion of female grade A staff by main field of science 2004 in the EU 25 Table 3.2 in European Commission 2006). The gender pay gap is also strong, the highest being in life science with health professionals (see Gender pay gap by selected occupations in private enterprise EU 25 2002 in Table 4.1 in European Commission 2006).

1.3.4 Industry, Private Research and Women and Biotechnology

As Donna Dickenson, Emeritus Professor of Medical Ethics, University of London points out, the relationship between industrial research (in biotech and for profit companies) and public funded research in universities is 'an extremely secretive and pernicious one'. Private research now dominates national and international biotechnology agendas and industry money often determines what is developed and what is not. There is a complex relationship between knowledge of scientists and ethics. It is important to find the right balance.

There are very complex economic social and ethical issues involved in looking at how industry determines private research agendas, and ultimately the whole development of the biotechnology field given the amount of money involved. Some critics call it a free market eugenics in relation to human germline manipulations and cloning and reproduction for profit pushing the genetic technologies. But as Dickenson says, the agribusiness and pharmaceutical companies' agendas that determine not only research interests but what type of food, bodies, medicines and environments we live in are hard to track and are bound up in economic and political agendas.

It is critical to note that many women are engaged in the biotechnology industry as scientists and business managers and entrepreneurs. There are more visible and more self-acknowledged women scientists in academia but the largest part of the research system is in industry where the figures on women in biotech are still very hard to find. Now that biotechnology is expanding as a proportion of the total research iceberg it is important to start to try and collect the figures of where women are working in biotechnology in industry as well as in research.

Life sciences attract women, or at least there are spaces allowed for women to enter the field, even though there are not proportional numbers of managers to workers in the biosciences. In India, women constitute 80% of the work force in agriculture and biotechnology. The significance of the number of women farmers and women working in life science, human and plant biotechnology has not been lost on Indian industry; forward-looking companies are aware and encourage women 'entrepreneurs' to engage. Several biotechnology companies encourage women to join start-ups, promoting biotech as particularly suitable for women (Women's Technology Cluster). King (2000) as well as claiming that biotech crops are particularly beneficial to women in the developing world goes on to say that 'as scientists around the world come to recognize women's role in agriculture, more research is taking into account their concerns and knowledge (…) in the future this research will yield tremendous benefits for women and the world'.[9]

From *She Figures* we see that there is a low proportion of women researchers in companies: for example women comprise only 28% of industrial science researchers in Ireland and 8% in Austria.

The European-wide study on Women in Industrial Research (European Commission 2003) looked at good practices in companies across Europe in order to illustrate the results of gender policies. They profiled several biotech companies that have gender policies: I contacted them to ask a little more about what gender policies in biotech industry meant. It is interesting to note that in the list of Best European companies four out the first top ten are biotech and pharma (out of 1,000 European companies). What is evident from this brief sample of both large and small companies' 'good practice' is that women are attracted to research in life sciences because of the subject area but also it is important that the working environment deals with the need to balance family and work needs and in some cases gender tensions. Whereas the percentage of women researchers seems balanced, women in management positions are few, even when the chief executive officer is a woman. Whether this is due to women employees' personal choice (for whatever reason) or gender bias, the glass ceiling for the majority of women working in biotech industry is clearly in place.

AstraZeneca pharmaceuticals, a large multinational with a turnover of over 16 billion euros with research and development laboratories in the UK (also in Sweden, North America and India) has been featured regularly as one of the top European best companies in the last five years. According to the (woman) diversity director the research policy sees women as a key constituency to recruit and maintain. Out of the 58,000 employed, 10,000 are researchers and 50% are women researchers. Of these, 29% of women are in senior positions and 21% of patented innovations in the last year were by women.

The company actively seeks to provide a friendly and conducive work environment with an innovative programme to help break down gender and racial discrimination

[9] Quoted from unpublished paper by Monsanto's senior vice president of public affairs Sarah Hull addressing a US Business Network Summit in Washington, DC, February 2006.

through interactive theatre where employees are encouraged to discuss sex discrimination and bullying, working conditions, and personal changes including family life. The company offers flexible working hours suitable for women with families, with part-time work, and parental leave of one year per child. In the company there is gender parity until you reach the top level; according to the diversity director many women are reluctant to take up top jobs because of the time commitment it involves.

Bio Alliance Pharma, a small biotech company working in the field of drug resistance, has a small turnover of 600,000 euros per year, is based in France and was founded by a woman who is its CEO. Of the 30 researchers, 20 are women. The CEO states that the selection of employees is based on ability, that women proved to be more suitable to the job, and hence more of them were employed. She concedes that, in terms of working conditions that suit women employees, it may help that she is a woman; e.g. she recognised the need for and established flexible working and career arrangements and help with child care as one of the important factors for keeping women workers on board.

Another French company with a woman CEO, Imstar SA Biotechnology with a 1.5 million euros annual turnover, has 50% women researchers. The CEO designed automated imaging systems for biomedical research diagnostics and drug development. She offers flexible working hours and merit and creativity are rewarded. The CEO sees as a woman that perhaps it is 'easier for her to recognize that women scientists have the required qualities and skills', but she also said it has been hard to recruit qualified women in France and three of her top researchers are from the Moscow Academy of Science.

Biotechnol, another small company based in Portugal with an annual turnover of 224,000 euros, has 70% women researchers among the 20, with 43% women managers. In asking the director why so many women the response was that biology and life sciences are where women are making their mark in the life sciences. The major issue of all women workers of balancing work and family life is relatively easy for this company as in Portugal child care is available.

Schering AG Pharmaceuticals, which specializes in new medical solutions for special health problems, is a large multinational headquartered in Germany with 5,023 million euros annual turnover. Of the 480 researchers 29% are women and 17% of management are women. The company has promoted equal opportunities since 1973 and opened a policy unit for women in 1990. Since 1995 the company promoted an interactive programme to encourage open dialogue between male and female staff in order to help men and women work on demands of career and family. The company won the first EU gender equality in the work place award in 2003.

1.3.5 Opinions of Women Biotech Researchers and Scientists

Important as it is to measure progress in numbers, figures can only tell us so much, particularly as figures on women and biotechnology in Europe are not yet fully documented. In order to probe deeper into gender relations for women working in biotechnology in Europe I conducted a brief survey by phone and email in order to find some personal narratives. I interviewed 20 women scientists and social scientists working in biotechnology in Europe, taken from among the list of women invited to engage in the WONBIT conference from among 15 countries.[10] I asked them to respond on how they perceived their own working conditions as well as their ethical positioning on the controversies around biotechnology research and industry. The women I spoke to were working in academia, bioscience laboratories (red, green and white) and those working on gender mainstreaming in science in European policy. Their ages ranged from 33 to 67 and from researchers of five years working in short term appointments in research laboratories to distinguished Professors at the top of the University scale.

The responses to the interviews fell into three quite distinct positions:

1. Bioscientists who separated their work as scientists from issues of gender equality
2. Bioscientists interested in how to achieve gender equity in the workforce by bringing in more women in the policy making arena
3. Feminist scientists and social scientists working theoretically on a critique of biotechnology, understanding science as a practice that is imbued by values and power relations

1.3.5.1 Group One

The first (the least number in the survey) were bioscientists who strictly separated out their work as scientists from issues of gender equality. They saw the main concern for women scientists as a professional choice between family and work place. They did not see any need to question the ethics of biotechnology as a science, nor did they understand what being a woman had to do with being a good scientist. They did not call themselves feminists and were not interested in seeing the connections between feminism and science.

A clear case of this position is in the following responses:

> I don't really see myself as a feminist. I am not really sure what a feminist is. I am certainly committed to gender equality, but at this point I see the main bottleneck here as poor acceptance of men adopting traditionally women's roles, rather than access for women to traditionally male roles. In particular, until men are allowed to participate fully in child care, it will remain hard for women to participate more fully in the workplace.

And:

> I don't think my views on gender issues impact directly on my research, although of course they impact on my activities as a teacher, supervisor, and line manager etc.

---

[10] Although I contacted over 45 scientists in 15 countries, I received only 20 responses on which I could draw for this essay. The following quotations are taken from telephone and email interviews carried out in February and March 2007.

Another comment was:

> It is clear that fewer women choose to work in the physical sciences than the biosciences. This difference is evident very early in school where girls tend not to choose physical sciences at GSCEs and A levels. I have never understood why this is the case.

Some comments underlined the prejudice against women, and the need for support by male partners, one respondent gave full credit to her husband supporting her successful career:

> I have a husband and two children. My husband is a writer and works from home. He has been the main child care provider and he has been free to move when my career made moving a good idea. Because of the support provided by my husband, I have very much enjoyed balancing my work and family life, with each enhancing the other.

She went on to add based on her own (somewhat unique I suspect) experience:

> I don't think there should be a European policy on women in anything. But I do think gender-neutral parenting laws would be helpful for everyone- e.g. parental leave entitlements that can be shared between the mother and father, rather than maternity leave etc.

Another response was:

> The most helpful thing to do to support women in science (and other professions) is to support men in taking up child care responsibilities. Better child care provision always comes up as a way to help women in science, but this is a symptom of the problem not a solution to it. If child care became a gender-neutral activity, then there would be no problem because all parents, male or female, would have the same issues.

1.3.5.2 Group Two

The second group of mostly younger women did see themselves as feminists. Their interest was on how to achieve equity in the workforce. They felt that more women scientists in research and industry as well as in the policy-making arena would lead to a better design of biotechnological research that would better serve women. Their major concern is to correct biases in the theory and practice of women and biotechnology in order to include women and their needs and interests. Much of their interest is in the numbers of women working on biotechnology, equality in the workplace, resistance to the glass ceiling and gender mainstreaming in policy. Some argued for a reallocation of resources for women to participate and reshape the knowledge of women and biotechnology as scientists, workers and consumers.

There was considerable discussion particularly from younger women respondents in Italy, Spain and Poland on how difficult it was to balance family and work needs and for them to move up the scale in the laboratory. Though as one young woman researcher working as a biologist in a research based children's hospital commented:

> We are almost all women in my research lab but very few of us have expectations to move up to management level. There is a lot of competition that breaks down any team spirit or support, particularly for those of us who have children. I have moved jobs a lot, partly from choice partly because it is hard to find interesting work once you decide to have children.

Another comment in agreement was that if women wanted to succeed they had to choose between male and female career paths:

> In order to succeed you need to put in such long hours and with small children I could not do that. Also the need to move around to suit the company makes it very difficult to have a family. Sometimes I think men and women both think that women who have children are less good scientists.

Another comment worth quoting is on the general prejudice encountered:

> I think that there is still an underlying view, held by both women and men, that while women can have a career, men need a career; and that child care is primarily a woman's responsibility. These societal biases have subtle effects throughout the workplace.

A senior woman in the UK academy commented on those difficulties linked to women's personal esteem and the difficulty of surviving a highly competitive environment:

> The biosciences academic career path involves a PhD followed by a post-doctoral research apprenticeship, before the possibility of a research team leader position in academia. This latter transition is a particularly critical step since there are very few such positions in comparison to the number of post-doctoral researchers. Meanwhile, the post-doctoral researcher stage is not a long-term career option for a range of reasons. Because it is so competitive, those who succeed in making this transition are usually highly motivated and committed to academic research to the point that they view alternative paths as lesser. This means that supervisors of post-docs see non-academic careers as failure, and hence most post-docs will "fail". I think this atmosphere has a big impact on the retention of women in science, compared to that of men.

On average, men have more of their self-esteem invested in personal success at work. This makes them more reluctant to admit to insecurity in aspects of their work. In contrast women are on average more willing to discuss openly uncertainty in their opinions and/or abilities. Men are on average more anxious to be seen to succeed, whereas women are on average more anxious to avoid excessively competitive situations and are less concerned about being seen to succeed. For these reasons it is easier for women than for men to consider and follow, alternative career paths than the perceived high-status academic career path.

Further up the ladder, these same average differences result in fewer women and more men applying for promotions.

Another comment from Germany had another take on women's approach to science in the workplace:

> I think that women find it easier on average to conduct collaborative research than men. This is important as research becomes of necessity increasingly collaborative and interdisciplinary.

1.3.5.3 Group Three

A third category of responses were from feminist scientists and social scientists that were working theoretically on a critique of biotechnology understanding science as a practice (in research and industry and policy) that is imbued by values and power relations (cultural, economic and social). They were interested in questions of dominance and power as well as the cultural, economic and social implications of biotechnology for women. They were critical but also saw changes in technology and science as opening up possibilities for changes in gender power relations both in Europe and globally.

Some of the responses included a strong concern on biomedical ethics that concern commodification of women's tissue, such as payment for eggs. Bioethics meant in this category not just autonomy of choice for individual women or scientists but more broadly justice. An interesting point was raised about how bioscientific work has entered the public domain in complex ways:

> Like everything else, biotechnological knowledge needs to be considered in terms of ethics and rights. However, there will inevitably be different views on the ethical use of and rights associated with each aspect of biotechnological knowledge. Policies and laws are therefore needed that reflect as far as possible an average view, while respecting as far as possible minorities on either end of the spectrum of opinion.

On stem cells, one respondent commented:

> Stem cell scientists, for example, or some of the early cloning researchers, have been very concerned to colonise popular understanding of their scientific areas and to promote a popular belief that their technologies offer cure-alls. They have had a great deal of success, which undermines the idea that there is a split between popular and scientific understanding. In other areas, such as the use of umbilical cord blood for stem cell research, doctors (not so much scientists) have been more in the business of issuing warnings against wholesale acceptance of this new development, but the public (egged on by commercial interests) have begun to assume it can and will be delivered as a matter of course.

In response to a question about whether autonomy should be the main guide for biotechnology in relation to egg donation for research the comment was made that it is important to raise the potential harm to women much more in order to understand if it is economic need, altruism or some other issue that induces women to donate:

> Overcoming the simplistic autonomy-based argument that "if they choose to donate, despite the risks, that ends the debate". Really the debate should start there, not end. Why might they "choose" to donate? Are they under pressure from family members with conditions that stem cell research is supposed to be able to cure eventually? Will the private agencies that already pay Eastern European women for their eggs in IVF turn their attention to eggs for research? Is that exploitation? Trafficking? Or again, genuine choice?

The ethical context or what one respondent called a new form of colonialism by first world drug companies focused on the exploitation of third world women and also Eastern European women:

> I'm quite worried about a new form of colonialism in which First World drug companies use specialist brokers to arrange clinical trials in the cheaper Third World countries. That's been going on since the 1990s, and it's now starting to extend to Eastern Europe. I'm also concerned about whether Third World women might become egg suppliers for somatic cell nuclear transfer stem cell technologies. Since the ova are enucleated, the race of the donor doesn't matter, as it does for IVF, so I've predicted that researchers will seek out the cheapest suppliers. Finally, of course, there is the whole vast issue of genetic patenting and pharmacogenetics. I think that: peoples with unusual degrees of genetic homogeneity, real or

> imagined (as in the Icelandic database), will become the targets of European and North American companies wanting to mine genetic databases or trial drugs. Those trials would be cheaper to run if the control and experimental arms of the trial have more similar genomes than usual in the First World.

One plant biologist put it differently but very eloquently:

> I am working on a network of interacting plant hormones that regulate shoot branching. I am interested in this problem as a model to understand how environmental and genetic inputs are integrated to control plant development. In the long run I would like to understand how this network changes under natural selection in comparison with the artificial selection that drives domestication of plants for use as crops. ... The work will certainly be useful both generally in terms of understanding plants, and specifically for example in speeding up new domestications...

As 90% of the biomass on the planet is plants they are the foundation for every ecosystem and every agricultural system, converting light energy into chemical energy, which can enter the food chain, or be used as fuel etc. Understanding plant biology is therefore an imperative. In my opinion, applications derived from fundamental knowledge about plants have the potential to contribute more to society than medical science.

Also in relation to Plant biology one explained in her response:

> Publicly funded research has three main aims. The first is simply that it is a basic human endeavour to pursue understanding on the natural world. The second is to provide the fundamental science base needed for future applications to be developed in commercial companies. The third is to work toward those applications of benefit to society that are not profitable and so will not be carried out by industry.

Publicly funded research and industrial research are therefore linked at the transition from fundamental knowledge to its application. One of the main reasons biotechnology is controversial is because of the idea of redesigning life, however I think this is substantially because of different views of life and particularly of what 'natural' is about. For example, there are several key issues that have led to the popular rejection of GM food in the UK. Perhaps the main one is the current fashion for natural = good, synthetic/manmade = bad, particularly when it comes to food. I find this quite ironic, because it requires a very human-centric view of nature to work as a doctrine. Seen from the point of view of a plant, it is rather obvious that being eaten is not good. Through natural selection, plants have evolved a wide range of defences, making them indigestible and/or toxic. As a result, natural plants are by definition not good for you. The whole point of agriculture is forced unilateral disarmament of plants, to reverse natural selection through artificial selection, removing their natural defences, making them more nutritious and higher yielding etc. We have been doing this for 10,000 years. Even then, it has been estimated that 99.99% of the toxic chemicals we eat are natural constituents of the plants in our diets.

This is one example among many of how a biological training brings a different view of nature than the popular conception. It is important to see the tensions in these approaches both as an interdisciplinary discussion but also one that suggests there are quite wide divides in how the topic Women and Biotechnology is approached in Europe with implications for how we move forward to advocate for policy change in relation to women and biotechnology in the European context.

1.4 The Global Development Debate on Biotechnologies

Activists, researchers, independent academics and others from the environmental, social justice, consumer rights, farm and women's movements are raising fundamental questions about the course of development in the 21st century. Biotechnology and its commercial applications are at the centre of their concerns. Perhaps the protesters are reminding us that answers are being proposed before many of the most basic questions have been asked. Madeline Boscoe (Co Chair of the Canadian Federal Advisory Committee on Reproductive and Genetic Technologies).

There are many complex ethical, social, economic and political implications of biotechnology for women in a global context. In this section I try to draw out how I can link the debates around women and biotechnology in Europe to the highly contested global policy debate on biotechnologies.[11] I have selected four areas of debate which are of particular importance in the global South which had strong ramifications for the WONBIT discussion on women and biotechnology.

1.4.1 Agriculture, Biodiversity and Biotechnologies

Social scientists and ecofeminists working in the South have been engaging with biotechnologies from a critical viewpoint in relation to concerns around biodiversity and sustainable development. Helen Zweifel in her study on modern biotechnologies in agriculture (Zweifel 1995) points out the relationship among development, technology and gender is a very complex one, and posits that an analysis must begin from the question of who controls the technologies and for whose benefit; i.e., will modern biotechnology undermine or benefit women's autonomy. In her studies on cocoa and cassava in Ghana and Colombia she argues that modern biotechnology as a technological innovation from the outside could function well without harming women's interests but that the problem is that the control over modern biotechnology is concentrated in the hands of a few private companies and that research is directed towards the interests of industrialized countries and large-scale agriculture more than small-scale farms. A study by IDRC and

[11] I am basing much of the discussion on contributions to the journal I edit *Development, 49*, 4 (2006) on *New Technologies and Development* (www.sidint.org/development) together with the work on *What Next* by the Dag Hammarskjold Foundation, the Canadian Action Group on Erosion, Technology and Concentration (ETC) Group and the Working Group on Women, Health and the New Genetics, as well as the USA based network Committee on Women, Population and Environment.

the Food and Agriculture Organization (FAO) on *Women and Plants* (Howard 2003) argues that women's knowledge of plants and all forms of biodiversity and agriculture along with their livelihoods are being threatened by the growing interest of biotechnology industry in plant genetic resources.

Vandana Shiva, an internationally known scientist from India since 1986, has been critical of western science attempts to modify food genetically. She presents biotechnology as a science and industry that is stealing the knowledge of seeds from women farmers in the South in a process she labels biopiracy. She sees two world views: the 'logic of death' ruled by profits of the global agribusinesses and biotechnology corporates and the 'logic of life' based on caring for the Earth and sustaining family and community of women farmers. The work of philosophers such as Mary Midgeley among others are exploring the importance of the Gaia theory as a new framework to ensure that biodiversity is preserved and that biotechnology does not destroy ecological methods that are continued and not violated; that women-centred knowledge ensures food security and sustains local agriculture.

Vandana Shiva sees the development of agricultural biotechnology as akin to the closing of the commons. Biotechnology grants private rights to previously shared properties, in this case germ plasm and plants (rather than fields and forests) with devastating social consequences. GM foods have provoked wide spread resistance not only among consumers in the North, particularly Europe, but also among small-hold women farmers in the global South (Shiva 1997, 1998; Sharma 2000).

There is also an innovative engagement of anthropologists and other researchers working in development research and policy. This particular entry point to the debate on women and technology includes my own research on women and the politics of place (Harcourt and Escobar 2005) looks at how to reformulate the relationship between nature and culture that respects the body, human relations, and relationships to other living beings. These studies look at biotechnology as an interesting point of intersection among production and reproduction, modernity and tradition, ecology and feminism in the rapid global changes to human social life and what it means to be human. In these debates feminists working transnationally are interested in biotechnology as a science of life that is linked to changes in relation to cultural and social practices around the body, food, community, environment at local levels that are determined by broader global economic and political debates.

Women's environmental struggles have focused on corporate ownership and control of vital biological resources impacting livelihoods, food security and health as well as the marginalization of women's role in decision making in these areas. It is assumed that the debate is about women as primary producers or consumers of food often responsible for feeding and ensuring the health of their family.

The debate is an intense one. On the one side the argument is that in the South genetically enhanced crops can increase yields and help subsistence farmers provide more and better foods for their families or, in the case of revenue crops such as cotton, get more value from their land by increasing yield. Crops, it is also argued, can be enhanced through biotechnology to contain more nutrients and help minimize the effects of malnutrition. Most importantly, GM crops can help provide these benefits in a sustainable way, so that increasing populations can be fed without placing additional pressure on natural habitats by destroying them for farming uses.[12]

On the other side, critics like Shiva and the Erosion Technology and Concentration (ETC) group[13] raise alarms about the social use of biotechnology and how life science companies are appropriating and manipulating genes for food and medical purposes without enough public debate and consensus. Shiva states in no uncertain terms that corporate and technological control over reproduction in plants, animals and humans represents the ultimate patriarchal project.

The argument of the biotech industry is that we have to accept genetically engineered foods if we are to feed the poor in the global south and that biotech will create a world free of pesticides and increase yields. Such technological solutions to hunger are strongly criticized (Mittal 2006). The problem critics suggest is not overall scarcity but unequal access.

According to the FAO 'The existence of 780 million chronically hungry people in the developing world today shows that there is something fundamentally wrong in the distribution of food and the resources with which to access it' (FAO 2002).

Golden Rice is a striking example of the promise of a technological solution to hunger.[14] Golden Rice, Syngenta's new version of genetically engineered rice, supposedly has a tenfold higher content of beta-carotene, which could fight Vitamin A deficiencies that cause blindness among children and adults in developing countries. But as WHO has pointed out, nutritional deficits can be easily and cheaply corrected with a more varied diet. Green leafy vegetables, oranges, and red palm oil all are high in beta-carotene.

Such biotech research has profound implications for farmers (and fisher people and pastoralists) and for food sovereignty worldwide, impacting millions of poor women's livelihoods. Major agribusiness firms, such as Syngenta, BASF, Bayer and Monsanto are reformulating their pesticides at the nanoscale to make them more biologically active and to win new monopoly patents. It is estimated that over the next two decades, the impacts of nanoscale convergence on farmers and food will exceed that of farm mechanisation or of the Green Revolution (ETC Group 2004).

Just like the biotech industry, the nanotech industry promises to feed the world and eradicate poverty, despite the fact that many nanoproducts have not been tested in relation to both health and environmental threats from the use and disposal of nano-particles.

It is the small farmers and agricultural workers in the developing world, the majority of them women, who are set to be the most affected by nanotechnology.

---

[12] Retrieved October 20, 2007, from http://www.syngenta.com/en/about_syngenta/biotech_faq.aspx

[13] See their website for their latest publications: Green revolution 2.0 for Africa? Retrieved October 20, 2007, from http://www.etcgroup.org/_page24?pub_id=611, News release: Food sovereignty or Green Revolution 2.0? Retrieved October 20, 2007, from http://www.etcgroup.org/_page24?pub_id=613), Gambling with Gaia. Retrieved October 20, 2007, from http://www.etcgroup.org/_page24?pub_id=608

[14] Retrieved October 20, 2007, from http://www.gene-watch.org/genewatch/articles/18-3Mittal.html

The Canadian based research ETC Group notes that 'Poor farmers are seldom in a position to respond quickly to abrupt economic changes. Particularly at risk are farm communities and countries in the global South that depend on primary export commodities such as rubber and cotton – products that could be displaced by new nanotech materials' (ETC Group 2004; Hickey and Mittal 2003).[15]

The response by civil society organizations such as ETC Group are to advocate for bans on nanoscale formulations of agricultural products such as pesticides and fertilisers until a regulatory regime specifically designed to examine these nanoscale products finds them safe. Recently, the Sierra Club, the largest grassroots environmental organization in the U.S., together with the Sierra Club of Canada, sent a letter to the World Bank challenging its involvement, together with the Global Environmental Facility, in biotech 'harmonization' programmes in Africa and Latin America (Mittal 2006; Mousseau and Mittal 2005).

We need to be aware of these tensions and resistances to biotechnology interventions in the South, particularly in relation to women's control over their knowledge and livelihoods. The problem of patenting means that their knowledge of agriculture and biodiversity becomes a profitable business for big biotech companies (Golden Rice for example required 70 patents) which will, given the large-scale gender inequalities between northern based industries and poor women farmers, can only undermine their autonomy.

1.4.2 Biopiracy and Genes

The famous example of Iceland's genetic heritage being sold to genomics company Decode, which then sold the human data to Hoffman LaRoche Switzerland for US 200 million, marked the shift of genomics research to mainstream commercial ventures (Delahanty 2000). Some seven years later the ethical questions around the access to data on human genetic diversity and the implications of the completion of the human genome project remain thorny. The Human Genome project launched in 1990 began with health, pharmaceutical and fertility research on genes as the fundamental determinants of health, but ethical questions about the commercial use of genetic material are still not being adequately addressed.

One important aspect, brought to public popular attention by a book (Le Carre 2001) and a film based on it, is the misuse of poor people in the South as research subjects in large scale genetic studies. The questionable ethical behaviour of researchers towards poor research subjects in the global South would

[15] For example the sale of Monsanto GM cotton seeds to farmers led not to a miracle but disaster with 60% crop failure due to the failure of Monsanto Bt seeds (through bollworm pest and 'Lalya' or 'reddening', a disease unseen before) compounded by sliding global prices and dumping of cheap subsidized cotton from outside the country, which led to thousands of suicides of farmers. In the western state of Maharashtra, 4,100 farmers committed suicide in 2004 (Mittal 2006).

not be possible in Europe. The bottom line is that these studies are exploiting people's poverty when what they need is food security not gene therapy.

Burrows[16] (2006) in her analysis of the Human Genome Diversity Project (HDGP), argues that biotechnologies are burdened by persistent ethical dilemmas, including the problem of how to access resources and engage subjects in the research endeavour without co-opting their rights or harming them or their communities. She defines biopiracy as the access to or acquisition of biological material and/or traditional knowledge related to its use, without the prior informed consent of those whose biological material or traditional knowledge has been 'accessed' or 'acquired'. She profoundly questions both industry and academia in the pursuit of plants, microbes and traditional knowledge whether for potential commercial interest or justified in the name of Science (or Medicine).

The need here is for a complementary relationship of genuine respect between potential researchers and potential subjects. Any other method of involving people in research in which the only allowable outcome is 'consent' to the goals of one side, can only be viewed as a sham. A process leading to 'informed choice' may be cumbersome and time consuming but crucial to any ethical means.

As Burrows states, 'Whatever the ambitions for understanding human history and evolution and uncovering data with possible biomedical applications, the goals of the Human Genome Diversity Project are only a small sample from the set of human possibilities. For many and perhaps most of us, academic curiosity is simply not the pinnacle of our aspiration' (Burrows 2006).

1.4.3 Nanotechnology and Manufacturing

The implications of reverse-engineering Mother Nature's designs for our own technological devices will be most profound on the economies of manufacturing. When companies can cheaply and chemically assemble materials and devices in the same manner that beer, cheese, and wine are manufactured today, it spells disruption and dramatic shifts in supply and value chains. Lux Research, Inc. *The Nanotech Report* 2004.[17]

The promises of nanotechniques in biotechnology are huge: medical and engineering capabilities that will allow us to cure, fix or provide technological assistance. Mohamed Hassan, President of the Third World Academy of Sciences, and Gordon Conway, the U.K. Government's Chief Scientific Advisor on International

[16] She is director of the Edmonds Institute in the United States, a public interest, non-profit organization that was founded in part to examine issues such as those raised by the Human Genome Diversity Project.

[17] Lux Research is a consultant group that provides strategic advice and on-going intelligence for emerging technologies. Leaders in business, finance, and government rely on us to help them make informed strategic decisions. http://www.luxresearchinc.com/

Development – see enormous potential for nanotech to improve the conditions of poor people in the 'developing world'.[18]

Worldwide, industry and governments have invested more than US$10 billion in nanotechnology research and development in 2004. The European Union, Japan and the United States are the leading investors, at roughly equivalent levels. China's government spends more money on nanotech research 'at purchasing-power parity' than any other country except the United States. Approximately 60 countries have established national nanotech research programmes, many of which are in Europe. The United States Government's National Nanotechnology Initiative (NNI) has spent over US$5 billion on nanotech research since 2001. The National Science Foundation in the United States estimates that the nanotech market will surpass US$1 trillion by 2011 (Thomas 2006).

To begin to understand what nanotechnology will mean for women and men in the global South – we need to understand the challenge nanocommodities pose to existing commodities such as rubber, minerals, metals and fibres. One example is in cloth. As explained by Jim Thomas from the ETC Group (Thomas 2006) Nano-Tex, a California-based company, has licensed its nanotech 'fabric enhancements' to more than 80 textile mills worldwide – including India's two largest mills. The treatments, incorporated in clothing and furniture sold by more than 100 companies, make the fabrics stain and spill-resistant, without changing texture.

Cotton is a staple export of many southern economies, grown in over 100 countries, 35 of them in Africa. It is the main cash crop for small-scale women farmers in Zambia, with cotton production estimated to have reached a ten-year high in 2003/04. Cotton is also the main cash crop in the Central African Republic, where cash crop production is the principal source of income for most of the rural population. Cotton accounts for 39% of Burkina Faso's exports, 37% of Chad's and 33% of Benin's. With a global market value of US$24,000 million (in 2003) and over one million people engaged in cotton production worldwide, nanotech's potential impact on the textile and fibre sector is devastating for many poor women farmers.

Commodity dependent nations, often the poorest and most vulnerable, will likely face the greatest disruptions. Currently, nanotech innovations and intellectual property are being driven from the North including in Europe. As core commodities are replaced by the new technological improvements to cloth etc. commodity-obsolescence or a drop in prices will impact heavily on women and men workers in the Global South who do not have the economic flexibility to respond to sudden demands for new skills or different raw materials (ETC Group 2005).

Within the larger context of technology transfer and intellectual property there are also major issues that have to be flagged. Gene patents have raised ethical concerns about the ownership of life; the new nanopatents go below the level of life, claiming the molecular processes and elements of nature at this fundamental level.

[18] Retrieved October 20, 2007, from http://www.publications.parliament.uk/pa/cm200405/cmselect/cmsctech/487/5032303.htm

Thomas quotes a study conducted by the University of Arizona and the United States National Science Foundation which found that 8,630 nanotech-related patents were issued by the United States Patent & Trademark Office (US PTO) in 2003 alone, an increase of 50% between 2000 and 2003 (as compared to about 4% for patents in all technology fields). The top five countries represented were: the United States (5,228 patents), Japan (926), Germany (684), Canada (244) and France (183) (Thomas 2006).

In short with this amount of investment it is possible to imagine a completely reordered economic world if manufacturing becomes nanosized. Manufacturing jobs would become high tech jobs that could be done from anywhere with, as the global economy is now functioning, dire consequences for women workers in the Global South.

1.4.4 Biotechnologies, Nanomedicine and the Poor

The combined market for nano-enabled medicine (drug delivery, therapeutics and diagnostics) will jump from just over US$1 billion in 2005 to almost US$10 billion in 2010 (Thomas 2006). These innovations in nanomedicine are currently being driven from the North and are designed primarily for OECD markets. But 'nanotech's medical innovations while likely to benefit the pharmaceutical industry will do little to address health and poverty in marginalized communities'. Indeed, Brazilian researchers Noela Invernizzi and Guillermo Foladori (2006) argue that nanomedicine could widen still further the gap between haves and have-nots.

The application of nanotechnologies in the field of medicine according to Invernizzi and Foladori is concentrated in three areas.

The first is diagnostics using sensors on the nanoscale that can be placed within the human body. Using quantum dots it is also possible to search out diseased cells and mark them; thereby an illness could be detected in the first moments, 'anticipating' a disease even before the body itself shows any reaction.

Another promising area of nanomedicine is that of drug delivery with the possibility to target drugs directly onto the affected cells or organs. This avoids harmful effects on healthy cells and tissues that do not require the drug and by going directly to its target, it is possible to increase the dosage. The so-called smart drugs can release a certain antibiotic only when an infection is present.

The third promising field of research is that of implants and prostheses. This ranges from repairing bones and teeth to any type of tissue. Nanomaterial coatings could increase the durability and adhesion of implants. Biological nanostructures could improve tissue regeneration. Nanoscale devices could be implanted as sensors that monitor the environment, detect properties, and deliver medicines. Nanostructured material can also be used in artificial sensory organs such as electronic eyes and nerves, although many of these projects are still in their earliest stages.

On the whole, these technologies promise to lengthen the human lifespan and reverse the effects of ageing. But does nanotechnology in medicine offer hope for the poor in the Global South?

In some governmental reports there is the optimistic assumption nanotechnologies will improve the living conditions of the poor. Task Force on Science, Technology and Innovation of the U.N. Millennium Development Project (Juma and Yee-Cheong 2005) and The Canadian Joint Center for Bioethics argue that nanotechnology can be used to help achieve five of the eight Millennium Development Goals (Salamanca-Buentello et al. 2005) solving specific problems such as energy, drinking water or disease-diagnosis. It is merely a matter of political will and ethics.

Others see developing countries as being able to take a role in the promising global market for nanoproducts as indicated by the rapid and growing trend in research in nanosciences and nanotechnologies in developing countries such as India, Brazil and China. The president of the Third World Academy of Sciences, Mohamed Hassan writes that '(…) developing countries have no choice but to embrace nanoscience and nanotechnology if they hope to build successful economies in the long term' (Hassan 2005).

However when considering nanomedicine for the poor from a social perspective several problems arise. The new nanomedical diagnosis requires an individualized surveillance system, and in many cases depends on pharmacogenetics to determine a patient's predisposition to illnesses, requiring individual care for the patient. Most poor people would not have access to these benefits if it follows today's patterns where the commercial form of distribution of vaccines or antibiotics has marginalized the poor from their benefits. According to the World Health Organization (WHO), since 2002, 80% of the world drug market is concentrated in North America, Europe and Japan, a geographic area where only 19% of the world population lives, whereas 90% of the burden of disease is located in poor countries where patients do not have the purchasing capacity to buy medicine.

In fact, the interest of pharmaceutical corporations is in diseases of rich people. According to Forbes the world's ten best selling drugs in 2005 were all for rich patients (high cholesterol, heartburn, high blood pressure, schizophrenia and depression) (Herper and Kang 2006).

By requiring a highly technical individualizing treatment for the consumer, nanomedicine will widen further the gap between haves and have-nots and distribution will continue to be concentrated in rich countries.

Traditional, complementary and alternative medicines, often forming the knowledge base of women in their role as carer of the family, are based on a holistic concept of treatment. The technological trajectory of nanomedicine leans instead towards the 'magic bullet approach' in which one type of medication is applied for each illness. In situations of poverty with precarious sanitary conditions and little infrastructure of services, the eradication of a certain illness is rapidly supplanted by another. For many poor communities, nanomedicine is destined to stay out of reach and out of view.

In short, medical science is not neutral but is embedded in the social political framework of our global, immensely unequal, society; it is important both that we do not let multinational corporations set the agenda nor that we develop biotechnologies to service the rich in Europe at the expense of the poor in the Global South.

1.5 Conclusion

It is important to reflect on the political, ethical and social issues that surround the issue of women and biotechnology in European research, industry, policy as well as global advocacy and development debates. Such dialogue and debate needs to be informed both by transnational feminist debate and by the knowledge and insights of scientists working in the field. The complexity of the debate as I have sketched it here is because of the convergence of different types of knowledge informing it, all of which have their own truths and understanding of political reality, and all of which are grounded in a vastly complex and changing social, economic and political world. The challenge for us is to work together through the inevitable tensions that emerge around these different discourses which inform the institutional, ethical, political, economic and social practices of biotechnology. The challenge is to map out how this knowledge and these practices can come together in new patterns, shaping an innovative, gendered paradigm of biotechnology.

Some of questions we have raised here which continue to be critical to a debate on women and biotechnology in Europe and internationally include:

- What are the political and ethical and cultural consequences of biotechnology for individual European women and men given the dynamic nature of Europe itself, as 27 countries with different histories and different modes and approaches to science and indeed to gender equality?
- How and where does biotechnological research fit into questions of addressing gendered social determinants for health such as sound nutrition, exercise, healthier work environments and better housing and by reducing poverty and violence which affect the overall wellbeing of so many European women?
- Are there creative and effective ways to develop critical public awareness about the political economic and scientific drivers of biotechnology research and its applications? How do we debate the alternatives?
- What are the responsibilities of European citizens, as well as the responsibilities of European policy and industry to poor women in the South who are bearing the brunt of the unregulated and unethical practices of biotechnological research and industry?
- How can women scientists come together with feminists working on biotechnology in order to interact, react and learn together, respecting the different realities, different histories and approaches to science and to policy making?
- How can European based networks of scientists and advocates working on women and biotechnology reach out to southern scientists, feminists and civil society groups?

Acknowledgements I would like to acknowledge the valuable insights of Lydia Alpizar, Ewa Barnik, Donna Dickenson, Barbara Duden, Jeanette Edwards, Elisabeth Ettore, Sarah Franklin, Sandra Harding, Barbara Leda Kenny, Nancy J. Lane, Ottoline Leyser, Pawel Lukow, Anna Magliozzi, Wanda Nowicka, Bronwyn Parry, Elettra Ronchi, Hilary Rose, Claudio Sardoni,

Patricia Spallone, Hilary Standing, Gillian Youngs and Flavia Zucco in helping me write the paper. The responsibility for any mistakes is mine alone.

References

AWID (Association for Women's Rights in Development) (2004a). Why new technology is a women's rights issue. *Facts and issues, 7. Gender equality and new technologies*. Retrieved April 31, 2007, from http://www.awid.org/publications/primers/factsissues7.pdf

AWID (Association for Women's Rights in Development) (2004b). Facing the challenges of new reproductive issues. *Facts and issues, 8. Gender equality and new technologies, young women and leadership*. Retrieved April 31, 2007, from http://www.awid.org/publications/primers/factsissues7.pdf

AWID (Association for Women's Rights in Development) (2004c). Nanotechnology. *Fact sheet, 1. Gender equality and new technologies*. Retrieved April 31, 2007, from http://www.awid.org/publications/primers/nanotech_en.pdf

Boscoe, M., & Tudiver, S. (2000). Biotechnology and women's health: Redefining the questions. In F. Miller et al. (Eds.), *The gender of genetic futures* (pp. 86–91). NNEWH Working Paper Series. Toronto: York University.

Burrows, B. E. (2006). Colonialism and the research endeavour: Reflections on the human genome diversity project (p. 76). *Development, 49*(4), 73–77.

Carson, R. (1962). *Silent spring*. New York: Fawcett World Library.

Delahanty, J. (2000). Gender and the gene giants: Research and action on women and the new genetics. In F. Miller et al. (Eds.), *The gender of genetic futures* (pp. 92–100). NNEWH Working Paper Series. Toronto: York University.

ETC Group (2004). *Down on the farm: The impact of nano-scale technologies on food and agriculture*. Ottawa: ETC Group.

ETC Group (2005). The potential impacts of nano-scale technologies on commodity markets: The implications for commodity dependent Developing countries. *South Centre Trade Research Papers, 4*. Retrieved April 19, 2007, from http://www.southcentre.org/publications/researchpapers/ResearchPapers4.pdf

European Commission (2003). *Women in industrial research good practices in companies across Europe*. EC: Community Research Science and Society.

European Commission (2005). *Women and science excellence and innovation – gender equality in science*. EC: Commission Staff Working Document (pp. 11, 15–16). Retrieved April 31, 2007 from http://europa.eu.int/comm/research/science-society/pdf/documents_women_sec_en.pdf

European Commission (2006). She figures 2006: Women and science statistics and indicators. Retrieved April 31, 2007, from http://ec.europa.eu/research/science-society/pdf/she_figures_2006_en.pdf

European Platform of Women Scientists (2007). *EPWS Newsletter,* 8.

FAO (United Nations Food and Agricultural Association) (2002). *Reducing poverty and hunger: The critical role of financing for food, agriculture and development* (p. 9). Rome: FAO.

Franklin, S. (2007). *Dolly mixtures*. Durham: Duke University Press.

Gaskell, G. et al. (2006). *Europeans and biotechnology in 2005: Patterns and trends. Final report on Eurobarometer 64.3: Report to the European Commission's Directorate-General for Research* (pp. 5, 32, 45). Retrieved June 12, 2007, *from* http://ec.europa.eu/research/ biosociety public_understanding/eurobarometer_en.htm

Haraway, D. (1991). A cyborg manifesto: Science, technology, and socialist-feminism in the late twentieth century. In D. J. Haraway (Ed.), *Simians, cyborgs and women: The reinvention of nature*. New York: Routledge.

Harcourt, W., & Escobar, A. (Eds.) (2005). *Women and the politics of place*. Bloomfield, CT: Kumarian Press.

Hassan, M. (2005). Small things and big changes in the developing world (p. 66). *Science, Policy Forum, 309*, 65–66.
The Helsinki Group on Women and Science (2005). *National policies on women and science executive summary*. EC: Directorate General for Research.
Herper, M., & Kang, P. (2006). World's 10 best-selling drugs. Retrieved November 20, 2007, from: http://www.forbes.com/2006/03/21/pfizer-merck-amgen-cx_mh_pk_0321topdrugs.html
Hickey, E., & Mittal, A. (Eds.) (2003). *Voices from the south: The third world debunks corporate myths on genetically engineered crops*. Oakland: Food First Books.
Howard, P. (Ed.) (2003). *Women and plants gender relations in biodiversity management and conservation*. London: Zed Books.
Hurst, R. (2006). Disability, development and biotechnologies. *Development, 49*(4), 101–106.
IDRC (Institute for Development and Research Canada) (2005). *Understanding the gender dimensions of biotechnology research and development: Consultative expert workshops: IDRC File 1021 13-004: Final technical report*. Retrieved April 31, 2007, from https://idl-bnc.idrc.ca/dspace/bitstream/123456789/128/1/77019.pdf
Invernizzi, N., & Foladori, G. (2006). Nanomedicine, poverty and development. *Development, 49*(4), 114–118.
Jones, R. (2005). The feminist dialogue: Multidimensional identities and internal diversities. *Development, 48*(2), 53–56.
Juma, C., & Yee-Cheong, L. (Eds.) (2005). *Innovation: applying knowledge in development. UN millennium project. Task force on science, technology, and innovation*. London: Earthscan.
Kerr, J., Sprenger, E., & Symington, A. (2004). *The future of women's rights: Global visions and strategies*. London: Zed Books.
King, P. (2000). *Women entrepreneurs in biotechnology industry women in science and engineering* (pp. 189–190). New York: New York Academy of Science.
Le Carre, J. (2001). *The constant gardner*. London: Hodder & Stoughton.
Mittal, A. (2006). Food security: Empty promises of technological solutions. *Development*, *49*(4), 33–38.
Mousseau, F., & Mittal, A. (2005). *Inequity in international agricultural trade: The marginalization of developing countries and their small farmers*. Oakland: The Oakland Institute.
OECD (Organisation for Economic Cooperation and Development) (2006). *OECD workshop on women and science report*. Ottawa: OECD.
Rees, T. (2005). Reflections on the uneven development of gender mainstreaming in Europe. *International Feminist Journal of Politics, 7*(4), 555–574.
Rockhold, P. (2006). Technology, impairment and disability. *Development Volume, 49*(4), 107–113.
Salamanca-Buentello, F., Persad, D. L., Court, E. B., Martin, D. K., Daar, A. S., & Singer, P. A. (2005). Nanotechnology and the developing world. *PLoS Medicince, 2*(5): e97.
Sharma, N. (2000). The antiracist feminist political standpoint against biopiracy. In F. Miller et al. (Eds.), *The gender of genetic futures* (pp. 113–119). NNEWH Working Paper Series. Toronto: York University.
Shiva, V. (1997). *Biopiracy: The plunder of nature and knowledge*. Toronto: Between the Lines.
Shiva, V. (1998). *Monocultures, monopolies, myths and the masculinization of agriculture*. Washington, DC: Workshop Women's Knowledge, Biotechnology and International Trade.
Spallone, P. (1989). *Beyond conception: The new politics of reproduction*. London: Macmillan.
Thomas, J. (2006). An Introduction to nanotechnology: The next small big thing. *Development, 49*(4), 39–48.
Tyson Darling, M. (2006). Gender, new technologies and development. *Development, 49*(4), 23–27.
UNESCO (United Nations Educational Scientific and Cultural Organisation) (1999). *Women, science, biotechnology: What does the future hold for the Mediterranean?, Turin (Italy), 29–31 Januray 1999*. Retrieved April 31, 2007, from http://www.unesco.org/science/wcs/meetings/eur_turin_report_e_99.htm

Vargas, V. et al. (2006). *Second manifesto of the campaign for a convention on sexual rights and reproductive rights*. Cladem, Lima, Peru.

Zweifel, H. (1995). Modern biotechnologies in agriculture: Impact on women in the south. *Biotechnology and Development Monitor, 23*, 10–13.

Chapter 2
Seeking A Seat at the Policy Table: Engaging Women in Biotechnology Research and in Decision Making

Nancy Hawkins and Elettra Ronchi*(✉)

Abstract: The under-representation of women in science and technology is today well documented and the underlying causes of this problem extensively studied. Progress in bridging this gap has, however, been slow and many questions remain unanswered, notably regarding women in the biotechnology sector. While various surveys indicate that women account for more than 60% of life science graduates in most of the 30 OECD member countries, women make up less than 5% of senior academic staff in these same countries. Furthermore, their representation in public office remains considerably lower than that of men.

This paper intends to provide an overview of today's policy debates, of the initiatives and approaches applied to enable greater participation of women in science and technology. It offers perspectives on the status of academic women in one of the OECD member countries for which comprehensive statistics and a broad range of support measures are available, Canada.

Keywords: Biotechnology, OECD, women, workforce, science and technology

Nancy Hawkins
Department of Molecular Biology and Biochemistry, Simon Fraser University, 8888 University Dr. Burnaby, BC, V5A 1S6, Canada
nhawkins@sfu.ca

Elettra Ronchi
Organisation for Economic Cooperation and Development, Health Policy Division, 2 Rue Andre' Pascal, 75016 Paris, France
elettra.ronchi@oecd.org

***Disclaimer:** The opinions expressed in this paper are the sole responsibility of the authors and do not reflect those of the OECD or of the governments of its member countries.

2.1 Introduction

Established in 1961, the Organisation for Economic Cooperation and Development (OECD) is a unique forum where the governments of 30 countries work together to discuss and coordinate economic and social policy in areas as diverse as health care, national accounts, economic indicators, employment, education, and the environment. The OECD is also one of the world's largest and most reliable sources of comparable statistics, economic and social data. A key focus of OECD work is how science, technology, innovation and education policies can efficiently contribute to sustainable economic growth and employment creation. The work aims to provide policy advice on the challenges arising from developments in new science-based industries, including biotechnology.

Biotechnology has become increasingly important in a number of OECD activities, since the topic was first addressed over 20 years ago. To name just a few, these activities include: the development of policy options for science and technology infrastructure; the implications of progress in genomics and genetics, intellectual property rights and licensing; and consideration for human health, environmental and food safety. Recurring policy questions in all this work are: What evidence provides a solid basis or argument for government to invest in biotechnology? Can biotechnology enable increased productivity and sustainable economic growth? Can it address society's unmet needs? What policy frameworks are necessary to enable its equitable and safe application?

One of the key factors in the growth of the biotechnology sector is the continued availability of skilled human resources. Competitiveness in biotechnology relies on appropriate investments in human capital through higher education and professional training in key areas such as leading-edge scientific disciplines, executive management, and regulatory affairs.

Today, however, demand for talent is exceeding the available supply. This should worry anyone interested in the future of biotechnology. To devise solutions or remedies to the problem, policy makers need to understand the nature and causes of this shortage. They also need to understand and monitor the evolving occupational characteristics and needs of the biotechnology sector. The lack of good data on the biotechnology labour market is, however, impeding such evaluation. In particular, there are important gaps in the data on the place and roles of women in the biotechnology workforce. In the absence of gender-disaggregated data many important questions remain unanswered.

How many women are in the biotechnology workforce? Are they fully operational when they arrive on the labour market after many years of study? What positions do they occupy in biotechnology companies? What is their representation at upper levels of management and do they occupy senior research positions? How many leave research for other types of jobs? For what reasons? What is their employment pattern in terms of public vs. private sectors? Are they mobile across sectors and internationally?

Although the focus of the WONBIT conference was on women and biotechnology, given the aforementioned issues, the paper inevitably widens this discussion.

For the purposes of this paper, the emphasis is on the contribution of women engaged in the life sciences, one of the primary areas of study that would provide the education necessary for a career in the biotechnology industry. More specifically, the paper will consider the data available on how women fare today in academia, their career paths, and discuss measures which aim to facilitate their participation in decision making. Concrete examples will be drawn from one of the OECD member countries for which comprehensive statistics and a broad range of support measures are available, Canada.

The paper confirms that there is under-representation of women at the senior levels of academia as well as in decision-making positions in the top research intensive universities. This is true across the whole of the OECD area. These inequalities extend to all science and technology disciplines, also to those areas and sectors which are of critical importance for biotechnology.

The paper argues that while science and technology workforce shortages are the crisis that has captured policy makers' attention, the under-representation of women deserves increased consideration for several other reasons as well. Greater participation of women at senior levels in research and in decision-making positions is essential if women are to participate more fully and equally in setting a biotechnology research agenda responsive to their needs. Not only has gender equity to be achieved as far as numbers are concerned, but an increased participation of women must also lead to greater influence in shaping the major scientific questions of the moment, many of which impact directly on the lives of women. The paper, therefore, reaffirms for the biotechnology sector what was said in 1995 by the Platform for Action adopted at Beijing[1] which stated that 'women's equal participation in decision-making is not only a demand for justice or democracy but can also be seen as a necessary condition for women's interests to be taken into account' (United Nations 1996).

2.2 Biotechnology: A Critical Technology for Sustainable Economic Growth and Development

Science is progressing rapidly and appears full of opportunities. In particular, advancements in the biological sciences during the last few decades have resulted in an explosion of knowledge not previously seen in all of human history. These leaps in biological understanding have provided the impetus for the development and expansion of the biotechnology industry. In the most simplistic terms, biotechnology can be defined as the use of biological processes to generate products and services. Because biotechnology is a diverse and rapidly evolving sector it has been a challenge to develop a commonly accepted definition. However, a universally accepted definition is crucial to comparing biotechnology activities between countries. *It has recently been defined by the OECD as the application of science and*

[1] Clause 181

technology to living organisms, as well as parts, products and models thereof, to alter living or non-living materials for the production of knowledge, goods and services (Beuzekom 2001).

Biotechnology represents one of the technological revolutions of this century. Its application spans many sectors including health, agriculture, food processing, environment, natural resources and aquaculture. Biotechnology has profoundly influenced the production of therapeutic and diagnostics for human health, improved agricultural yield through genetic modification, and through bio-remediation and new industrial processing methods, has addressed environmental protection issues. It is considered a pillar of the knowledge economies, and a powerful driver of economic growth.

The impact of biotechnological innovation – both in strengthening existing industries and facilitating the growth of new businesses – is inescapable. There are over 5,000 biotechnology companies globally. In 2006, the United States had the largest number of 'dedicated' biotechnology firms,[2] (2,196) followed by Japan (804) and France (755). Canada currently has approximately 500 biotechnology companies (Beuzekom and Arundel 2006). The European Union (15) had a total of 3,154 biotechnology companies (Beuzekom and Arundel 2006). A recent report by the European Commission's in-house scientific service, DG Joint Research Centre, which maps the applications of biotechnology in Europe, showed that in 2006 the production and use of modern biotechnology supported the generation of up to 1.69% of the EU economy (Gross Value Added). In Europe, biotechnology's economic importance has become comparable to the largest industry sectors. In Canada, biotechnology revenues now exceed 4 billion dollars (Statistics Canada 2007).

2.3 Gender Dimensions of Biotechnology

In 2004, the Gender Advisory Board (GAB) of the United Nations Commission on Science and Technology for Development[3] reported that 'very little research existed on assessing the gendered patterns of benefit, use, development and effect of biotechnologies and that what existed had not been compiled or examined in such a way as to enable assessment of the status or depth of knowledge in the area, or understand what may be the most effective policy responses'. Essentially

[2] Dedicated biotechnology firm – is defined by the OECD as a biotechnology active firm whose predominant activity involves the application of biotechnology techniques to produce goods or services and/or the performance of biotechnology R&D.

[3] 'Understanding the Gender Dimensions of Biotechnology Research and Development-regional Consultative Expert Workshops', organised by the Gender Advisory Board, UN Commission on Science and Technology for Development in collaboration with the University of Pretoria and the Pakistan National Commission on Biotechnology in Pretoria and Islamabad, on November 26, 27 and 30–December 1, 2004.

this means that in the 30 years since the advent of biotechnology – marked by the *in vitro* production of a human protein (somatostatin) – the gender dimensions of biotechnology have remained relatively unexplored. Yet, there are clearly a few biotechnologies that have a significant and proven impact on gender equality and women's rights (Thomas 2003; Klinge and Maguire 2004).

As mentioned in other chapters of this same publication, many reproductive and genetic technologies can reinforce gender inequalities. For instance, ultrasound imaging, combined with genetic tests or conventional cytogenetic analysis, is routinely used for medical care purposes to determine whether an unborn child may be carrier of a genetic disease (e.g. Duchenne muscular dystrophy, which is linked to the X chromosome). The same mix of technologies can also be used for sex selection purposes, as has been the case in India and China, leading to dramatic imbalance in sex ratios (Chan et al. 2006). Currently, ratios of males to females in China are alarming: 100 females:117 males in cities and 100:130 in rural areas, where the vast majority of the Chinese population live. In 2002, Dr. Chan, Director of the Center for Behavioural Health in Hong Kong, reported that a common name for girls in China in the 1950s and 1960s was 'Die Di', meaning 'bring a young brother' (Chan et al. 2002). Needless to say, these and other reproductive technologies have been a traditional focus of gender equality advocates.

While not immediately obvious, the use of genetically modified (GM) crops may also differentially affect women compared to men. In many parts of the world, women comprise the majority of subsistence farmers. Repercussions or consequences from the use of GM crops may have a more direct impact on women. Women are also often the primary workers in the field to be exposed to pesticides, herbicides for weed management and insecticides (AWID 2004). Currently, the most outspoken critics of GM foods are environmental groups, though gender rights advocates have now entered the fray (AWID 2004; Pradesh 2007). Overall, these developments represent only a minor fraction of the biotechnology industry potential. There are undoubtedly many other applications that can and will have an impact on women's lives and new technologies are constantly emerging. Yet, there appears to have been few or no attempts to systematically apply gender mainstreaming, in other words apply a gender 'lens', to the formulation and implementation of policy decisions and projects in biotechnology. Although it is obvious that gender dimensions relevant to biotechnology development and application exist, little systematic research has been done to date to understand their nature.

There is clearly a gap between intention and practice to engage gender mainstreaming analysis and tools in biotechnology. It is unclear if this gap is a result of a lack of knowledge or understanding about how gender mainstreaming works and can be used with biotechnology, or a lack of the necessary organizational will, or a mixture of the two. How would gender mainstreaming really look like for biotechnology? How do we develop a context that truly and systematically takes into account gender differences in biotechnology applications, tries to identify developments that can make a difference to the lives of women, and strives for gender equality?

In addition to lack of analysis regarding the impact of biotechnologies on women, even less attention has been paid to the representation of women within the industry itself. This raises the question – Why should we care? There is of course the most obvious answer: gender equality is a fundamental human right and a question of social justice. The promotion of gender equality is instrumental and key to achieving internationally agreed development goals. However – within the narrower context of biotechnology – there are other socio-economic reasons for why it is imperative that attention be paid to this issue. First, if women are active participants in the development of biotechnologies, they can influence policy decisions that ultimately could affect women's lives. Women bring in different perspectives and research interests and as such can contribute in steering the direction and quality of research. This has implications beyond just protecting women's rights. Also, the success of the biotechnology industry requires both public acceptance and accountability, and as women constitute half of the public, they must be actively engaged in the debate.

A second critical argument underlying the importance of ascertaining the level of participation of women in the biotechnology sector relates to human resource issues. As mentioned in the introduction, one of the most critical determinants of growth and success of the biotechnology sector is access to a highly skilled labour force. To be globally competitive, countries cannot afford to lose highly talented individuals because of gender.

2.4 Economic Success of Biotechnology is Not Inevitable

While the success of biotechnology as an enabling tool to advance scientific knowledge and discovery appears inevitable, economic success of biotechnology also requires public acceptance and accountability. Today many people recognize that when used discerningly, biotechnology appears to offer new possibilities to improve the quality of life. There is nonetheless also evidence of dissatisfaction with a number of recent applications of biotechnology research, specifically in health and agriculture; and there are calls for a better 'match' between the scientific research agenda(s) and societal needs. In Europe, in particular, the most recent Eurobarometer (Gaskell et al. 2006) reports that support is strong for medical applications of biotechnology as long as there are clear benefits to human health. For example, while the public generally perceives recent developments in pharmacogenetics as useful and worthwhile pursuing, there are significant mixed reactions to gene therapy. As the judged usefulness of technologies declines there is also an increase in perceived – risk. The crisis of confidence on the directions and applications of biotechnology-which has regularly been monitored by the Eurobarometer – is of great importance both to society at large and for scientific and technological progress. It helps, therefore, to take a step back and briefly review some of the arguments and considerations underlying the public debate surrounding biotechnology that reveal factors that may restrain innovation and commercialization in biotechnology.

2.5 Public Acceptance and Accountability

Since its very inception biotechnology has raised ethical issues, challenged traditional social and normative institutions and has elicited strong sentiments across civil society. Given its intrinsic power to manipulate life and life forms and the extent to which it can affect all aspects of human life and human choice, this could not be otherwise. Indeed a commonly heard comment is that biotechnology has transformed scientists into 'God-like creators', capable of dismantling the integrity of all species, including humans. Examples of some recent experimentation give a sense of the potential for transgression: mice engrafted with human cartilage (ears) on their backs; baboons with transgenic pig hearts transplanted into their necks (Chen et al. 2000); genetically modified salmons that grow twice as fast as their wild relatives (Fletcher et al. 1999) fruit-flies where eyes are induced to grow in unusual parts of their body (Halder et al. 1995).

These research applications have clearly captured the imaginations of journalists and the public. Yet, western science is not new to transgression or to the deconstruction of human essence. For example, the view of the human body as a set of parts that can be manipulated, analysed, transformed and enhanced has its origin in medical practice and can be traced back to 1772 with Robert Boyle and his mechanical view of the disfunctional body (Hunter and Davis 1999–2000). And it is not by chance that Mary Shelley's famous character, Frankenstein, has often been associated with the products of new biotechnologies. Shelley's visionary inspiration, as reported in her own words in the preface to Frankenstein: 'I have thus endeavoured to preserve the truth of the elementary principles of human nature, while I have not scrupled to innovate upon their combinations' appears strikingly foretelling of today's genetic engineering research. More recently, anthropologist Margaret Lock has described the scientific view of the human body as 'isolated, decontextualised and abstracted from real time and social space' (Lock 1993). It is indeed in the context of advances in molecular biology, human genetics, genomics, and tissue bioengineering that the concept of the human body has undergone the most significant transformation.

Today the human body is seen as a depository of genes, or in the words of biologist Thomas Eisner, as a loose-leaf book, whose individual pages, the genes, are available for selective transfer and modification (Eisner 1982). With the complete human genome now sequenced, the human body has become a resource to be mined, ultimately an object whose most intrinsic subjective properties are – as all other chemical substances – part of an unbroken continuum with the rest of the universe and can be used and modified at will.[4] The fact that some bodily parts or genetic characteristics may be moved about between species, including from animals to humans (e.g., through xenotransplantation or gene therapy) inspires in many people a sense of disarray, fear, perhaps of disgust or of 'yuk' (Midgley 2000).

[4] See: The Icelandic Health Sector Database in: Genetic Testing, 'Policy Issues for the New Millennium' (2000) – OECD

This has resulted in a widespread preoccupation with the actual risks of using biotechnology and in a public outcry for more and better information and for more direct ways of influencing decision-making on scientific and technological issues, on the directions that innovation should take. Not surprisingly for many governments, this represents a formidable task and at the same time an opportunity. Policies prepared after long intensive study, negotiation and compromise have recently been called into question by adverse public reactions. The controversy over the appearance of transgenic foods in Europe remains a poignant example.

Consumers today demand to know how innovations come about. They have gained an enhanced understanding of the potential adverse consequences of human intervention, particularly if it is guided by misplaced commercial interests. Hence, they are calling for more rigorous review, greater accountability and more open debate on the risks and benefits of genetically modified plants and animals, new medical tests and therapies and other applications of biotechnology.

The responses of governmental and legal institutions to public controversy and to the emergence of new biotechnologies have consistently held the potential to make or break the commercial prospects of the sector. A great majority of governments have therefore come to recognize the need to develop more effective mechanisms for involving all relevant stakeholders in discussion on risk and in the formulation and implementation of policy decisions. Clearly, for women's voices to be heard they must be present in those institutions that count in this debate, in research and decision-making positions. Yet, as discussed in the next sections, women make up only a minority of senior academic staff in the majority of countries. Furthermore, their representation in 'official and managerial positions' in industry and in public office remains considerably lower than that of men. As a result, women are often under-represented in advisory boards or in consultations where 'eminent international experts in senior positions' are called to share opinions, even when their interests are directly at stake.[5]

2.6 Sustaining a Highly Skilled Science and Technology Workforce

The main economic drivers of biotechnology are embodied in the skills, knowledge and potential for innovation of highly trained personnel. Success in the biotechnology industry requires companies to recruit and retain top talent. Demand for this talent is, however, rapidly exceeding the available supply. In the case of the European Union, the 'Gago report', a strategic report drafted by Portugal's former science

[5] As an example, the 2005 report The Top 10 Biotechnologies for Improving Health in Developing Countries, by the University of Toronto, provides a significant and valuable contribution to the fierce debate about the potential for good that biotechnology holds for the global community. It suggests a process for priority setting. Regretfully only five of the twenty eight contributing experts were women.

minister, Jose Mariano Gago in 2005, estimated that at least 500,000 more researchers would be needed to reach the EU's Lisbon Strategy objectives, i.e. – to become the world's leading knowledge-based economy – by 2010. In 2005, the OECD Global Science Forum reported that there is a general declining interest in science studies among young people and a significant decrease in doctorate holders (OECD 2006a). The study found that the relative share of graduates in science and technology disciplines had actually declined over the period 1993–2003 in 10 out of the 16 OECD countries studied and that the trends were even more negative at the doctorate levels in all except three countries (Japan, Turkey, Korea). Furthermore, the study found that science and technology disciplines were affected differently by the decline, for instance, enrolments decreased in absolute and relative terms for physics or mathematics in several countries, while the opposite occurred in computing studies. These trends have significant implications for the vitality and sustainability of the biotechnology sector where expertise is needed in a range of disciplines including biology, medicine, chemistry, engineering, physics, law and business.

In November 2006, Canada's federal government released *Advantage Canada*, an economic plan to make Canada a world leader for current and future generations (Canada 2006). *Advantage Canada* recognizes that Canada can and must do more to turn ideas into innovations that provide solutions to environment, health, and other important challenges, and to improve Canadian economic competitiveness. The plan advocates creating five Canadian 'advantages' that will help Canada prosper in the world economy. One of these is to strive for a knowledge advantage. However, currently in Canada it has been noted that there is a 'lack of supply of skilled and educated workers required to become a global leader in biotechnology' (Munn-Venn and Mitchell 2005). Too few Canadian students choose to pursue advanced S&T degrees. Compared to the OECD average, Canada has fewer natural science and engineering degree students within the total student population and fewer Ph.D. holders among young Canadians (Industry Canada 2007).

2.7 Participation of Women in Biotechnology

OECD countries cannot remain inactive with regard to the need to build human potential in the field of science and technology. With over 5,000 biotechnology companies worldwide, economic success clearly depends on talent and human resources. Women and other minorities are the most obvious source for increasing the Science & Technology (S&T) workforce. But what do the statistics say about the participation of women in biotechnology?

In 2000, the OECD initiated a major effort to outline a statistical framework to guide the measurement of biotechnology activity and to develop conventions for measuring and comparing biotechnology activities across OECD countries. A first compendium of biotechnology statistics was released in 2001, but was based on diverse sources, mainly from private unofficial providers (Beuzekom 2001). The second (Devlin 2003) and the more recent third editions (Beuzekom and Arundel 2006) are based to a much a larger degree than the first version on official data. Despite these considerable efforts we still have little information to draw upon for an analysis of women in biotechnology. First, differences in the definitions and methods that countries use to compile their data create problems for comparing developments in the biotechnology economy across countries, including employment. Second, national labour force data is often not sufficiently disaggregated by gender. As a result, many unanswered questions remain also at the level of individual countries.

We will focus here on a review of the information available for Canada. Statistics Canada has conducted surveys that provide a comprehensive analysis of the biotechnology industry since the mid 1990s. The first survey, the Biotechnology Use Survey in 1996, looked at the use of biotechnologies in a number of industries (Statistics Canada 1998). This was followed by a second survey in 1997, the Biotechnology Firm Survey, which focused on core biotechnology firms (Traore 2001). Subsequently, the elements of the first two surveys were combined in the 1999 Biotechnology Use and Development Survey (BUDS) (McNiven 2001). This survey has since been conducted every two years (2001, 2003, 2005) and provides a detailed analysis of the biotechnology industry sector in Canada, including a breakdown of human resources. Although detailed information is available with respect to biotechnology employment by position type, by sector, and by size of companies, none of this information is disaggregated by gender. Separate from the biotechnology surveys, Statistics Canada has recently published a detailed gender-based statistical analysis, the fifth in a series that began in 1985. Although it provides data on the representation of women in different occupational groups, it does not specifically focus on the biotechnology sector. However, two occupational classifications would include a subset of professionals who are employed in the biotechnology sector. These include managerial positions and professionals in natural sciences, engineering and mathematics. As of 2004 women comprised only 21% of professionals in natural sciences, engineering and mathematics. This has changed little since 1987 when they accounted for 19.5% of employees in these fields. Women also only make up a minority (24%) of senior level managers in Canada.

2.8 Which Fields of Science and Technology Do Women Choose?

The OECD and the European Commission's statistical office, Eurostat, have worked closely to improve the collection of gender disaggregated statistics in a number of Science and Technology (S&T) fields of studies relevant to biotechnology: *life sciences, mathematics and statistics, physical sciences, computing sciences, and engineering*. The data for 2005 show that the number of female students in tertiary education has increased more rapidly than that of males. But the proportion of women choosing science and technology studies still remains below 40% in most OECD countries and the choice of discipline is still highly gender-dependant. Fields of study such as engineering or computing sciences are largely male-dominated.

In Canada, in 2004–05, women comprised the majority of students (58%) enrolled full time in university at the undergraduate level (Canadian Association of University Teachers, CAUT 2007). Similar to what was observed for most OECD countries, women are still significantly under-represented in the mathematics and computer sciences (25%), engineering and related technologies (21%). Women undergraduates are still a minority in the traditionally male dominated discipline of physics. They exceed men in the majority of the life sciences study areas including biochemistry, biophysics and molecular biology, biological and physical sciences, general biology, genetics, biotechnology, plant biology, ecology, microbiology, pharmacology and toxicology, physiology and zoology. It is only in the relatively new field of bioinformatics, which can be loosely defined as the use of computers and computer science to study biological questions, that women account for only 32% of the undergraduate enrollment.

Recent data shows, however, that, when compared to the numbers at the undergraduate level, fewer women choose to continue graduate level life science studies (CAUT 2007). In 2004–05 women accounted for 45.2% of Ph.D. candidates in the life sciences and received 43.1% of doctoral degrees. There are, nonetheless, a subset of disciplines in which currently just over half of the Ph.D. candidates are women, including botany, genetics, ecology/evolution/population biology and physiology/pathology.

It will be interesting to revisit these numbers in a few years to observe if the increase in enrollment at the undergraduate level translates into an increase in the representation of women in different Ph.D. programs.

2.9 Women Academic Career Paths

During the past few decades, scholarly dialogue on the topic of women academic career paths has increased. The existence of a 'glass ceiling' or 'sticky floor' for women trying to progress to senior positions is well documented and affects all occupational sectors, even those which are dominated by women.

To examine the representation of women throughout their academic career we will focus on Canada. In Canada, detailed statistics exists regarding the representation of women at the different tiers of faculty (CAUT 2007). In 2004–05, 35.8% of assistant professors in the agricultural and biological sciences were women. This figure declines to only 30.9% at the associate professor level. At the most senior academic level women are significantly under-represented. They account for only 17.9% of full professors specifically in the agricultural and biological sciences. This under-representation of women at the highest academic rank is not restricted to the life sciences. The total percentage of women full professors in Canada, regardless of discipline, is only 18.8%. The only academic rank in which the representation of women exceeds that of men in the agricultural and biological sciences is classified as 'other'. This category includes lecturer and instructor positions, most of which are non-tenure track. These are less desirable positions offering lower pay and less job stability. Women occupy 63.9% of these positions.

A recent study by the CAUT reveals that this situation is not unique to Canada (CAUT 2006). CAUT compared the percentage of women at the highest academic rank from selected English speaking countries. Both the United Kingdom and New Zealand fare slightly worse than Canada, with just 13% and 14% of full professors being women. By comparison, in the United States 28% of full professorships are held by women. In most European countries, the percentage of women in the top grades of academia is below 20% (European Commission She Figures 2006).

2.10 Why Are Women Under-Represented in Senior Academic Ranks?

A variety of explanations have been proposed to account for the under-representation of women in senior academic ranks. A discussion of all the possible factors influencing a women's progression along the academic ranks is beyond the scope of this paper. A comprehensive analysis can be found in the 2006 report titled 'Beyond Bias and Barriers: Fulfilling the potential of women in academic science and engineering', by the US National Academy of Sciences. We will focus here briefly on the 'leaking pipeline'-metaphor,[6] and the broader issue of reconciling work and family life.

The progression from undergraduate to full professor is often referred to as the pipeline and the factors leading to leakage along this pipeline have come under intense scrutiny. Most researchers start as Ph.D. students, proceed thereafter into post doctoral positions, researchers or assistant professors; then become associate professors and some eventually achieve the rank of full professor. If the proportion of women differs from one level to the next it is assumed that the pipeline 'leaks'. In other words, if there is an increase in the percentage of women entering 'the S&T pipeline' and obtaining Ph.Ds., unless the pipeline is 'leaky', one would also predict a significant increase in the gender ratio at the end of the pipeline, i.e., of full professors. This assumption can be tested against the data available for Canada. In Canada, between 1991–1992, women obtained 28.5% of earned doctorates in the agricultural and biological sciences. Sufficient time has elapsed for these women to have entered the most senior academic ranks. We have previously reported that the total percentage of women full professors in Canada, regardless of discipline, is only 18.8% (CAUT 2006). These numbers would suggest the existence of a 'leaking pipeline', however, it is likely not the only explanation.

The OECD has recently reviewed national policies for reconciling work and family life in its series Babies and Bosses. Volume 4, published in 2005, covers Canada (in particular the province of Québec), Finland, Sweden and the United Kingdom (OECD 2005). This series found that child care is more readily available today in many countries than in the past, though many tax and benefit systems still discriminate against women in some countries. Part-time work for men and women

[6] For a reader-friendly description of the leaky pipeline see Saunders (2002).

is, at least in some countries, also an effective way of reconciling earning and caring. But overall working life is still experienced very differently by men and women (OECD 2006b). In particular, tension between work and family can inhibit women's progression through the academic ranks. Mary Anne Mason, the first woman Dean of Graduate Studies at the University of California Berkeley, is a strong advocate of this theory. She has analyzed extensively data from the Survey of Doctorate Recipients (SDR) conducted by the National Science Foundation in the United States. She discovered that women who had children within five years post Ph.D. were less likely to achieve tenure than men who also had children during the same period (Mason and Goulden 2002). This is a very critical period in the lifetime of a scientist. For life scientists who choose to pursue an academic research career, additional training after receiving a PhD, is almost always required. This postdoctoral position can last on average three to five years. During this period, postdoctoral fellows develop their own research interests and gain prominence through the publication of their research results. It appears, therefore, that the biological clock is on a direct collision course with the tenure clock.

In addition to the most senior academic ranks, women are also under-represented in decision-making positions at research-intensive Canadian universities (WISEST 2002). An interim study released in 2002 analyzed the number of women in senior administrative positions in 12 top medical/doctoral universities throughout Canada. These positions include President, Vice-President, Associate Vice-President, Dean of Faculties and Chairs of Departments. The data from this preliminary study are striking. Overall, women only occupy 17.1% of these key positions. These are influential positions. Individuals in these positions have input into decisions about institutional funds for research, staff, strategic planning etc. They ultimately can shape the future of an institution.

2.11 Positive Trends

A positive indicator of the status of Canadian academic women is the narrowing of the gender income gap. In 1995, the average salary of women professors in the agricultural and biological sciences was only 82.3% of that of their male colleagues. By 2005, this gap had closed to 91.7%. The picture is even more encouraging when the data is broken down by academic rank. While a gender pay gap exists at the level of full professor, at the level of assistant professor women's salaries are now on par with men (CAUT 2007).

However, in Canada the picture is less rosy outside of academia. Although data is not available specifically for the biotechnology industry, in general women working full time in the private sector earn consistently less than their male counterparts (Statistics Canada 2006). The average annual income for women with university degrees employed full time is slightly less than 70% of that of men. This picture does not improve significantly when the average income of women and men is compared according to the field of occupation. Women professionals employed full time in the natural sciences earn 83.1% of the average male income for the same position. In managerial positions the differential increases. The earnings drop to 67.4% of that of men (Auriol 2007). Thus, all the available data strongly suggests that if gender disaggregated statistics existed for the biotechnology industry, a gender pay gap would be apparent.

The OECD has recently published a first analysis of available statistics on doctorate holders in nine countries, including the United States and Canada. The report shows that in the United States, as may be expected, women are less well paid than men. For research positions the differential is larger in the United States (–27%) than in Canada (–20%). The differences are more pronounced in business and less marked in higher education. There is an exception to this in postdoctoral positions, where women receive higher salaries than men. One should also note that post-doctoral salary levels are much lower than other research position salary levels (Kergroach and Cervantes 2006) and salaries for early stage researchers are rather low in many countries relative to per capita GDP.

Another measure of success for academic scientists is their ability to obtain grant funding, an absolute necessity for conducting research in the life sciences. One of the major granting agencies in Canada is the Natural Sciences and Engineering Research Council of Canada (NSERC), a federal government agency. The Discovery Grant is the largest NSERC granting program and provides essential funding for many laboratories in Canada. In 2004–05, 15% of Discovery Grants were awarded to women (Women in Science and Engineering in Canada 2006). While at first glance this appears to be very low, it must be pointed out that in addition to the life sciences, Discovery Grants are also awarded in other disciplines such as engineering and physics, fields of study that have a more significant shortage of women faculty. A more reliable measure is the success rate of male versus female applicants. By this measure there has been little difference in the success rates of men and women over the last several years. In addition, another positive indicator is the longevity of females in the Discovery Grant program. After 10 years in the program the percentage of women still receiving a grant are on par with their male colleagues. Finally, there is no appreciable difference in the average grant size. Thus, at least for NSERC, once women achieve a faculty position they are as successful as men in obtaining and maintaining grant funding.

It must be remembered, however, that women are in a minority in many of the key subjects for which research funding is sought, so equal rates of success still disguise continuing differences in numbers. Further analysis would also be needed to understand application patterns, since the rate at which men and women apply is also important.

2.12 Direct Support Measures

Most OECD countries have specific programmes in place which aim to achieve a better gender balance in science education and research (e.g. improved childcare, measures to balance work and family responsibilities, mentoring programmes).

Such programmes are very important at the level of individual institutions. Most of these instruments and measures are geared to the universities and public sector research.

For example, as mentioned previously, specific measures include grants to support positions for women at universities. In Canada, NSERC has acknowledged that women face barriers in becoming faculty members and has actively developed several programs to increase the participation of women. Thus, in 1998 NSERC launched the University Faculty Award (UFA) program.[7] The goal of this program is to promote the recruitment and early career progression of women and aboriginal people in tenure-track faculty positions in Canadian universities. UFA award holders receive a portion of their salary for five years. As a result, it is expected they have a reduced teaching and administrative loads during this time so they can devote more time to establishing a competitive research program. This award encourages universities to appoint women (or ethnic minorities) researchers to tenure-track positions. Because this is a relatively new program its impact on increasing the representation of women at senior academic levels has yet to be evaluated.

Another program specifically designed by NSERC to increase the participation of women in science and engineering is the Chairs for Women in Science and Engineering.[8] These Chairs are specially funded positions in which the NSERC funding is matched by contributions from corporate sponsors. These prestigious positions are held by successful women academics that have demonstrated research excellence. These women serve as role models for women considering careers in science and engineering and as part of the Chairs responsibilities they are actively engaged in developing, implementing and communicating strategies to increase the participation of women in these fields.

Recent research suggests that efforts to close the gender gap in science must begin at the earliest levels of schooling. Thus, NSERC has also developed the PromoScience program.[9] This program provides grant funding to a wide variety of organizations, such as non-profit groups, which aim to foster and develop an interest in science and engineering in young Canadians from elementary school through to high school.

In 1999, the federal government of Canada launched a new program, the Canadian Research Chairs (CRC) program,[10] which was designed to help universities recruit and retain talented researchers and thereby strengthen research excellence in Canada. To achieve this goal this program involved the creation of 2,000 research professorships by the year 2008. These chairs are awarded in three different disciplines; natural sciences and engineering, health sciences, social sciences and humanities. They are awarded for five to seven years, depending on the stage of career, and are renewable. Each eligible university is individually responsible for

[7] http://www.nserc.gc.ca/sf_e.asp?nav = sfnav&lbi = c7

[8] http://www.nserc.gc.ca/programs/wise_e.htm

[9] http://www.nserc.gc.ca/promoscience/index_e.htm

[10] http://www.chairs.gc.ca/web/home_e.asp

nominating researchers to fill the chair positions. Nominations undergo an outside peer review process.

It quickly became apparent two years into the program that there was an under representation of women. Shortly after its inception in 2000–01 only 14.5% of CRC chairs had been awarded to women. Thus, CRC officials commissioned a gender-based evaluation of the program (Begin-Heick and Associates Inc. 2002). The evaluation revealed that the under-representation of women in the CRC program lay with the universities' failure to nominate female candidates. On the other hand, once nominated, women were granted the chair position at the same rate as men, suggesting that the approval process did not discriminate against women.

As a result of these findings, the CRC committee took several courses of actions. In particular, the committee asked universities to update their strategic research plans and to address the issue of gender representation. Progress has, however, been slow and in 2005 the steering committee decided to increase data collection and monitoring and to hold universities accountable for meeting gender distribution targets. This sterner approach appears to have yielded tangible results. The percentage of women awarded CRC chairs has increased from 14% in 2000 to 32% in 2006.[11] Continued monitoring is necessary to ensure that the proportion of women CRC chairs adequately reflects the available pool of women faculty.

2.13 The Importance of Mentoring

While there is no magic bullet to ensure representation of women in science, the importance of mentoring cannot be underestimated. One of the recommendations from the conference 'Women & Science: Celebrating Achievements, Charting Challenges', organized in 1997 by the US National Science Foundation, was to recognize the importance of mentoring. According to Shirley Tilghman, President of Princeton University in the United States, 'mentoring is important and it is important at every level' (Tilghman 2004).

Mentoring can be simply defined as a relationship in which a more experience person shares its knowledge and insight with a less experienced person. These exchanges can be very informal in nature. For instance, a faculty member can have an 'open door policy' where students are encouraged to drop in when they have questions or need advice. Mentoring can also take on more formal arrangements involving organized events and activities.

One organization in Canada with an active mentorship role is the Society for Canadian Women in Science and Technology (SCWIST).[12] SCWIST is a non-profit voluntary organization located in Vancouver, Canada. It was established in 1981 to

[11] http://www.chairs.gc.ca/web/about/stats/may2007.pdf

[12] http://www.scwist.ca/

promote, encourage and empower women working in science and technology. The goals of SCWIST are to:

- Promote public awareness of the opportunities for women in science and technology;
- Encourage the full participation of girls and women in all aspects of science and technology education; and
- Increase the representation, retention and status of women in the science and technology workplace.

SCWIST aims to support girls and women at all stages of their education and careers, from young girls in elementary school to women established in their careers. In SCWIST's 26 year history, the organization has implemented and coordinated many exciting, entertaining and interesting programs to achieve these goals. Financial support for these programs have come from a variety of sources, the most notable of which was a generous donation from the late Nobel Prize-winning chemist Dr. Michael Smith.

SCWIST has long recognized the importance of encouraging young girls to pursue math and science. In 1990, the organisation developed an educational outreach program called *ms infinity* that aims to motivate girls to continue with their math and science education. The program consists of four main components: the *ms infinity* conferences, Quantum Leaps conferences, Telementoring and Science Days for Girl Guides. *The ms infinity* conferences, as well as the Science Days, allow young girls to learn about science and participate in hands-on science workshops led by enthusiastic, successful and engaging women that work in different science and technology fields. The program also aims to dispel myths and stereotypes regarding women in science. Quantum Leaps conferences and telementoring provide opportunities for girls to interact with women in scientific careers. One key to the continued support and success of the *ms infinity* program is the participation of dedicated women that serve as mentors. These are dynamic women who are passionate about their careers, have rewarding personal lives and outside interests. Overall, this program provides an excellent model for organizations that would like to develop programs to motivate girls to consider careers in science and technology.

2.14 Conclusions

Much has been said and tried to advance women in science and technology in the last 12 years, since the Fourth World Conference on Women in Beijing (United Nations 1996). Yet, in spite of remarkable progress towards equal opportunity, occupational segregation persists. Women are less favoured than men on the S&T labour market, in terms of unemployment, time to employment, type of contract, access to research positions and salary. Women are also still significantly under-represented in senior and decision-making posts. With varying degrees, this is true for all OECD countries. These inequalities extend to all science and technology disciplines and also to those areas and sectors which are of critical importance for biotechnology.

To gain insights on why this is the case, better data on women employment and career paths are needed. In particular, there is an imperative need to fill the current data gap about women's participation in the biotechnology sector and in the creation of science-based start-ups.

Although simple statistics are not sufficient for policy makers to devise solutions or remedies, gender disaggregated data can improve the general knowledge of what's going on, and, even more importantly, of what is not going on. In the absence of data, the need for change may not be fully understood by those who are in a position to make a difference. Furthermore, data collection enables measurement of progress and, as the examples presented in this paper show, a lot can be gained by monitoring over time the impacts of the implementation of new policies and strategies.

The examples presented in this paper also show that the 'toolkit' can be quite rich and varied. Many different support measures can be used today to directly impact recruitment and retention of women in science and technology. Most of these measures are geared to the public education and research sectors. There are no quick fixes, though, and there is no best single formula - even at the level of an individual country. Strategies need to adapt to the different local contexts. And to succeed, there needs to be a coherent package of support measures 'at every level of education and career'.

These 'rules of thumb' could also surely apply to the biotechnology sector. However, there is no hard evidence to speak about since the gender dimensions of biotechnology have remained relatively unexplored and many questions remain unanswered. We listed in the introduction a number of the most obvious questions, particularly those that relate to women in the biotechnology work force. There are however, the less obvious ones too. Not only has gender equity to be achieved as far as numbers are concerned, but an increased participation of women must also lead to greater influence in shaping the major scientific questions of the moment, many of which impact directly on the lives of women. Women bring in different perspectives and research interests and as such should contribute in steering the direction and quality of research. It is therefore necessary to understand how women can influence this direction and the impacts of the research. How are research agendas set today? For whose benefit? In science departments are the policy priorities targeting women and men equitably? How can women's participation in decision-making be assured?

Left unattended, the absence of policy debate in these areas, together with the scarcity of women in senior scientific and policy positions, inevitably means that women individual and collective opinions are less likely to be voiced in policy and decision making processes and equally in setting a biotechnology research agenda responsive to their needs.

References

Auriol, L. (2007). Labour market characteristics and gender mobility of doctorate holders: Results for seven countries. Retrieved October 20, 2007, from http://www.oecd.org/dataoecd/17/57/38055153.pdf

AWID (Association for Women's Rights in Development) (2004). Why new technology is a women's rights issue. *Facts and Issues, 7. Gender Equality and New Technologies*. Retrieved April 31, 2007, from http://www.awid.org/publications/primers/factsissues7.pdf

Begin-Heick, N. & Associates Inc. (2002). Gender-based analysis of the Canada Research Chairs program.Retrieved October 20, 2007, from http://www.chairs.gc.ca/web/about/publications/gender_e.pdf

Beuzekom, B. van (2001). *Biotechnology statistics in OECD member countries: Compendium of existing national statistics*, STI Working Papers 2001/6. Paris: OECD. Retrieved October 20, 2007, from http://www.olis.oecd.org/olis/2001doc.nsf/43bb6130e5e86e5fc12569fa005d004c/c1256985004c66e3c1256ac600350f21/$FILE/JT00112476.PDF

Beuzekom, B. van & Arundel, A. (2006). OECD Biotechnology statistics-2006. Retrieved from http://www.oecd.org/dataoecd/51/59/36760212.pdf

Canada (2006). Advantage Canada: Building a strong economy for Canadians. Retrieved October 20, 2007, from http://www.fin.gc.ca/ec2006/pdf/plane.pdf

CAUT (Canadian Association of University Teachers) (2006). Women in the academic workforce. *CAUT Education Review, 8*(1), 1–6.

CAUT (Canadian Association of University Teachers) (2007). *CAUT Almanac of post-secondary education in Canada 2007*. Retrieved October 20, 2007, from http://www.caut.ca/en/publications/almanac/2007_CAUT_Almanac.PDF

Chan, C. L., Blyth, E., & Chan, C. H. Y. (2006). Attitudes to and practices regarding sex selection in China. *Prenatal Diagnosis, 26*, 610–613.

Chan, C. L. W., Yip, P. S. F., Ng, E. U. H., Chan, C. H. Y., & Au, J. S. K. (2002). Gender selection in China: Its meanings and implications. *Journal of Assisted Reproduction and Genetics, 19*(9), 426–430.

Chen, R. H., Kadner, A., Mitchell, R. N., & Adams, D. H. (2000). Mechanism of delayed rejection in transgenic pig-to-primate cardiac xenotransplantation. *The Journal of surgical research, 90*(2), 119–25.

Devlin, A. (2003). An overview of biotechnology statistics in selected countries: Statistical analysis of Science, Technology and Industry, STI Working Papers 2003/13. Retrieved October 20, 2007, from http://www.olis.oecd.org/olis/2003doc.nsf/43bb6130e5e86e5fc12569fa005d004c/7524fee7966a86c2c1256de50049f989/$FILE/JT00154671.PDF

Eisner, T. (1982). Chemical ecology and genetic engineering: the prospects for plant protection and the need for plant habitat conservation. Abstract. Paper presented at the Symposium on Tropical Biology and Agriculture, St. Louis, MO: Monsanto Company, July 15, 1985).

European Commission (2006). *She Figures 2006. Women and Science Statistics and Indicators.* Retrieved April 31, 2007, from http://ec.europa.eu/research/science-society/pdf/

Fletcher, G. L., Shears, M. A., Goddard, S. V., Alderson, R., Chin-Dixon, E. A., & Hew, C. L. (1999). Transgenic fish for sustainable aquaculture. In N. Svennevig, H. Reinertsen, & M. New (Eds.), *Sustainable aquaculture: Food for the future?* (pp. 193–201). Rotterdam: AA Balkema.

Gaskell, G. et al. (2006). Europeans and biotechnology in 2005: Patterns and trends, final report on eurobarometer 64.3: Report to the European Commission's Directorate-General for research. Retrieved June06, 2007, from http://ec.europa.eu/research/biosociety/public_understanding/eurobarometer_en.htm

Halder, G., Callaerts, P., & Gehring, W. J. (1995). Induction of ectopic eyes by targeted expression of the eyeless gene in Drosophila. *Science, 267*(5205), 1788–1792.
Hunter, M. & Davis, E. B. (Eds.) (1999–2000). *The works of Robert Boyle*. London: Pickering & Chatto.
Industry Canada (2007). Mobilizing science and technology to Canada's advantage. Retrieved October 20, 2007, from http://www.ic.gc.ca/cmb/welcomeic.nsf 532340a8523f33718525649d 006b119d/ 69c0cd4e782234b5852572de00503b8f!OpenDocument
Kergroach, S., & Cervantes, M. (2006). *Complete Results of the SFRI Questionnaire on the Working Conditions of Researchers in the Universities and Public Research Organisations, DSTI/STP/SFRI(2006)1*, OECD.
Klinge, I., & Maguire, P. (2004). The policy implications of gender mainstreaming for healthcare research in the EU. *PharmacoEconomics, 22*(Suppl. 2), 87–94.
Lock, M. (1993). *Encounters with aging* (pp. 370–371). Berkley, CA: University of California Press.
Mason, M. A., & Goulden, M. (2002). Do babies matter? The effect of family formation on the lifelong careers of academic men and women. *Academe: Bulletin of the AAUP, 88*(6), 21–27.
McNiven, C. (2001). Biotechnology use and development: 1999. Retrieved October 20, 2007, from http://www.statcan.ca/english/research/88F0006XIE/88F0006XIB2001007.pdf
Midgley, M. (2000). Biotechnology and monstrosity: Why we should pay attention to the "yuk factor". *Hastings Center Report, 30*(5), 7–15.
Munn-Venn, T., & Mitchell, P. (2005). Biotechnology in Canada: A technology platform for growth. The Conference Board of Canada. Retrieved October 20, 2007, from http://www.agwest.sk.ca/biotech/documents/115-06-Biotechnology%20in%20Canada.pdf
OECD (2005). *Babies and bosses: Reconciling work and family life. Vol. 4: Canada, Finland, Sweden and the United Kingdom*. Paris: OECD.
OECD (2006a). Evolution of student interest in Science and Technology Studies Policy Report. Retrieved October 20, 2007, from http://www.oecd.org/dataoecd/16/30/36645825.pdf
OECD (2006b). *Women in scientific careers: Unleashing the potential*. Paris: OECD.
Pradesh, A. (2007). *Plea for ban on GM crops*. The Hindu. Retrieved from http://www.hinduonnet.com/thehindu/thscrip/print.pl?file = 2007041115900500.htm&date = 2007/04/11/&prd = th&
Saunders, D. M. (2002). *This was never meant to be a career: Gender and monopsony in academic labor markets*. Paper presented at Eastern Economics Association Meetings, Boston, MA, March 2002.
Statistics Canada (1998). Biotechnology use by Canadian industry: 1996. Retrieved October 20, 2007, from http://www.statcan.ca/english/research/88F0006XIE/88F0006XIB1998005.pdf
Statistics Canada (2006). Women in Canada: A gender-based statistical report. 5th ed. Retrieved October 20, 2007, from http://www.statcan.ca/english/freepub/89-503-XIE/0010589-503-XIE.pdf
Statistics Canada (2007). *Innovation Analysis Bulletin, 9*, 1.
Thomas, S. (2003). *Critical issues pertaining to the gender dimension of biotechnology*. A Paper for the Gender Advisory Board, United Nations Commission on Science and Technology for Development. Retrieved from http://gab.wigsat.org/gdrbiotech.pdf
Tilghman, S. M. (2004), Lonmo C. (2007). Ensuring the future participation of women in science, mathematics, and engineering. In G. Reinhart (Ed.), *The Markey Scholars Conference: Proceedings* (pp. 7–12) Washington, DC: The National Academic Press.
Traore, N. (2001). Canadian biotechnology industrial activities: Features from the 1997 biotechnology survey. Retrieved October 20, 2007, from http://www.statcan.ca/english/research/88F0006XIE/88F0006XIB2001012.pdf
United Nations (1996). *Platform for action and the Beijing declaration*. New York: United Nations Department of Public Information.

WISEST (Women in Scholarship, Engineering, Science and Technology) (2002). *The interim report: Preliminary research findings 2002*. Edmonton, AB: University of Alberta. Retrieved October 20, 2007, from http://www.uofaweb.ualberta.ca/wisest//pdfs/InterimReport.pdf

Women in Science and Engineering in Canada (2006). Produced by the Policy and International Relations Division, Natural Science and Engineering Research Council of Canada, Ottawa, Canada.

Chapter 3
Perception and Attitudes Towards Biotechnology in Hungary

Judit Acsády(✉) and Zoltán Ferencz

Abstract: Surveys on science and technology indicate that in general, women are less interested in, and less supportive of science and technology than are men. However, differences between women and men highly depend on the specific technology, and amongst the more educated groups gender differences tend to be rather small.

According to recent empirical data, Hungarian attitudes towards biotechnology are not significantly determined by gender. Women as a social group do not identify as supporters of or opponents to these technologies, although certain gender-specific concerns are expressed, as for example in the case of consumption of genetically modified (GM) food.

This paper first presents some background information followed by data collected by international and national surveys related to science-society, biotechnology and gender issues. Next, it introduces two examples for biotechnology-related legislation: (i) the special legislation introduced in Hungary concerning production of MON 810 corn, and (ii) the law regulating assisted procreation. The paper concludes that although feminism on the part of civil organisations has not overtly appeared in processes leading to the above laws, women play an important role in expert bodies influencing the regulation of biotechnology.

Keywords: Biotechnology, Hungary, science, assisted procreation, GM food

Judit Acsády
Sociology Research Institute, 1014. Budapest, Úri u.49, Hungary
acsady@socio.mta.hu

Zoltán Ferencz
Institute of Sociology, 1014. Budapest, Úri u.49, Hungary
ferencz@socio.mta.hu

F. Molfino, F. Zucco (eds.) *Women in Biotechnology,*

3.1 Introduction

With the exceptions of the consumption and production of gene manipulated food production and assisted procreation, questions about technology and the dilemmas they present are rarely discussed in Hungary. Even though gender consciousness appears in the body of experts who legislate, do research, or participate in civil society initiatives, feminist points of view on social issues are represented by very few in public life. Thus, we can assume that these efforts have rather weak impact on how people formulate their views on biotechnology.

Our inquiry looks at how biotechnology is perceived by lay or interest groups and professionals in Hungary. We are interested in the underlying values, ethical and other considerations governing these perceptions and attitudes. On the one hand, the question can be raised of what kind of society and what kind of human values or needs are assumed behind the promotions of these scientific developments. On the other hand, it can be asked in what way people are aware of biotechnology's impact or how it influences their everyday life. How are these articulated in connection with gender issues? In what way do gender or gender-conscious views influence the formulation and articulation of values and attitudes and how relevant civil organisations raise their voices in these questions?

The problem is subsumed within the broader issue of the public's understanding of science and their perceptions of humankind's ability and willingness to interfere with nature.

Earlier studies show that for most people in Hungary, the lack of information makes it difficult to formulate relevant opinions about scientific achievements (Tamás et al. 2000). Yet, in everyday acts of living, men and women are becoming aware of, e.g., the issues of safe food (Vári 2005). Data shows that the weight of civil organisations in Hungary (within which women's organisations have an even weaker position) is not significant enough in policy and decision making. In spite of this, the civil sector can make certain issues visible and can sometimes initiate debate.

These observations will be supported by analyses based on interviews with representatives of biotechnological research and with representatives of relevant civil organisations (e.g., greens, women's organisations, consumer protection groups) to illustrate the formulation and gender-based differences of perceptions and attitudes toward biotechnology.

3.2 Methodology

Our statements are based on the mixed methodology of the empirical social sciences. We analysed some databases from external sources: Eurobarometer (Gaskell et al. 2006, Jouhette and Romans 2006) and from internal sources: Hungarian Academy of Sciences, Institute of Sociology (2006a, b). Besides this we organised interviews with significant participants in research on biotechnology, representatives of NGOs and other experts.

3.3 Gender Differences and Opinions

If we want to examine how gender determines opinions about biotechnology in a comparative international context, first will be useful to highlight a few relevant data about gender relations in Hungary.

Even though Article 66 of the Constitution states a general prohibition of discrimination between men and women by declaring: 'The Republic of Hungary shall ensure the equality of men and women in all civil, political, economic and cultural rights' (Paragraph (1)), yet in the period of the transition to a constitutional state, gender inequalities became reformulated in the spheres of the labour market, political representation, cultural life and in the everyday practice of domestic responsibilities.

In this period of transition, employment declined drastically after decades of a policy of full employment that existed in the state-socialist system until 1990. Though in recent times there has been a world-wide tendency of growth in women's participation in the labour market, this participation remained low in the Central-Eastern European Countries, and especially in Hungary, where after a dramatic decline it stabilized at a very low level, less than the European Union's average. It is surprisingly true that Hungary is still among the few countries where female unemployment is lower than male unemployment. The reasons lie both in the uneven gender distribution among the economic sectors and in the extremely high rate of economic inactivity of the working-age population (Nagy 2003). In the Hungarian context we can observe traditional attitudes concerning women's social roles, but at the same time due to their education an increasing number of women have achieved managerial positions up to the middle level. On the grounds of the latest survey we find that the glass ceiling is all too evident between the middle and top levels of management (Bálint 2003).

Even though there is a tendency of slightly declining differences between earnings of men and women, yet in 2004 the average gross earnings of women were 86% of those of men in a total of all occupations (Bukodi et al. 2005, 124).

In earlier research,[1] members of the Hungarian elite groups were asked about their views on these differences of income between men and women. Their perception was that such differences would continue to increase. The majority of male respondents agreed, yet women were more optimistic on this point. All of the respondents found the situation to be a basis of potential conflict and that resolution would require state intervention (Table 3.1).

Inequality in the sphere of political representation is even more significant. Recently, as a result of the last parliamentary elections in 2006, only 11% of the MPs are women.

[1] The results of the research are available: Hungarian Academy of Sciences, Institute of Sociology (2006).

Table 3.1 Opinion of Hungarian elite groups on the future: 'The differencies of income between men and women will increase more' (Hungarian Academy of Sciences, Institute of Sociology 2006)

	Female	Male
Probability of the occurence	51	42
Potential of the conflict	65	54
Need for the state-intervention	67	56

Average on 100 degree scale

Traditional gender role expectations largely determine women's position in public and private life. Time use statistics show that, compared to men, women devote almost three times more hours to household responsibilities including domestic work and caring for children or elderly family members (Bukodi et al. 2005, 76–77).

3.4 Overview on Interest in Science and Technology

The portrait of European citizens painted by the 2005 survey (Gaskell et al. 2006) on biotechnology shows them to be increasingly optimistic about biotechnology, more informed and more trusting of the biotechnology system. The European public is not risk-averse about technological innovations that are seen to promise tangible benefits. While the majority are willing to delegate responsibility on new technologies to experts, making decisions on the basis of scientific evidence, a substantial minority would like to see greater weight given to moral and ethical considerations in decision taking about science and technology, as well as to the voices of the public. There is widespread support for medical (red) and industrial (white) biotechnologies, but general opposition to agricultural (green) biotechnologies in all but a few countries.

Although more women in Europe are employed, more are entering higher education and more are going into science than in the past, it is still the case that women are under-represented in careers in many areas of science.

While such human resource issues are beyond the scope of the Eurobarometer on biotechnology, in the following paragraphs we take a brief look at gender differences as evidenced in this survey (Fig. 3.1).

This figure (Fig. 3.1) shows that, in the EU, there are clear differences between men and women in their levels of interest in matters of science and technology. Compared to women, a greater percentage of men say they are 'sometimes' or 'often' interested in science and technology. In relation to knowledge about biology and genetics, there is a small average difference between men and women.

The findings from the Eurobarometer suggest that we must be cautious about generalisations on gender differences. On five of the eight technologies, women are almost as optimistic as men that these technologies will improve our way of life. While men are generally more knowledgeable about biology and genetics, women

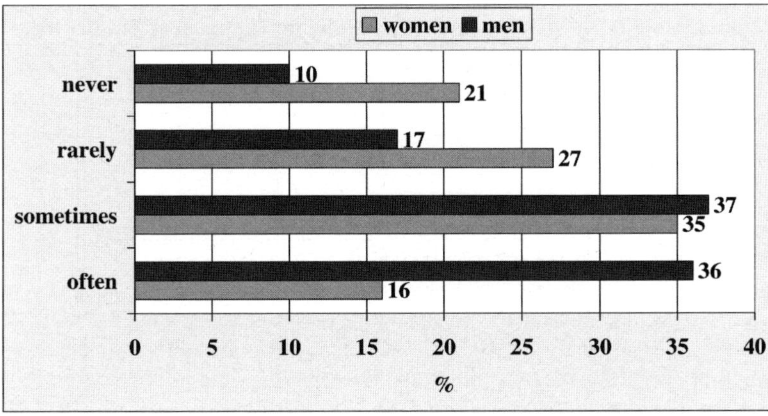

Fig. 3.1 Gender and interest in science and technology (Gaskell et al. 2006)

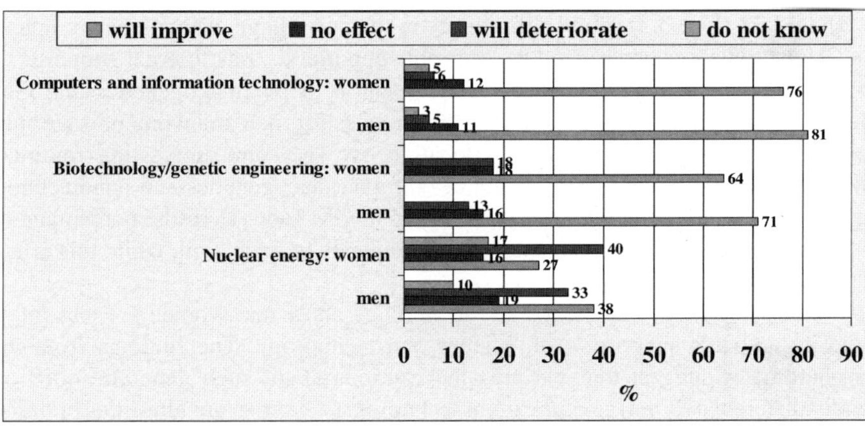

Fig. 3.2 Gender and optimism about technology (Gaskell et al. 2006)

out-score men on questions about pregnancy – an issue of direct concern to them. On approval for nanotechnology, gene therapy and pharmacogenetics, differences between women and men are not blatant and amongst the more educated women the gender difference is much smaller.

The figure (Fig. 3.2) shows the average levels of optimism and pessimism about various technologies for men and women in the EU. The percentage of 'will improve' answers is higher for men across all technologies and the percentage of 'will deteriorate' answers is greater for women, except in the case of 'wind energy'. The proportion of 'don't know' answers is always higher for women than it is for men, reaching almost half of those women surveyed (49%) in the case of nanotechnology.

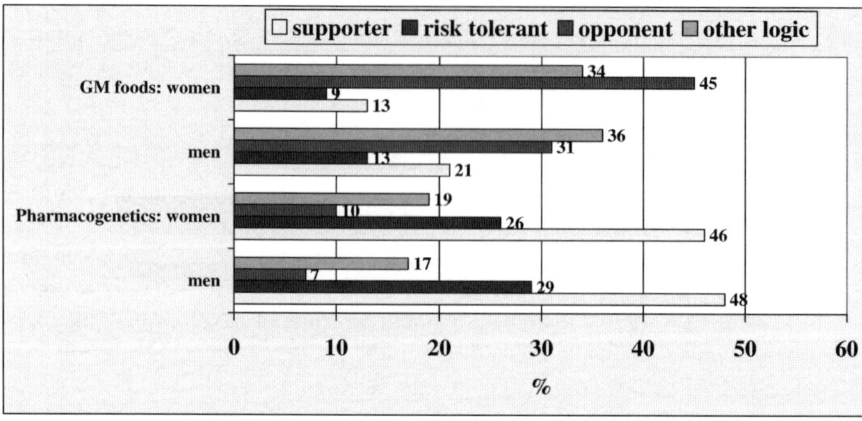

Fig. 3.3 Gender and logic of support for special technologies (Gaskell et al. 2006)

The figure (Fig. 3.3) shows differences in the percentage of women and men in the EU, who were characterized as 'outright supporters', 'risk tolerant supporters', 'opponents' and having 'other logics' in relation to pharmacogenetics and GM food. While we can observe some clear differences for men and women, if we split the sample according to levels of education, we see some interesting findings. There is a similar pattern in the cases of both pharmacogenetics and nanotechnology. But the situation is reversed with regard to GM food. Here the percentage of men who are 'supporters' increases with more years of education, while this effect is not seen in women.

Typically, surveys on science and technology show that women are less interested in, and less supportive of, science and technology. The findings from the Eurobarometer suggest that we must be cautious about such generalisations on gender differences. On five of the eight technologies, women are almost as optimistic as men that they will improve our way of life; while men are generally more knowledgeable about biology and genetics, women out-score men on questions around pregnancy – an issue of direct concern to them; on approval for nanotechnology, gene therapy and pharmacogenetics differences between women and men are not so evident and amongst more educated women the gender difference is much smaller.

However, women with higher education are less likely to show an attentive or an active interest in biotechnology. Is this more likely to be a consequence of the traditional division of labour in European households, rather than an intrinsic lack of interest among women?

In 2005, ten new Member States joined the European Union. In addition to the Mediterranean islands of Malta and Cyprus, the other eight – Estonia, Latvia, Lithuania, Hungary, Czech Republic, Poland, Slovenia and Slovakia – had experienced around 50 years of socialism. Through the lens of the Eurobarometer we investigate the 'science cultures' that these new Member States have brought to the EU.

Generally speaking, these are countries that are in the industrial phase of development (in comparison to some of the EU15 countries that are moving towards the post-industrial phase). Typically, in the industrial phase of development, while science has made only limited penetration into the public sphere, it is idealised as the preferred route to social and economic progress. However, this industrialisation hypothesis may not hold for the eight ex-communist countries. Could 50 years of communism have suppressed cultural differences, and/or led to a quite different science culture? The ten new countries respectively are just about as heterogeneous as are the former EU15 countries, seen by this set of indicators of science culture. All of the ten are in the industrial stage of development; they share some common features that were also seen in other 'new entrants' to the EU in the past. As such, the New EU10 are somewhat different from the EU15 countries in 2005. First, by comparison to EU15, science has not achieved much penetration in public awareness in the new Accession States. Second, the publics in these countries are relatively more optimistic about the contribution of technology to society, and are just as supportive of medical, industrial and agricultural biotechnologies. They also have greater trust in actors and institutions involved in science and technology. But, as has been seen in other EU Member States, such views can be subject to dramatic changes.

In Hungary, as we have seen in the indexes of the Eurobarometer, the level of optimism and trust is significantly higher than in the EU15 countries. On the contrary, we found a lower level of knowledge in case of science and technology. The results of a public opinion survey in Hungary show that people recognize the importance of science and technology. In Fig. 3.4, when we asked people if 'Science and technology make work more interesting', a majority of them agreed.

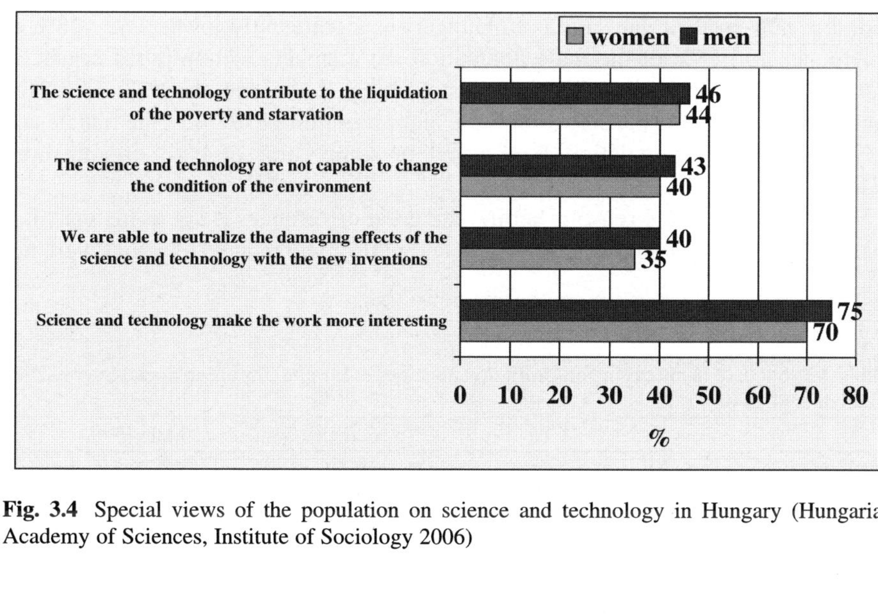

Fig. 3.4 Special views of the population on science and technology in Hungary (Hungarian Academy of Sciences, Institute of Sociology 2006)

When we asked them about the details and special influences, the knowledge level decreased considerably.

The members of Hungarian elite groups were more familiar but confused on this topic. They thought that in the period to 2012 'Consumers will be able to recognize the GMOs', because this problem contained a high level of conflict and would need strong state intervention. The pessimistic view from another aspect said that the statement: 'the results of gene-technology can decrease the problems of agriculture' is not possible in the short time period assumed. Nevertheless the situation contains a high level of conflict and a need for state intervention.

3.5 Interests and Strategies

The Hungarian public trusts scientific achievements, yet is not well informed about issues of biotechnology in general. As will be pointed out in the following, even though there is a lack of knowledge and information, the public is only concerned about certain aspects of biotechnology, such as for example safe food, therapeutic stem cell practices, or the safety of different procedures of assisted procreation. The question is if dilemmas emerge and different points of view are formulated on the basis of different value systems, beliefs and interests, what strategies do given social groups have to represent and defend their particular interests. It is a very crucial problem in a society like Hungary where in a market-oriented public life, non-governmental and non-profit organisations of civil society are weak and these might not have much impact on decision-making processes. Thus too much responsibility is transferred to the state and governmental agencies to balance and defend civil interests. In comparison with the European average, participation of the population in non-profit organisations in Hungary is significantly lower. Only 18% of the female and 28% of the male population are members of non-profit organizations. The difference is even more astonishing if we see the data from the Netherlands where the participation is the highest in Europe (78% of the female and 85% of the male population is engaged in some form of civil organization) (Table 3.2).

However one of the reasons behind the great differences is socio-historical: it was illegal to form social organizations and participate in activities that were not

Table 3.2 Share of members in non-profit organizations in Europe (European Social Survey 2003 [Bukodi et al. 2005])

	Female (%)	Male (%)
Hungary	18	28
EU average	50	59
The Netherlands	78	84

controlled by the Communist Party-led authorities during the four decades of state-socialism. As a result of the transition, a new Association Law was accepted in 1990 that permitted the foundation of civil groups. By now the number of registered organizations has become about 50,000, but many of them are not active and only exist formally. From the point of view of our inquiry about the perceptions and attitudes towards biotechnology, the relevant examples of civil organisations are: the Association of Consumer's Rights, Green's and Women's Organizations. In the case of the representatives of the former two (Association of Conscious Consumers and Greenpeace) *sustainability* serves as the base of arguments concerning the consumption and production of GM food. They argue that consumers should by all means be fully informed about the ingredients and the way of production of a given food product. Yet, the question arises of who is responsible for presenting this information. Is it the producer, the distributor, the state or civil society? The above mentioned organizations argue that informed European consumers are very critical towards GM food and if the market follows the needs of consumption there will not necessarily be a broadening of the market for GM products in Europe, at least not until the long term effects of these products are demonstrated by trustworthy experiments.[2] This concern was relatively strongly represented in the so-called GM Round Table that organized a successful Open Discussion Day on 22 November 2006 in Budapest at the Hungarian Parliament with the participation of representatives of the Agricultural, Health Care, Environmental, Economy and Information Committees of the Parliament, experts, researchers, embodying pro- and counter arguments against gene-technology and also representatives of the GM seed market (Darvas 2007).[3] The Round Table itself is a civil society initiative founded by experts and activists who were interested in making the issue of GM food public and in preparing the work of the Hungarian lobby team in Brussels that managed to defend the local interest by means of the Safe Guard Clause 2001/18 concerning the growing of MON 810 corn.[4] According to the decision made in January 2007, on the basis of the risk assessments of the Pannon Region, planting of MON 810 corn is prohibited in Hungary. The leader of the lobby team was Katalin Rodics from the Ministry of Environment. So, this time the interest of the local market, that is the local non-GM corn producers and the GM- skeptics, won. Yet, before being able to formulate the Hungarian standpoint at the European level, several opposing interests crushed locally. The Ministry of Environment and the Hungarian Academy of Sciences financed research on the effect of introducing GM plants containing toxic elements into

[2]Based on interviews with Vera Móra, Ökotárs, Budapest (16 April 2007) and Noémi Nemes, Greenpeace, Budapest (14 May 2007)

[3]The GM Round Table proceedings is a rich source of information about biotechnology in Hungary (Darvas 2007).

[4]MON 810 is an insect resistant corn line produced by Monsanto.

the environment.[5] Yet, these findings did not persuade the Ministry of Economy who supported the planting of GM corn and represented the interests of the biotechnology lobby, including local representation of the multinational company, Monsanto which keeps on promoting its products and formed an Association in Hungary to defend its interests. The Ministry of Agriculture changed its position during the process, first supporting the side of the biotechnology lobby to authorize the planting of MON 810 corn in Hungary but later, realizing the local producer's economic interest (in defending their position in the market for non-GM plants), took the side of prohibition. This side was strongly influenced by Greenpeace Budapest office and also local environmental groups. The GM Round Table mediated in this debate. The process shows that, with respect to growing MON 810 corn, it was not necessarily the civil society that formulated a platform against the state or multinationals but the opposers and the supporters of biotechnology who are divided along value systems and beliefs and their approach to sustainability (Fig. 3.5).

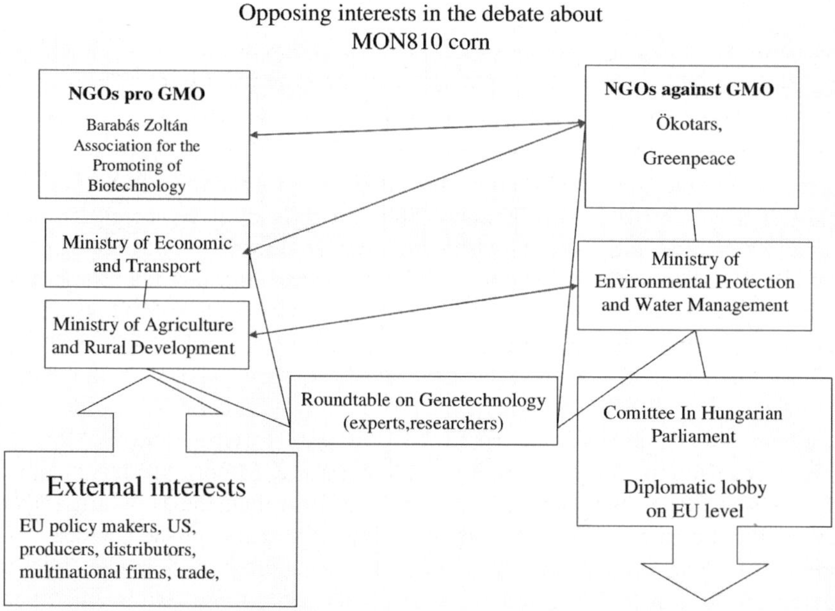

Fig. 3.5 Opposing interests in the debate about MON810 corn

[5] As a result of this research it was determined that the toxic element does not deteriorate as soon as it was expected and even eight months after it could still be tested in the field. The test showed damages to the worms of some protected species of butterflies living on the given territory. Also, the biological activity of the soil changed as a result of planting of the GM plants.

As it was pointed out earlier in this debate, neither side included feminist criticism in their arguments. Furthermore, the civil society organizations that were engaged in this diplomatic struggle did not express much concern towards other aspects of biotechnology.[6] Neither did women's organizations in Hungary appear to have the problems and dilemmas of biotechnology on their agenda.

3.6 Gender Aspects Without Feminism

Partly as a result of their being very few in number, the impact of women's organisations in Hungarian society is relatively weak. Even the number of registered women's organizations is very few (their number is about 100, and out of these a mere two dozen are active and appear in public life) compared to the total number of non-profit organizations in Hungary mentioned above.

Following new challenges in the 1990s, in the period of the transition in gender relations these organizations have recently become preoccupied with equal opportunities in the labour market or improvements in women's position in politics and in leadership. Because of Hungarian traditional patterns of behaviour, domestic violence and violence against women in general are also important issues to tackle, as are the unsolved problems of legislation concerning prostitution and control of the illegal sex industry. Also among the most relevant questions are to what extent the state can (or can motivate the employers to) support families in harmonizing work and family life. How can these supports be structured to be available and attractive for both fathers and mothers (suggesting new forms of parenting) given a rather stereotypical approach to gender roles in Hungary?

Gendered attitudes and stereotypical views are reflected in statements concerning different aspects of biotechnology as well. 'Women are much more concerned about buying safe food, as they do the shopping mostly' – claimed the leader of the Association of Conscious Consumers in Budapest, being a young woman herself, pointing out the potential move towards more conscious consumers. At the same time her account demonstrates one of the very typical misunderstandings about progressive women's movements in Hungary: 'I am not sure if progressive women's organizations would support the issue of conscious consumption, as feminism is against the traditional gender roles' (Kata Újhelyi, Association of Conscious Consumers).[7] According to Réka Matolay's research and observations concerning

[6]It is worth mentioning here that, as far as it was possible to discover this attitude during the process of preparing for this presentation, other aspects, such as, e.g., therapeutic stem cell research and the practice of bone marrow transplantation have been accepted by the public. Yet cloning is not much debated and is not even very evident in public awareness, except for media coverage of the results of the successful researcher, András Dinnyés who managed to produce several cloned animals such as mice and rabbits in his laboratory in Gödöllö, Hungary.

[7]Interview with Kata Újhelyi (Association of Conscious Consumers), Budapest, 3 May 2007.

strategies of the agri-biotech companies in Hungary: 'women represent the responsible side and men are mostly concerned with innovation and distribution'.[8]

A further example illustrates very well that social activity, even in such a new technology-determined field as test-tube baby programs, is organized along classical gender role patterns. According to Judit Sándor's[9] observation, mothers or mothers-to-be set up internet networks to share information and experiences and feelings about in vitro processes. In Sándor's view these networks can play a key role in dealing with the emotional aspects and psychological challenges the participants of such programs have to face. However it is known that these participants are not committed to discussing or dealing with the ethical, philosophical or legal dimensions of biotechnology, but are much more focused on the details of the program they are involved in. That is, their concern is that, e.g., the test-tube baby program they are participating in should be safe, effective and easily accessible. The public debates that emerge around this subject and the media representation of this field also tend to focus on given individual cases when these practical considerations are at issue.

Yet, unique gender aspects were recognised in the law regulating assisted procreation. According to Judit Sándor, who took part in the preparation of the law, legislation favours women because it was made by taking into account aspects of human rights, dignity and the principle of non-discrimination. As a result, under Article 1666 § (1) of the Health Care Act accepted in 1997: 'Right to the continuation of the infertility treatment (….) can be enforced exclusively by women who became widows or divorced after the medically assisted fertilisation had been achieved. Hungarian law recognises the different nature of the contribution by woman (invasive and potentially risky medical interventions) and man (sperm donation) in the assisted procreation treatment'. Also, couples can decide about donating surplus embryos for research purposes (Sándor 2000).

As we contended earlier, even though feminism from the side of civil organizations does not overtly appear in social and legal processes concerning biotechnology in Hungary, the examples above illustrate that women have significant influence on those organizations in their regulation of this field.

3.7 Conclusion – Diverse Visions

Attitudes towards biotechnology in Hungary are not significally determined by gender. Women as a social group do not identify either as supporters or as opponents to these technologies, however do express certain gender-determined concerns such as for

[8]Personal communication by Réka Matolay, Ph.D. candidate at Corvinus University, Budapest. For further information see Matolay (2007).

[9]Interview with Judit Sándor, Dr. Professor at the CEU Budapest. Leading expert and consultant in legal issues concerning human reproduction, 30 May 2007, Budapest.

example in the case of consumption of GM food. In general, critical views are formulated much more on the base of sustainability or on ethical, religious concerns. Among the promoters of biotechnological research and practices some are formulating their arguments by neglecting the potential risks. They even see biotechnology and gene manipulation as the solution to overcome climate change, global warming.[10] Cloning is 'natural' as András Dinnyés pointed out expressing optimism and enthusiasm for the results of experiments (Dinnyés 2007).

Yet, a question emerges as biotechnology, for the first time in the history of humankind, makes human intervention into the process of evolution possible. How far should scientists be supported in their work without being made aware of the consequences of their experiments? In what ways can biotechnology be governed by regulations, especially in societies where the possible base of the legislation, that is, the 'ethical consensus of the society' (Judit Sándor) is not expressed? The 'researcher is a Homo ludens – and he/she might not be able to take the responsibility and to recognize that not all aspects of biotechnology are compatible with sustainability', says Emília Madarász, biologist, a leading expert and consultant in stem cell research of the Hungarian Academy of Sciences, Institute of Experimental Medicine (personal communication). Given that not all scientists reflect on their work in as responsible way as she does, the responsibility for keeping society informed about the advantages and dangers of biotechnology should be shared among scientists, civil society, intellectuals, and other decision makers so as to formulate a consensus for regulations and practices.

References

Bálint, Z. (2003). *Analysis of equal opportunities of the male and female managers in the Hungarian business sphere*. MSc dissertation, Corvinus University, Budapest.

Bukodi, E., Mészáros, J., Polónyi, K., & Tallér, A. (Eds.) (2005). *Women & men in Hungary, 2004*. Budapest: Ministry of Youth, Family, Social Affairs and Equal Opportunities.

Darvas, B. (Ed.) (2007). *Mezögazdasági géntechnológia: Elsögenerációs gm-növények: Országgyülési nyílt nap. 2006 nov22*. Budapest: Magyar Országygülés Mezögazdasági Bizottsága. (Genetechnology in agriculture: first generation of gene-modifified plants: Parliamentary open forum 22 November 2006).

Dinnyés, A. (2007). *State-of-the-art: The clones are here*. Paper presented at Perfect Copy? Comparative and Interdisciplinary Approaches to Reproductive Cloning and Stem Cell Research, workshop organized by the Center for Ethics and Law in Biomedicine, Budapest, March 1–2, 2007.

Gaskell, G. et al. (2006). Europeans and biotechnology in 2005: Patterns and trends, Final report on Eurobarometer 64.3: Report to the European Commission's Directorate-General for

[10]In the interview with Noémi Nemes (Greenpeace and member of GM Roundtable) she referred to the point of view of the biologist, Dénes Dudits, Szeged, who was the founder of the pro-GM NGO.

Research Retrieved June06, 2007, from http://ec.europa.eu/research/biosociety/public_understanding/eurobarometer_en.htm

Hungarian Academy of Sciences, Institute of Sociology (2006a). *Delphi-type opinion survey on Hungarian elite groups*. Database.

Hungarian Academy of Sciences, Institute of Sociology (2006b). *Public opinion survey on the future visions of the Hungarian population*. Database.

Jouhette, S., & Romans, F. (2006). EU Labour Force Survey: principal results 2005. In *Statistics in Focus. Population and Social Conditions, 13.*

Matolay, R. (2007). *Legitimacy Strategies of Agri-biotech Companies in Budapest*. Dissertation, Vállalatgazdaságtan Intézet, Döntéselméleti tanszék, Budapest.

Nagy, B. (2003). Women in the economic elite. In M. Domsch, D. Ladwig, & E. Tenten (Eds.), *Gender equality in Central and Eastern European countries* (pp. 151–168). Frankfurt am Main: Peter Lang.

Sándor, J. (2000). Reproductive rights in the Hungarian law: A new right to assisted procreation? *Health and Human Rights, 4*(2), 196–218.

Tamás, P., Schmidt, A., & Ferencz, Z. (2000). Innovation clusters in the Hungarian economy. In G. Papanek (Ed.), *The Hungarian innovation system* (pp. 56–62). Budapest: OMFB.

Vári, A. (2005). Food Safety and Governance: Questions after the EU Connection. In Gáthy V. (Ed.) *Életminöség: holisztikus szemlélettel: Nádasdy Akadémia szimpóziumok 2005-ben: a Nádasdy Alapítvány rendezvényei: angol összefoglalókkal* (pp. 128–131). Budapest: Nádasdy Alapítvány (The Quality of Life: A Holistic View).

Chapter 4
Contribution of Bulgarian Women to Plant Biotechnology: Institute of Genetics Case

Georgina Kosturkova(✉)

Abstract: Plant biotechnology emerged in the 1950s and had an immediate positive impact on many fields of science, agriculture and industry. Since 1970, in Bulgaria more than 20 research units of *in vitro* culture were established utilizing a wide range of methods. Allocation of resources for plant biotechnology is still increasing in most bioscience institutes. Concerning gender – work with *in vitro* cultures in Bulgaria attracts predominantly women, who represent more than a half of the staff in two big units (in the Institute of Genetics, BAS and the University of Plovdiv) or all the personnel in small laboratories. Regarding level of positions, female researchers predominantly hire female assistants, while leadership depends on personal abilities, rather than gender. The specificity of biotechnology work needs certain attributes traditionally found in women, such as skilled hands, quick fingers, gentle action, patience and creativity. The Department of Plant Biotechnology of the Institute of Genetics has joined in the world tendencies of development and application of plant biotechnology in science, agriculture and industry. Its main achievements are in the area of induced organogenesis and androgenesis, somaclonal variation and modification of plant genomes, modeling of stress and micropropagation of valuable and endangered species, contributing to enrichment of genetic diversity and protection of biodiversity.

Keywords: Plant biotechnology, Bulgaria, Institute of Genetics, women in science

4.1 Introduction

Biotechnology is a complex of techniques, methods and approaches which can be applied in many industries. Plant biotechnology emerged in the 1950s, based on the possibility to cultivate isolated plant tissues and cells. The ability of plant cells to

Georgina Kosturkova
Department of Plant Biotechnology, Institute of Genetics,
Bulgarian Academy of Sciences, 1113 Sofia, Bulgaria
gkosturkova@yahoo.com

express their entire genetic information and regenerate whole plants (i.e., totipotence), is the basis of their development in controlled *in vitro* conditions and to the establishment of biotechnological techniques and methods for genetic manipulations (Griga et al. 2001). In vitro cultures created in the spheres of science merged very quickly to the spheres of production and at the end of the twentieth century plant biotechnology was already a multi-billion industry in agriculture, food processing, health care and environment protection.

4.2 Plant Biotechnology in Bulgaria

Plant biotechnology history in Bulgaria started during the 1970s. For 30 years more than 20 research units working on plant *in vitro* cultures were established in research institutes and universities. The oldest one, existing since 1974, is the laboratory of tissue cultures of the Institute of Genetics, Bulgarian Academy of Sciences (BAS). There are four more institutes (Institute of Botany, Institute of Microbiology, Institute of Plant Physiology, Institute of Forestry) of the Bulgarian Academy of Sciences carrying on research on *in vitro* cultures. The greatest number of experimental units appears in the National Center for Agricultural Sciences (NCAS) where *in vitro* techniques are applied in nearly half of the research institutes: AgroBioInstitute (ABI), Dobrudja Agricultural Institute (DAI), Plant Genetic Resources Institute (PGRI), Maize Research Institute (MRI), Cotton & Durum Wheat Research Institute (CDWRI), Agricultural Institute of Karnobat (Karnobat AI), Agricultural Institute of Shoumen (Shoumen AI), Vegetable Crops Research Institute (VCRI), Fruit Growing Research Institute (FGRI), Viticulture & Enology Institute (VEI), Agricultural Institute of Kyustendil (Kyustendil AI), Regional center for Extension Services in Septemvri (Septemvri RCES). There are several universities providing teaching and research in different aspects of plant biotechnology: University of Sofia, University of Plovdiv, University of Agriculture, University of Forestry (detailed data in Kosturkova 2006).

A wide range of *in vitro* methods is utilized in the different organizations depending on the targets and scope of their research. Table 4.1 illustrates the most commonly used techniques which are better represented in institutes with wider area of investigations (IGen, ABI, IPGR, University of Plovdiv, University of Agriculture), as well as institutes concentrated on specific crops (DAI, VCRI, CDWRI).

Biotechnology importance and development varies in the different units as is illustrated by the allocation of resources to the major areas for crop improvement: plant biotechnology, germplasm enhancement, line development and evaluation (for more details see Kosturkova 2006). Table 4.2 shows, as well, the tendency for application of plant biotechnology during the last decade. Allocation of resources is increasing in most of the institutes under survey (7 out of 11). The number is greater if newly formed units like that in the Institute of Botany and in the University of Forestry are taken into consideration.

Table 4.1 Application of *in vitro* techniques

Unit	Tissue cultures/ micro-propagation	Haploids, embryo rescue	Muta-genesis, selection	Genetic engineering	Molecular characteri-zation
Institute of Genetics (BAS)	•	•	•	•	•
Institute of Botany (BAS)	•				
Institute of Microbiology (BAS)	•		•		
Institute of Forestry (BAS)	•			•	
University of Plovdiv	•			•	•
University of Agriculture	•		•		•
University of Sofia	•			•	
University of Forestry	•				
AgroBioInstitute (NCAS)	•	•		•	•
Institute of Plant Genetic Resources (NCAS)	•	•			•
Dobrudja AI (NCAS)	•	•		•	
Vegetable Crops RI (NCAS)	•	•		•	•
Cotton & Durum Wheat RI (NCAS)	•	•		•	
Shoumen AI (NCAS)	•	•			
Fruit Growing Institute (NCAS)	•	•			
Kyustendil AI (NCAS)	•	•			

BAS – Bulgarian Academy of Sciences; NCAS – National Center for Agricultural Sciences; AI – Agricultural Institute; RI – Research Institute

Table 4.2 Budget distribution for: plant biotechnology (PB), germplasm enhancement (GE), line development and evaluation (LDE)

	Allocation of resources (%) to					
	Plant biotechnology in year		Germplasm development in year		Line enhancement evaluation in year	
	1995	2004	1995	2004	1995	2004
Institute of Genetics	12	20	28	22	60	58
AgroBioInstitute	40	60	40	20	20	20
Dobroudja Agricultural Institute	30	20	30	20	60	70
Plant Genetic Resources Research Institute	14	15	56	60	30	25
Cotton & Durum Wheat Research Institute	10	25	20	25	70	50
Maize Research Institute	12	18	15	81	83	
Agricultural Institute of Shoumen	20	20	10	10	70	70
Vegetable Crops Research Institute	30	35	35	35	35	30
Agricultural Institute of Kyustendil	17	19	16	11	67	70
Viticulture and Enology Research Institute	20	10	30	40	50	50
Agricultural University, Genetics & Breeding Department	10	10	15	10	75	80

The importance of plant biotechnology for plant improvement can be evaluated by the relatively big ratio (1:3,7) of biotechnologists to plant breeders (Table 4.3).

Following growth to a significant number of scientists in both spheres over the last 20 years, there is currently a decreasing tendency in that number (because of economical reasons). However, the share of those with higher education (Ph.D. holders) in plant biotechnology is increasing, indicating the great interest generated by modern scientific developments.

Concerning gender – work with *in vitro* cultures in Bulgaria attracts predominantly women, who represent more than half of the staff in two big units – the Department of Plant Biotechnology of the Institute of Genetics (see Table 4.4) and the Department of Plant Physiology and Molecular Biology of the University of Plovdiv (see Table 4.5).

For the last 20 years in the Institute of Genetics the tendency has been an increase in the female share from 77% to 86%, 89% and 83%. There are laboratories exhibiting such numbers in the Institute of Botany and in the Cotton and Durum Wheat Research Institute which employ only women personnel. However, in the University of Plovdiv the number of women has fluctuated above 50% but recently decreasing from 75% to 57%. When comparing the positions occupied by women and men, it is notable that there are no male technicians and that female researchers have predominantly female assistants. Concerning leadership in the Department of Plant Biotechnology – the position of Head has been occupied equally, but project leaders have been predominantly women.

One may ask the question: ‘Why women in biotechnology?’. The answer after 30 years of experience in most of the spheres of *in vitro* and DNA techniques could be the following – biotechnology work is specific and needs: skilled hands and quick fingers; gentle action and tenderness; patience and persistence; curiosity and watchfulness; imagination and creativity; dedication and love. In addition, it requires the ability to manage several things simultaneously. All these are traditionally attributes of women.

Table 4.3 Total number of scientists in plant biotechnology and plant breeding and their educational level

Educational level of scientists in	1985	1990	1995	2000	2004	Total for the period
Biotechnology						
B.Sc.	4	13	12	14	16	59
M.Sc.	34	36	29	22	22	143
Ph.D.	21	29	31	32	34	147
Total number	59	78	72	68	72	349
Plant Breeding						
B.Sc.	16	22	14	9	8	69
M.Sc.	172	164	127	100	96	659
Ph.D.	120	131	119	108	102	580
Total number	308	317	260	217	206	1,308

Table 4.4 Educational level and positions occupied by women and men in the Department of Plant Biotechnology, Institute of Genetics

	Period of investigation									
	1980–1984		1985–1989		1990–1994		1995–1999		2000–2007	
Position/education	Women	Men	Women	Men	Women	Men	Women	Men	Women	Men
Technicians	2	0	2	0	0	0	0	0	0	0
M.Sc. & B.Sc.	3	1	7	0	8	0	4		4	0
Ph.D. students	2	0	3	0	2	0	0	0	1	0
Researchers	2	0	4	2	4	1	2	1	3	2
Senior researchers	1	2	3	1	2	1	2	0	2	0
Total number	10	3	19	3	16	2	8	1	10	2
%	77		86		89		89		83	
Degree holders										
Ph.D.	2	1	7	0	6	0	3	0	2	0
D.Sc.	0	0	0	0	0	0	1	0	1	0
Total number	2	1	7	0	6		4	0	3	0

Table 4.5 Educational level and positions occupied by women and men working on biotechnology in the Department of Plant Physiology and Molecular Biology, University of Plovdiv (kindly provided by Professor I. Minkov, Head of the Department)

	Period of investigation									
	1980–1984		1985–1989		1990–1994		1995–1999		2000–2007	
Position/education	Women	Men	Women	Men	Women	Men	Women	Men	Women	Men
Technicians	1	–	1	–	1	–	1	–	1	–
M.Sc. & B.Sc.	2	1	1	1	3	1	2	1	5	4
Ph.D. students	1	–	1	1	3	2	3	2	1	2
Researchers	2	1	2	1	4	1	3	3	4	3
Senior researchers	3	1	3	2	1	1	2	1	2	1
Total number	9	3	8	5	12	5	11	7	13	10
%	75		62		71		61		57	
Degree holders										
Ph.D.	3	1	3	2	3	2	3	2	5	4
D.Sc.	–	–	–	–	–	1	–	1	–	1
Total number	3	1	3	2	3	3	3	3	5	5

4.3 Department of Plant Biotechnology of the Institute of Genetics

The first Bulgarian laboratory for plant *in vitro* cultures was established in 1974 at the Institute of Genetics. The Institute of Genetics 'Acad. Doncho Kostoff', Bulgarian Academy of Sciences, was established in 1910 as a Central Agricultural Research Institute and has undergone several transformations. Nowadays complex genetic studies are carried on organism, cell, chromosome and molecular level. The research scope of the Institute has always been of multiple character – along with theoretical problems and methodological approaches, numerous topics closely related to practical crop breeding were elaborated. Its objects of research have always been of wide range, covering the most important crops for the country: tobacco, tomato, sweet pepper, wheat, barley, triticale, maize, sunflower, soybean, pea, beans, alfalfa, medicinal and ornamental plants. Major results were achieved in the theory of gene expression, regulation and inheritance; induced variability *in vivo* and *in vitro*; heterosis; genetics of resistance; induced mutagenesis and radio-protection.

The Department of Plant Biotechnology in its 35 years history has participated in world tendencies, focusing its research on establishment and development of *in vitro* techniques and their use in theoretical and applied science for crop improvement, enrichment and reservation of plant biodiversity, environment protection, and production of valuable substances.

Its main topics of research could be formulated as:

- Genetic and physiological mechanisms of induced organogenesis, somatic embryogenesis, androgenesis and plant regeneration in callus, cell and protoplast cultures from generative and somatic tissues.
- Micropropagation of medicinal, ornamental, rare and endangered plant species

- Reconstruction and modification of plant cell genomes using different approaches: (1) genetic transformation, (2) somatic hybridization, (3) *in vitro* mutagenesis, (4) somaclonal and gametoclonal variation.
- Modeling of biotic and abiotic stress *in vitro* using pathogenic fungal filtrates, high osmotic substances, extreme temperatures and heavy metals. Development of tests for both screening of genotypes with higher tolerance and systems for cell selection for higher resistance.

The main objects of research are economically important species from *Nicotiana*, *Lycopersicon*, *Capsicum*, *Triticum*, *Hordeum*, *Pisum*, *Glycine*, *Helianthus*, *Medicago*, *Rubus*, *Arnica*, *Galantus*, *Gentiana*, *Rhodiola*, *Schizandra* and some other species depending on the demands of our collaborators.

The Department of Plant Biotechnology conducts mutual research with the other institute departments (Molecular Genetics, Biochemical Genetics, Cytogenetics, Plant Disease Resistance, and Applied Genetics), as well as with various institutes from the Bulgarian Academy of Sciences, and other organizations in Bulgaria and abroad (e.g., Czech Republic, France, Greece, Hungary, India, Italy, The Netherlands, Poland, Russia, Sweden, Turkey, UK, Ukraine, USA).

The main outputs are original *in vitro* techniques, methods and approaches; new plant forms and lines with improved agronomic traits being a basis for breeding programs and fundamental science.

4.4 Some Outcomes of the Department of Plant Biotechnology

The totipotence of the plant cell is realized by formation of callus tissue (consisting mainly of nondifferentiated cells) and/or of organogenic/embryogenic cultures leading to regeneration of plants. Establishment of efficient and reliable protocols for *in vitro* cultures is the first step in plant biotechnology and in many cases this is a crucial point for application of more sophisticated *in vitro* manipulations (e.g., mutagenesis and cell selection, genetic transformation and somatic hybridization, metabolic engineering, and other frontier investigations like nanotechnology). However, induction of cell division and growth, followed by morphogenesis and development of plant regenerants is a complicated process depending on many factors.

Research on the effect of various factors (physical, chemical, physiological, genetic etc.) for initiation of *in vitro* cultures has been carried out since establishment of the *in vitro* laboratory of the Institute of Genetics. The important role of genotype of the donor plant, the type of explant, media composition and consistency, culture conditions, stress factors, etc. and their interaction was observed and used for successful establishment of *in vitro* cultures in a large number of crops (e.g., tobacco, tomato, pepper, wheat, barley, triticale, sunflower, pea, soybean, alfalfa, aronia, raspberry, datura, leucojum, arnica, gentiana and some others). Accumulated knowledge helped in revealing mechanisms for induction of morphogenesis in tissue cultures from somatic and generative cells.

The importance of these studies is very illustrative in the cases of sophisticated *in vitro* cultures like protoplast cultures, or these of reculcitrant species (like some cereals and grain legumes), or those initiated from generative tissues.

In this respect outstanding work has been done on development of efficient protoplast systems for large scale isolation of protoplasts from *Medicago* species using various sources (callus, cell suspension, mesophyll and cotyledons). Alternating media composition and culture conditions, morphogenic pathways could be directed to organogenesis, direct or indirect embryogenesis. Stimulation of direct embryogenesis and its sustainability by protoplast electrotreatment was observed for the first time. (Antonova-Kosturkova 1987; Antonova-Kosturkova and Davey 1990; Kosturkova 1998a).

Original work has been done on the importance of genetic factors for callus and morphogenesis induction in cereal cultures. Efficiency of these processes and their dependence on the genotype was demonstrated when wheat lines with disomic addition of rye chromosomes were used as initial material. Substitution of chromosomes gives information on the loci responsible for the processes from differentiation to regeneration. Morphogenic response of tissue cultures from triticale depended on the ploidity level (tetra-, hexa-, octoploid) of callus cells and on the duration of *in vitro* cultivation (Dryanova and Ganeva 1995, 1996/1997a, 1996/1997b; Dryanova et al. 1998).

In more recent years, knowledge of the interaction of various factors was crucial in regeneration of recalcitrant grain legume species. Induced direct and indirect organogenesis was achieved in all tested *Pisum* and *Glycine* genotypes when using special combinations of culture media and explants (immature embryos or cotyledonary nodes) (Kosturkova 2005; Kosturkova et al. 1997, 2005a). Regeneration of plants by setting flowers and pods in test tubes (a rare phenomenon) was stimulated by low-dose irradiation, as well as by treatment with 'Humostim' – a natural organic fertilizer. Efficient (index of multiplication up to 50) and long-term (more than 4 years) *in vitro* organogenic pea cultures were established allowing *in vitro* genetic manipulations (Kosturkova et al. 2003a, 2005b).

In the above mentioned examples mechanisms of *in vitro* development were studied in somatic cells of the plant. However, generative cells were the object of investigations for many years and induced androgenesis and ginogenesis for production of haploid and dihaploid plants was another important area of research. The role of genotype, different physical and chemical factors effecting androgenesis were studied in economically important species from different genera (*Lycopersicon*, *Nicotiana*, *Capsicum*, *Medicago*, *Datura*, *Rubus*). The efficiency of androgenesis, for example, in tomato depended on donor plant growth conditions, anther size, developmental stage of the microspore, anther treatment prior to culture (Shtereva et al. 1998). Among the outstanding results are haploid regeneration in alfalfa and flowering haploid and dihaploid tomato regenerants reported for the first time by Zagorska (Zagorska et al. 1990, 1998).

Regenerants of tomato differed phenotypically, citologically and biochemically from the donor plant and among themselves, some of them having valuable traits like higher disease resistance, higher content of dry matter, sugars and vitamin C (Zagorska et al. 2004). This is a good example of gametoclonal variation which is a successful addition to previous achievements in the field of somaclonal variation of tobacco, tomato, wheat, and triticale (Karadimova 1993; Nedkovska 1993; Zagorska 1993). Somaclones as new plant forms are a good source for enrichment of genetic diversity and were included in breeding programs.

In most cases the established *in vitro* cultures were used as a step to other investigations and more sophisticated manipulations.

The *in vitro* approach and different methods (tissue cultures, embryo rescue, protoplast fusion) were applied for overcoming the barriers of incompatibility between remote species in the genera of *Nicotiana*, *Capsicum*, *Licopersicon*, *Phaseolus*, *Brassica*, and *Helianthus*. One of the most applied techniques is the use of embryocultures which was successful in producing interspecific hybrids of tobacco, sweet pepper, tomato and common bean (Zagorska 1993).

Tissue cultures are intensively used to overcome hybrid incompatibility and obtain cytoplasmic male sterility (CMS) forms in *Nicotiana*. A series of CMS forms were created crossing *N. tabacum* with various *Nicotiana* species like *N. africana*, *N. plumboginifolia*, *N. longiflora* which are valuable breeding material (Nikova et al. 1991, 2001, 2004).

Somatic hybridization was also used to avoid the barriers of sexual crosses with wild species. Isolated protoplasts from the cultivated *N. tabacum* L. were fused with protoplasts from the wild species *N. megalosiphon* and *N. rotundifolia* and somatic hybrids were selected (Ilcheva et al. 1997, 1998).

Protoplast cultures from *Medicago* species were used in genetic studies and for genetic manipulations. Models to study the effect of γ-irradiation, antibiotics and pathotoxins on protoplast development were created. Plant regenerants and protoplast clones resistant to tetracycline and ampicillin were selected after spontaneous and induced mutagenesis and resistant lines were obtained. (Kosturkova 1993a). Efficient DNA transfer and transient gene expression (90–96% chloramphenicol acetyltransferase (CAT) activity) was achieved after protoplast electroporation with multiple DC electric pulses. Kanamycine resistant transformed plants were recovered (Kosturkova 1993b; Jones et al. 1993).

Genetic transformation by other techniques is also of interest. Transgenic tobacco and tomato plants were selected after *Agrobacterium tumefaciens* mediated transformation with genes for carotenoid biosynthesis obtaining information for carotenoid pathways biosynthesis and its regulation (Corona et al. 1996; Kosturkova 1998b). Particle bombardment was used for β-glucoronidase (GUS) genetic transformation of haploid wheat and triticale calli (Dryanova et al. 2000). Experiments are in progress for *Agrobacterium* mediated transformation of medicinal plants.

Recently, application of some biotechnology methods is under discussion and under consideration, especially those concerning genetically modified organisms (GMOs) as a food source. However, the problems of plant damages by pathogen attacks, drought, salinity, pollution and improvement of plant performance remain. Recently the tendency of directing science to more environmental friendly technologies grows. The requirements for investigations of biotic and abiotic stress have intensified. *In vitro* cultures could be used to develop models to study stress factors which can be applied in *in vitro* conditions or imitated in these conditions.

Procedures for *in vitro* selection for disease, herbicide, drought and salinity resistance were established for pea and soybean using culture filtrates of pathogen fungi, a herbicide, polyethylene glycol (PEG) causing water shock, and NaCl. Regenerants with higher stress tolerance were selected. (Kosturkova et al. 2001, 2003b, 2006). Models for comparative *in vivo* and *in vitro* assessment of different pea and soybean genotypes tolerance to fungal pathogens *Ascochyta pisi* and *Sclerotium bataticola*, as well as to drought conditions, were created proving the efficiency of *in vitro* procedures, important for shortening the breeding processes for selection for resistance (Kosturkova et al. 2004a; Rodeva et al. 2001, 2005). These systems and similar approaches could be applied to other species and stress factors.

A plant biotechnology approach has been successfully used in different cases of environment protection. One of the applications of *in vitro* cultures is to study the effect of pollutants or protectors. Mutagenic effect of the herbicide Stomp 330 was observed in triticale calli (Dryanova and Dimitrov 2000). Radioprotective effect and bioregulatory activity was observed using *in vitro* cultures for several new substances of artificial or natural origin (Kosturkova et al. 2004b; Mehandjiev et al. 2002; Noveva et al. 2006).

Micropropagation is another strong area of any department with concern for environment protection. It was successfully developed for a number of crops, ornamental and medicinal plants. Recently, after the pioneer work of *in vitro* cultivation of *Leucojum aestivum* Mill, important for preparation of poliomyelitis remedies (Zagorska et al. 1997) research has focused on establishment of *in vitro* procedures for micropropagation of the endangered medicinal species *Gentiana lutea* L., *Rhodiola rosea, and Arnica montana*. Targeted objectives are both multiplication of the extinct species and recovery of their areals, and substitution methods for drug production (Petrova et al. 2006a, b; Tasheva et al. 2003, 2005).

4.5 Conclusions

Plant biotechnology in Bulgaria has been developed with the active participation of women in both scientific and administrative areas. Most of the women along with their research careers have managed a good accommodation in their family life and in society.

In this paper only a part of the Plant Biotechnology Department results was presented to illustrate its research in the past and in the present. Our research has been guided, usually, by world tendencies in plant biotechnology, working on both areas – fundamental and applied. An illustrative example is the observation of Dr. Zagorska that cells cultured *in vitro* differ cytologically from the initial mother plant cells, which was a scientific contribution to the theory of *in vitro* cultures, but became a basis for somaclonal variation which has practical importance for obtaining new plant genetic forms, thus creating new genetic diversity. Another example is the research on cell mechanisms of induced androgenesis and the factors influencing it, which is in the area of pure science, but has practical importance when used for haploid and dihaploid plant production. Another example for successful complementation of theory and practice is the research of Dr. Kosturkova on the effect of different stress factors on the cell (respectively protoplast) level and on modeling of biotic and abiotic stress *in vitro* for the development of tests and schemes for selection of resistant genotypes in aid of plant breeding of grain legumes (pea and soybean). A good illustration of the possibilities for extending knowledge in one sphere by utilization of new methods is also the pioneer research on the use of alfalfa protoplast cultures in induced mutagenesis and DNA transfer by electroporation, as well as application of *in vitro* cultures in radiobiology, antimutagenesis and environment protection. Concerning nature, a good illustration of a practical application of tissue cultures is the work of the young scientists Petkova and Tasheva on micropropagation of endangered medicinal plants to avoid their extinction.

Future investigations will continue on overcoming problems related to biotic and abiotic stress, and poor plant production. New methods concerning nature protection, metabolic engineering, and nanotechnologies are in process of development and/or in a stage of feasibility studies.

The philosophy of plant biotechnology development should be towards its application to sustainable agriculture and environment protection, while gaining acceptance from modern industry. Wider support of this tendency could contribute to easier acceptance of plant biotechnology today, when some opposing opinions exist on the extent and limits of the new techniques. In this respect the greatest fears of society are raised by the concept of genetic engineering (respectively genetically modified organisms – GMOs) which for some people could 'create monsters', but for others could 'make miracles' overcoming any limits of nature. However, theoretical predictions cannot always be fully realized (e.g., in genetic transformation some genes could remain silent, being not expressed in the new cell or in somatic hybridization after fusion of protoplasts, especially from remote species, part of the genetic material could be eliminated) and the new ideas, methods and techniques may have to undergo changes and improvement. The case with plant biotechnologies is similar when scientists have taken into consideration social and ethical issues and tried to make these technologies more widely acceptable in the developed and developing world. However, genetic engineering is only a part of the huge area of plant biotechnology which for its half century history has proven its positive impact on the development of different spheres of science (molecular biology, genetics, bioinformatics, physiology, biochemistry, botany, plant breeding, forestry, ecology, etc.) and has contributed to emergence of new areas like proteomics and nucleomics. Plant biotechnology participates in agriculture, floriculture, food and pharmaceutical industries where a lot of women are engaged to provide food, better living and beauty. Women give life and are in charge of continuation of life, hence, new technologies are not likely to be misused in women's hands.

Acknowledgements The author thanks Fondazione Brodolini, Women and Science Association (Associazione Donne e Scienza), Italy and the European Commission.

Preparation of this paper was possible with the support of biotechnologists from the Institute of Genetics (Professor Zagorska, Dr. Dryanova, M. Petrova, K. Tasheva, M. Dimitrova) and the other organizations cited in the paper. Professor Minkov is particularly thanked for providing

University of Plovdiv data. Research was possible with the financial support of the National Scientific Fund to more than ten projects (CC1201 still active).

References

Antonova-Kosturkova, G. (1987). Morphogenesis in calli clones derived from alfalfa protoplasts. *Genetics and Breeding, 20*(2), 171–175.

Antonova-Kosturkova, G., & Davey, M. R. (1990). Somatic embryogenesis in cotyledon protoplast cultures of alfalfa. *Proceedings of the Vth International Conference of Genetics*, Sept., Varna, BAS Publishing, 277–230.

Corona, V., Aracri, B., Kosturkova, G., Bartley, G. E., Pitto, L., Giorgetti, L, Scolnik, P. A., & Giulianno, G. (1996). Regulation of carotenoid biosynthesis gene promoter during plant development. *The Plant Journal, 9*(4), 505–512.

Dryanova, A., & Ganeva, G. (1995). Morphogenetic response of tetra-, hexa- and octoploid triticale calli depending on the duration of cultivation period. In A. Börner and A.J. Worland (Eds.), *European wheat aneuploid co-operative newsletter* (pp. 36–37).

Dryanova, A., & Ganeva, G. (1996/1997a). Callus induction and plant regeneration of wheat lines with disomic addition of rye chromosomes. *Genetics & Breeding (Sf), 28*(3), 3–9.

Dryanova, A., & Ganeva, G. (1996/1997b). In vitro regeneration system from immature embryos of different ploid genotypes triticale. *Genetics & Breeding (Sf), 28*(2), 36–42.

Dryanova, A., & Dimitrov, B. (2000). Influence of the herbicide Stomp 330 on morphogenetic response of triticale callus cultures. Cytological evidences for its mutagenic action. *Cytologia, 65*(1), 17–23.

Dryanova, A., Stoinova, J, Vladova, R., & Ganeva, G. (1998). Genome analysis of tetraploid triticale forms. *Comptes rendus de l'Academie bulgare des Sciences, 51*(10), 107–110.

Dryanova, A., Zagorska, N., Beltcev, I., Pantchev, I., Ganeva, G., & Gözükirmizi, N. (2000). Biolistic transformation in haploid wheat and triticale callus cultures. *Proceedings of the Second Balkan Botanical Congress*, May, 2000, Istanbul, Turkey, vol. II, 235–239.

Griga, M., Kosturkova, G., Kuchuk, N., & Ilieva-Stoilova, M. (2001). Biotechnology. In C. L. Hedley (Ed.), *Carbohydrates in grain legume seeds: Improving nutritional quality and agronomic characteristics* (pp. 145–207). Wallingford, UK: CABI.

Ilcheva, V., San, L. H., Dimitrov, B., & Mehandjiev A. (1998). Genetic constitution of somatic hybrids obtained after protoplast fusion between *N. tabacum* L. and *N. megalosiphon. Comptes rendus de l'Academie bulgare des Sciences, 50(7/8)*, 87–91.

Ilcheva, V., San, L. H., Dimitrov, B., & Zagorska, N. (1998). Morphological and cytological characteristics of somatic hybrids between *N. tabacum* L. and *N. megalosiphon* Heurk. et Müll. *CORESTA Bulletin, 3/4*, 11–12.

Jones, B., Antonova-Kosturkova, G. P, Vieira, M. L. C., Rech, E. L. Power, J. B., & Davey, M. R. (1993). High transient gene expression, with conserved viability, in electroporated protoplasts of *Glycine*, *Medicago* and *Stylosanthes* species. *Plant Tissue Culture, 3*(2), 59–65.

Karadimova, M. (1993). Somaclonal variation in wheat. *Biotechnology & Biotechnological Equipment, 2*, 26–29.

Kosturkova, G. P. (1993a). Protoplast cultures of alfalfa and their application in genetic manipulations in vitro. *Biotechnology & Biotechnological Equipment, 2*, 40–42

Kosturkova, G. P. (1993b). Foreign gene expression following electroporation of *Medicago* protoplasts. *Biotechnology & Biotechnological Equipment, 2*, 43–46.

Kosturkova, G. P. (1998a). Induced somatic embryogenesis in protoplast cultures of *Medicago* species and some factors influencing it. *Annuaire de l'Universite de Sofia "St. Kliment Ohridski", 88*(4), 286–290.

Kosturkova, G. P. (1998b). Scheme for *in vitro* regeneration for *Lycopersicon esculentum* convenient for genetic transformation with *Agrobacterium tumefaciens. Annuaire de l'Universite de Sofia "St. Kliment Ohridski", 88*(4), 282–285.

Kosturkova G. P. (2005). In vitro development of various soybean (*Glycine max*) explants from mature seeds. *Breeding and Technological Aspects in Production and Processing of Soybean and Other Legume Crops: Scientific Reports of the Jubilee Scientific conference*, 8–9 Sept., Pavlikeni, 94–99.

Kosturkova G. P. (2006). Report on *Plant breeding and biotechnology capacity. Survey in Bulgaria* Rome, Italy: FAO. http://apps3.fao.org/wiews/wiews.jsp.

Kosturkova, G., Mehandjiev, A. D., Dobreva, I., & Tsvetkova, V. (1997). Regeneration systems from immature embryos of Bulgarian pea genotypes. *Plant Cell, Tissue and Organ Cultures, 48*, 139–142.

Kosturkova, G., Rodeva, R., & Mehandjiev, A. (2001). *In vitro* selection systems for *Ascochyta pisi* and herbicide higher tolerance of pea. In Association Européenne de Recherche sur les Protéagineux (AEP) (Ed.), *Towards the sustainable production of healthy food feed and novel products,* 158–159.

Kosturkova, G., Tasheva, K., & Mehandjiev, A. (2003a). *In vitro* callogenesis and organogenesis in pea (*Pisum sativum*). *Scientific Publications "TEHNOMAT and INFOTEL 2003", 4*(2), 121–127.

Kosturkova, G., Angelov, G., Rodeva, R., Tchorbadjieva, M., & Mehandjiev, A. (2003b). *In Vitro* modelling of biotic stress - higher resistance of pea cultures to *Phoma medicaginis* var. *pinodella* culture filtrates. *Proceedings of the Vth International Symposium "BioProcesses"*, Sofia, Oct., 186–189.

Kosturkova, G., Todorova, R., & Mehandjiev, A. (2004a). Development of *in vitro* tests to study soybean drought tolerance. In Association Européenne de Recherche sur les Protéagineux, (AEP) (Ed.), *Legumes for the benefit of agriculture nutrition and the environment: Their agronomics, their products, and their environment,* p. 301.

Kosturkova, G., Dimitrova, M., & Kintia, P. (2004b). Effect of the natural bioregulator "Moldstim" on pea in vitro cultures. *Phytochemistry and Application of Plant Saponins, Proceedings of the International Conference on Saponins*, Sept., Pulawy, Poland, 88.

Kosturkova, G., Dimitrova, M., & Mehandjiev, A. (2005a). *In vitro* organogenic potential of new mutant lines of pea (*Pisum sativum*). *Plant Science (Sf), 42*(3), 222–225.

Kosturkova, G., Mehandjiev, A., Tasheva, K., Ditrova, M., Rodeva, R., & Mihov, M. (2005b). Establishment of long-term organogenic cultures of pea (*Pisum sativum*) for crop improvement programs. *Proceedings of the COST Action 843 final conference*, June, Stara Lesna, Slovakia, 66–68.

Kosturkova, G., Nedev, T., & Dimitrova, M. (2006). Application of callus cultures of soybean (*Glycine max*) to study abiotic stress factors. *Field Crops Studies, 3*(2), 245–249.

Mehandjiev, A., Kosturkova, G. P., Vasilev, G., & Noveva, S. (2002). Radioprotective effect of novel disubstituted thioureas on pea (*Pisum sativum*) development. *Radiation Biology and Radioecology, 42*(6), 649–658.

Nedkovska, M. (1993). Cell selection for developing of herbicide tolerance in tobacco. *Biotechnology & Biotechnological Equipment, 2*, 30–33.

Nikova, V., Zagorska, N., & Pundeva, R. (1991). Development of four tobacco cytoplasmic male sterile sources using in vitro techniques. *Plant Cell Tissue and Organ Cultures, 27*, 289–295.

Nikova, V., Pundeva, R., Vladova, R., & Dimitrova, A. (2001). A new tobacco cytoplasmic male sterile source from hybrid combination *Nicotiana longiflora* Cav. and *N. tabacum* L. using *in vitro* techniques. *Israel Journal of Plant Sciences, 49*, 9–13.

Nikova, V., Pundeva, R., Vladova, R., & Dimitrova, A. (2004). Application of tissue cultures to overcome complete sterility *Nicotiana plumboginifolia* Viviani x *Nicotiana tabacum* L. F_1 hybrid. *Israel Journal of Plant Sciences, 58*(1), 67–72.

Noveva, S., Lazarova, N., Kosturkova, G. P., & Mehandjiev, A. (2006). Study of toxicity of heavy metals in pea (*Pisum sativum* L.) using different methods. *Field Crops Studies, 3*(3), 405–413.

Petrova M., Zagorska, N., Tasheva, K., & Evstatieva, L. (2006a). *In vitro* propagation of *Gentiana lutea* L. *Genetics and Breeding (Bg), 35*(1/2), 63–68.

Petrova, M., Stoilova, Ts., & Zagorska, N. (2006b). Isoenzyme and protein patterns of *in vitro* micropropagated plantlets of *Gentiana lutea* L. after application of various growth regulators. *Biotechnology & Biotechnogical Equipment, 20*(1), 15–19.

Rodeva, R., Kosturkova, G. P., Georgieva, I., & Mehandjiev, A. (2001). Evaluation of pea genotypes for ascochyta blight resistance *in vivo* and *in vitro. Beitrage zur Zuchtungsforschung, 7*(1), 47–51.

Rodeva, R., Kosturkova, G. P., & Mehandjiev, A. (2005). Searching for resistance to ascochyta blight in pea (*Pisum sativum*) by conventional and biotechnological methods and approaches. *Plant Science (Sf), 42*(3), 226–230.

Shtereva, L., Zagorska, N., Dimitrov, B., Kruleva, M., & Oanh, H. (1998). Induced androgenesis in tomatoes (*Lycopersicon esculentum* Mill.) II. Factors affecting induction of androgenesis. *Plant Cell Reports, 18*, 312–317.

Tasheva, K., Zagorska, N., Dimitrov, B., & Evstatieva, L. (2003). *In vitro* cultivation of *Rhodiola rosea* L. (2003). *Proceedings of the International Scientific Conference "75 years Institute of Forestry"*, BAS, Oct., 161–165.

Tasheva, K., Petrova, M., Zagorska, N., & Georgieva, E. (2005). Micropropagation *in vitro* of *Rhodiola rosea* L. *Proceedings of the COST Action 843 final conference*, June, Stara Lesna, Slovakia, 69–71.

Zagorska, N. (1993). Some aspects of development and application of plant biotechnology in Bulgaria. *Biotechnology & Biotechnological Equipment, 2*, 3–8.

Zagorska, N., Schtereva, L., & Robeva, P. (1990). Alfalfa (*Medicago* spp.): *In vitro* production of haploids. In Y. Bajaj (Ed.), *Biotechnology in agriculture and forestry* (pp. 458–471). Berlin: Springer.

Zagorska, N., Stanilova, M., Ilcheva, V., & Gadeva, P. (1997). Micropropagation of Leucojum aestivum. In Y. Bajaj (Ed.), *Biotechnology in agriculture and forestry* (pp. 178–193). Berlin: Springer.

Zagorska, N., Shtereva, L., Dimitrov, B., & Kruleva, M. (1998). Induced androgenesis in tomatoes (*L. esculentum* Mill.) I. Influence of genotype on androgenetic potentiality. *Plant Cell Reports, 17*, 968–973.

Zagorska, N., Shtereva, L., Dimitrov, B., Sotirova, V., Baralieva, D., & Kruleva, M. (2004). Induced androgenesis in tomatoes (*L. esculentum* Mill.) III. Characterization of plant obtained by anther cultures. *Plant Cell Reports*, 22, 449–456.

Part II
Bodies, Cultures and Scientific Metaphors

Chapter 5
Embryo Transfer: A View from the United Kingdom

Sarah Franklin(✉)

Abstract: Although the rise of human embryonic stem cell research has been almost entirely enabled by the prior birth of In Vitro Fertilisation (IVF) in the late 1970s, the relationship between these two fields remains somewhat under-examined. From a feminist perspective the relationship of IVF to human Embryonic Stem cell (hES) derivation is understandably a matter of significant concern. Moreover, this relationship is one factor accounting for widespread national differences in the regulation of hES derivation. This paper examines these, and other, aspects of what is described as the 'IVF-Stem Cell Interface', with an emphasis on work in the UK. The paper also explores the work of the group Human Embryonic Stem Cell Coordinators (HESCCO): a recent UK initiative aimed to enable greater ethical oversight of the IVF-Stem Cell interface.

Keywords: IVF-stem cell interface, human embryonic stem cell coordinators, egg donation, consent form, feminist perspective

5.1 Introduction

The birth of Louise Brown in England in the summer of 1978 was an event surrounded by uncertainty from many quarters: on the one hand this new miracle of scientific progress was seen to offer hope for thousands, if not millions, of childless couples, while also confirming the power of scientific and technological innovation. On the other hand, the very term test-tube baby conjured up more negative images of scientific innovation gone too far, and in particular too deep into the biological frontier of human reproduction (Challoner 1999; Henig 2004; Mulkay

---

Sarah Franklin
Department of Socialogy, BIOS Centre, London Schl of Economics, Houghton Street, London WC2A 2AE, UK
s.franklin@lse.ac.uk

F. Molfino, F. Zucco (eds.) *Women in Biotechnology,*

1997). Not surprisingly, this ambivalence was reflected in the feminist literature as well: while some feminists called for a complete ban on IVF, others lobbied for its improvement – and it is in many ways an ambivalence about technological assistance to reproduction that is its most conspicuous social legacy (Kerr and Franklin 2006; Throsby 2004).

Much has happened since the first successful clinical application of IVF nearly 30 years ago, and this remarkable period of social history has, in some respects, only just begun to be examined properly by scientists, clinicians, journalists, social scientists, and historians.[1] Three trends are nonetheless clear. One is the rapid acceptance of IVF technology to the point it is all but an everyday occurrence to have a test-tube baby. Second is the vast scale of this new form of human reproductive assistance: more than three million children have been born of IVF worldwide, and in some countries, such as Israel and Denmark, it accounts for as much as 3% of the birth-rate. Lastly, but of equal significance to the first two trends (which are complementary to one another) is the striking discrepancy in what we might call the national cultures of IVF. Hence, for example, even in countries side-by-side in Europe, such as Italy, Germany, and Switzerland, the profile of IVF in terms of its practice varies enormously. Moreover, while these variations can be put down to influential factors, such as religion, such factors alone cannot explain the almost opposite principles through which IVF is regulated and practiced in otherwise religiously-similar countries such as France, Austria, Spain, Ireland, and Portugal – all of which are predominantly Catholic, but none of which have very similar situations concerning IVF. The reverse is also true, in that several countries which in some ways have rather similar IVF profiles do so for very different reasons, as in the case of Germany, Ireland and Austria.

The advent in the 1990s of the high-profile post-IVF technologies of cloning by somatic cell nuclear replacement (CNR) and human Embryonic Stem cell (hEs) derivation has brought increasing attention to the vexing question of how to achieve a consensual, over-arching, and unified mechanism for the regulation of human embryo research and hES derivation – processes which are inextricable from IVF. In other words, a number of new issues that are as complex technologically as they are ethically or politically have arisen at the interface between IVF and human embryonic stem cell derivation which move far beyond the familiar (and rather hackneyed) questions about the 'moral status of the embryo', such as the rights and wrongs of various means of embryo procurement (altruistic donation? paid donation? egg sharing for research?), as well as both specific and general questions about commercialisation, anonymization, screening, and also issues of Good Manufacturing Practice (GMP), standardisation, and safety. In this chapter I argue that the synthetic historical relationship between IVF and hES derivation needs to

[1]For a selection see Alberda et al. (1995), Brown and Brown (1998), Clarke (1998), Corea (1985), Edwards (2001, 2004, 2005), Fishel and Symonds (1986), Kannegiesser (1988), Pfeffer (1993).

be more fully understood in order to make sense of the current controversy about stem cell science – including its gendered dimensions.

To begin with it is important to acknowledge that it is the close relationship between human embryo research and IVF that enabled the latter to become a clinical reality in the late 1970s. IVF and human embryonic stem cell derivation have an identical pedigree in the research of scientists such as Richard Gardner, one of Bob Edward's PhD students at Cambridge, who conducted the first successful biopsy of a rabbit egg tophectoderm in the 1960s, using unprecedented methods of embryo surgery, and successfully achieving a 'technologically-assisted conception' (and birth) with very obvious implications for both human reproduction and clinical genetics. Thus, while the birth of Louise Brown was met with a level of public debate that is today often forgotten, the rapid international acceptance of IVF as a form of technological assistance to those in search of a 'miracle baby' quickly cemented into place the rationales for IVF that are all but taken for granted in contemporary society.

In retrospect, then, it is not only clear that public acceptance of IVF also meant implicit acceptance of human embryo research, but the inevitable accumulation of embryos in storage, and the moral imperative to continue to pursue human embryo research – in particular because IVF has never had a particularly high success rate to begin with (although, in no small part because of embryo research, it has been improved dramatically and is now effective approximately 50% of the time in the contexts of best practice using single blastocyst transfer).

The interlinked features of the introduction and recent history of IVF are – that it:

(a) Came from embryo research;
(b) Would by definition involve in the creation of more human embryos than could be used clinically;
(c) Would necessarily involve further embryo research to continue to improve the clinical outcome of IVF.

These are the backdrop to much of today's controversy about what kind of relationship between, and ethical governance over, hES derivation and clinical ART – or the IVF-stem cell interface should exist. This question, of embryo supply, and its source in IVF, thus comprises a crucial dimension of hES cell derivation, and its highly politicised ethics, which are increasingly focussed on the sourcing, distribution, and ethical oversight of human embryos that could potentially be donated to research. Indeed, one way to condense the main question posed of human embryonic stem cell research by its various interested and often anxious publics is simply to ask what reciprocities, responsibilities or obligations will return from these donations – to individual women or couples, as goods and services, as hope or cure, as improved IVF services for the future, or in the form of a more diffuse, non-specific and generalised idea of 'common good' (Waldby and Mitchell 2006). In sum, this interface, between the context of assisted conception and human embryonic stem cell derivation, is yet another microcosm of wider issues we often understand in terms of science and its publics.

The controversy surrounding the allegedly fraudulent conduct of South Korea's Woo Suk Hwang has, paradoxically, increased pressure on the scientific community to demonstrate the very highest standards of ethical practice in the context of hES cell derivation, while simultaneously illustrating how easily scientific evidence, materials, and documents can be forged. At the same time, then, that issues such as informed consent have become more complex in the context of embryos donated for hES derivation (e.g., their amplification and potential perpetuity), the science too is under new pressures, such as international competition, that place new burdens on its traditional mechanisms of ethical governance (i.e. editorial oversight and peer review). The very same extensive informed consent documentation that made Hwang's research *appear to comply* to the highest ethical standards, but which was revealed to have been faked, demonstrates that documentation alone is never a sufficient guarantee of veracity: as Hwang's paper also showed, highly technical DNA scans can also be faked almost without effort simply by substituting a 'before' for an 'after' image. A paper trail may be only that.

These issues of what are commonly referred to as accountability and transparency point to the need for robust ethical protocols for hES derivation, such as those set out in the UK Stem Cell Bank's Code of Practice, which is in turn partly modelled on Britain's notoriously strict Human Fertilisation Embryology Authority (HFEA) requirements for ART, and even stricter requirements for animal research. Procedures such as informed consent for egg and embryo donation to stem cell research must be, and be seen to be, more than paper-tight. But what will these consist of? Informed consent was originally developed for procedures such as surgical interventions that posed a risk to the life or health of the individual, and was intended to protect patients against the pressures of medical paternalism. Over time, informed consent has been expanded to cover everything from liability protection for clinicians to organ donation. Unsurprisingly, the expansion of informed consent has been accompanied by criticisms of its 'empty ethics' from those who claim it has become more of a panacea than an actual form of protection, and more an exercise in 'box ticking' than actual communication between doctor and patient (Corrigan 2003).

In the context of embryo donation, informed consent is complicated by several features, such as the fact that two people must give their consent as a couple if the embryo is to be donated to research (whereas informed consent was initially envisaged as individual). There is a growing literature on embryo donation in which some of the factors that influence couples perceptions of this practice have begun to be studied, albeit largely inconclusively.[2] Studies show wide divergences in couples' willingness to donate their embryos to research in general (Burton and Sanders

[2] See Bangsboll et al. (2004), Bjuresten and Hovatta (2003), Burton and Sanders (2004), Choudhary et al. (2004), Newton et al. (2003), Kovacs et al. (2003), Laruelle and Englert (1995), Lornage et al. (1995), McMahon et al. (2003), Söderström-Anttila et al. (2001), Svanberg et al. (2001), Westlander et al. (1998), and for over-arching bioethical discussions see Robertson (1995), Magnus and Cho (2005).

2004; Choudhary et al. 2004), and this finding has been replicated in more recent research on donation to stem cell research in particular (Bjuresten and Hovatta 2003; Parry 2005). This is not surprising given that embryos produced in the context of IVF are complex embodiments of reproductive hope, reproductive labour, and often reproductive loss (Kitzinger and Williams 2005). Moreover, as already mentioned above, donation to hES derivation raises new and unfamiliar issues for informed consent, such as how much information should be taken from donors, how practically should reversible anonymity be organised, how should patients' questions about the production or use of artificial gametes derived from their reproductive cells be answered truthfully, and so forth.

These questions, and many others, although obvious, and increasingly widespread, have only just begun to be studied empirically, in particular by social scientists – whose insights are more often drawn from the real-world contexts of actual practice than the more ethereal realms of speculation common to bioethics. Although comparatively little has been published about the precise dynamics of the IVF-stem cell interface, some social scientists and clinicians have begun to examine some of the key factors and influences that define this context, such as patient perceptions of embryo donation to stem cell research (Choudhary et al. 2004; Parry 2005), media representations and public concern about embryonic stem cells (Doring and Zinken 2005; Kitzinger and Williams 2005; Williams et al. 2003), and the wider social implications of the issues raised by stem cell research (Holm 2002; Ganchoff 2004; Hogle 2003; Liddell and Wallace 2005; Parry 2005; Rapp 2003; Sperling 2004; Waldby 2002; Waldby and Squier 2003; see also Franklin 1999a, b, c, 2001, 2003b; Franklin et al. 2005).[3]

5.2 The UK Stem Cell Initiative

In the second half of this chapter, I focus on the experience of the UK, where ethical oversight of embryo donation to stem cell research comprises a significant component of a coordinated programme of hES derivation and banking. This system of 'human embryonic stem cell coordination' is seen not only to offer the most efficient and cost-effective route to successful translation of hES products, but also greater accountability and transparency, and ethical integrity, in the fast-paced and often controversial field of bioscientific innovation. One of the 'lessons to be learned' from the UK, I suggest, is that while debate about ethical governance (and

[3] Other scholars who have examined the 'economic' aspects of donation of embryos, and other tissue, such as blood, organs, genes, or other body parts, by examining how they become invested with particular kinds of value. As numerous scholars have pointed out, such forms of 'trade' (Parry 2005, 2006) raise wider sociological, ethical, and political issues that require further definition and analysis (Liddell and Wallace 2005; Lock 2002; Rose 2001, 2006; Thompson 2005; Waldby 2002; and see esp. Tutton 2002; Corrigan 2003).

especially informed consent) is often rightly criticised as empty (Corrigan 2003) or as limited or as too narrowly, and individualistically, defined, it is, from another point of view, an increasingly important context of active deliberation and dialogue, and one in which patient input and public participation may play a surprisingly important role. Using a case study from the UK that reveals the complexity of negotiations that are occurring in the context of standardising informed consent protocols at the IVF-stem cell interface, and based on my own fieldwork for the past five years as part of one of the UK's leading stem cell teams, I suggest this context of unique, explicit, and unprecedented deliberation has much of value to tell us about the promise and politics of stem cell research.

The question of what is at stake in the context of hES cell derivation and banking is of particular significance in the UK, which is widely acknowledged to have one of the most extensive regulatory frameworks for governance of assisted reproduction and embryo research anywhere in the world (Gunning 2000; Mulkay 1997). It is also often noted that this widely emulated system, while highly bureaucratic (it operates through a licensing body), and strict (it is backed up by criminal law), is also one of the most tolerant, progressive, and liberal towards embryo research, and, more recently, hES cell derivation. In the latter context, the UK is also advantaged by both a national healthcare system, and a comparatively joined up climate of academic research, funded in large part by the national research councils, and the Wellcome Trust, the world's second largest medical charity and a major donor of research income to UK universities. Embryo research in the UK is thus protected by a stable regulatory environment and a strongly positive government policy of support, which in turn are directly linked to promoting UK leadership in the 'knowledge economy' of the biosciences, where scientific achievement and research investment are considered to be vital matters of national, commercial, and political self-interest (Franklin 2001, 2003a, b, c).[4]

Since the late 1990s, it has been the official policy of the UK government to promote hES cell research, and in order to further this priority the government has both increased research funding to this area, and supported the commissioning of the world's first National Stem Cell Bank to make of the UK a world hub of hES cell science. As in many other European countries, both the UK government and the general population share a positive evaluation of the potential benefits of stem cell research, and support its development, including the necessity for embryo

[4]Following in the wake of a number of distinctive UK 'firsts' in the fields of reproductive biomedicine, embryology, developmental biology, and genetics, including the discovery of the structure of the double helix (1953), the first embryo transfer and biopsy in rabbits (1967); the first test-tube baby (1978), the first immortalised cell lines in mice (1981); the first successful preimplantation genetic diagnosis (PGD) (1990); and the first successful cloning of a mammal from an adult cell (1996); it was clear that the UK would enjoy an 'indigenous' advantage in the field of hES cell derivation – the first major post-genomic industry to galvanise into a global 'race' for clinical and commercial success since the completion of the draft sequence of the human genome map in 2001 (Edwards 2001, 2005).

research.[5] The strategy of the UK government under the leadership of Prime Minister Tony Blair, for whom stem cell research was a personal priority, has been to promote stem cell research through public funds as well as public-private partnerships, and to align the governance and regulation of this new field with existing legislation on embryo research, tissue banking, and quality assurance. Also key to the national science strategy under the Blair government has been more attention to issues of public support for, and participation within, scientific innovation – exemplified by the UK's GM Nation initiative to stimulate public dialogue in the wake of the infamous Monsanto debacle of the late 1990s. The efforts to regulate stem cell research have consequently aimed to promote and to protect UK stem cell science by ensuring public confidence in the ethical governance of this new field, while attempting to produce a stable environment for successful research and development, translating these potential clinical applications, and stimulating commercial investment while attempting to avoid the more divisive and unpopular consequences of a culture of privatised research.

As recommended by the Blair government, and in accordance with the conclusions of a Department of Health (DH) consultation paper into human embryonic stem cell research, published in 2000, new regulations were introduced into Parliament to amend the 1990 Human Fertilisation and Embryology Act in order to widen the criteria for research on human embryos, including the creation of embryos through cell nuclear replacement (CNR) for research purposes. Under the Human Fertilisation and Embryology (Research Purposes) Regulations (2001), three new criteria for embryo research were added to the existing Act to widen the possibility for understanding the cellular bases of serious diseases, and to permit research that might lead to their treatment – including through the method known as 'therapeutic cloning' (i.e. CNR).

In 2002, following the recommendations of the DH and also those contained in a special Report of the House of Lords Select Committee on Science and Technology (2002), the UK Stem Cell bank was commissioned and publicly funded as a joint endeavour between the Medical Research Council (MRC) and the Biotechnology and Biological Sciences Research Council (BBSRC, together £2.3 million). 2002 also saw the first licenses granted for hES cell derivation from the Human Fertilisation and Embryology Authority (HFEA), the statutory body which oversees all research concerning human embryos in the UK, and in 2003 the first UK-derived lines were announced by King's College in London (Pickering et al. 2003). In this same year the UK stem cell bank was constructed and began its rigorous process of quality accreditation by the MHRA (Medicines and Healthcare products Regulatory Agency). In the meantime, its Steering Committee (initially chaired by Professor Genevra Richardson and later by Lord Naren Patel) began the equally arduous process of devising means to fulfil the aims and goals of the bank in a manner that would be simultaneously consistent

[5] For full results of these polls see http://www.ipsos-mori.com/polls/2003/amrc.shtml, and http://www.yougov.com/archives/pdf/TEL050101042_1.pdf

with existing UK and EU legislation, including the guidelines of its national research councils (e.g. MRC, BBSRC), the practices of other public tissue-banking organizations such as the National Blood Service (NBS), international cGMP and ISO requirements, MHRA accreditation criteria, and the licensing criteria of the HFEA. By the end of 2003, the HFEA had granted eight further licences for hES cell derivation nationwide. The bank was accredited and opened in 2004, the same year the UK Human Tissue Act (HTA) came into effect, under which regulation the first UK lines were banked in 2005. The simultaneous adoption in 2004 of the EU Directive on Setting Standards of Quality and Safety for the Donation, Procurement, Testing, Processing, Storage, and Distribution of Human Tissues and Cells (widely known as the EU TD) increased the necessity for the bank, in coordination with the UK's increasingly numerous hES cell derivation laboratories, to reach agreed-upon standards and protocols for all aspects of derivation – from cGMP to informed consent.[6]

The mission statement of the UK stem cell bank is to 'work with the scientific and clinical community to assure the quality of human stem cell lines used in research and therapy'. Its four main aims are to:

- Produce, test and release well-characterised seed stocks of adult, foetal, and embryonic stem cell lines within a stringent quality framework
- Promote basic research in the UK and abroad through the provision of 'Research Grade' cell banks
- Provide stringently tested, safe 'Clinical Grade' cell banks under EU cGMP conditions as starting material for therapeutic uses
- Work with the scientific and clinical communities, commercial organisations and regulatory agencies to assure the quality of human stem cell lines used in research and clinical therapy and disseminate best practice

The bank is a public, non-commercial, facility, which will provide a major resource for the global biotechnology community, including the commercial sector. The cell lines that are banked and exchanged through the new facility are intended to comprise a unique population that acts as a vital repository for an emergent form of reproductive value that literally fuses social technologies of regulation and consent together with biological techniques for cellular control and management, and medical and social aspirations for scientific progress. In addition to the quality and stability of lines that are adequately characterised to be banked, the facility is responsible to guarantee the provenance of any lines it accepts – a procedure that has required devising new protocols for informed consent to donate embryos to stem cell research (as we shall see further below).

---

[6] 2004 also saw the first HFEA licenses granted to two UK clinics for the deliberate creation of embryos for research using cell nuclear replacement (CNR). A blastocyst derived from this method was announced by researchers at Newcastle University in the spring of 2005. The legality of creating human-animal (hybrid) embryos for research was affirmed by the UK Parliament in May 2008.

With the bank as its hub, a national system of human embryonic stem cell derivation, characterisation, and banking has been developing rapidly since the first UK lines were successfully derived at King's in 2003 (Pickering et al. 2003). Two of the most important components of the UK stem cell network are the connections between and amongst the various centres (the added value of national scale), and the interface between stem cell derivation laboratories and IVF clinics (which are in many cases literally being merged into a single facility).

Successful clinical introduction of stem cells and the emergence of international stem cell exchange and trade, as well as commercialisation, will thus involve a combination of several components: they must be safe, regulated, and standardised to an acceptable GMP standard; they must retain public confidence and support to become viable products on a significant scale; they have to work and be successful enough to be either clinically or commercially viable; and they must be seen to be ethically-derived and subject to appropriate legislation. In sum, good governance of hES cell derivation, banking, and manufacture will be essential to their potential clinical, commercial, and social value (Banchoff 2004).

5.3 Human Embryonic Stem Cell Coordinators (HESCCO)

The UK is currently one of the only countries to begin to attempt to develop national standards and regulation not only for hES cell derivation and banking, but also the ethical protocols for informed consent in the context of IVF, where the embryos necessary for hES cell cultivation are sourced.[7] An important UK initiative launched in 2003 established a network of HESCCO, which is currently in the process of updating, improving, piloting, and semi-standardising a national protocol for patient consent and information materials for donation of embryos to stem cell research, as well as GMP and other derivation criteria.[8] This ongoing British initiative is an unprecedented practical intervention into the practices and procedures that structure the 'embryo transfers' linking hES derivation to IVF, and thus offers a unique opportunity to sketch out some preliminary reflections on how these are beginning to be managed, standardised, and regulated.

The necessity for greater national coordination of UK hES cell line derivation prompted the MRC and the BBSRC to make available strategic research funding towards development of cooperation between centres capable of deriving hES cell,

[7] Informed consent is thus becoming more important for stem cell research just as its legitimacy is increasingly questioned by some commentators, see Corrigan (2003).

[8] The first network of its kind, HESCCO, the human embryonic stem cell coordinators group, was founded in June 2005 in Leeds with approximately 25 members.

and assisted conception units (ACUs), who could provide spare or surplus eggs and embryos produced in the context of In Vitro Fertilisation (IVF) treatment for infertility. Seven such centres were funded around the UK (Newcastle, York (Leeds), London/Leeds (the Bridge Centre), Sheffield, King's College London (Guy's Hospital), Roslin Centre (with Manchester), and the North of Scotland Initiative (Aberdeen/Dundee/Edinburgh). A number of these appointed stem cell coordinators to facilitate recruitment of patient donors for stem cell projects. In an effort to facilitate ethical sourcing of eggs and embryos, and to devise semi-standardized national protocols for patient information and consent for embryo donation to stem cell research, funding was sought and approved to bring the coordinators and stem cell scientists together biannually specifically for national stem cell coordination, a primary benefit of which would be increased national cooperation and more effective and efficient use of available embryos for stem cell research. In addition to the seven MRC supported units or networks, other assisted conception or stem cell units also joined the initiative at its outset: Manchester, Birmingham, Oxford, Bourn Hall Cambridge, Nottingham.

Through this nationwide programme of cooperation, centres encountering similar experiences, be they in meeting GMP or drafting patient information and consent forms, could share experiences and essential information. Efficiency of scale would also be gained to achieve more rapid compliance with existing and emergent legislation, and in particular the 2004 EU Tissue Directive, as well as the quality control requirements of GMP, ISO and other accreditation measures. In sum, national coordination would offer the additional potential to increase the accountability and transparency of stem cell derivation and banking, by ensuring the credibility and rigour of its ethical governance, thus enhancing successful scientific innovation with the potential to improve public health.

Aims of HESCCO are:

- To establish a nationwide programme of coordination
- To increase levels of standardization
- To maximise use of potentially available eggs and embryos
- To improve patient communication and consent procedures
- To comply with UK, EU, & international regulatory policy
- To contribute to the process of creating such policy
- To coordinate and share quality control experiences

Besides the individual 'stem cell coordinators', clinicians, scientists, and embryologists participated in HESCCO meetings as well as two social scientists, and invited representatives of national regulatory bodies, including the HFEA, and organizations with relevant tissue banking expertise, such as the national blood service (NBS). Members of the bank (UKSCB) also attend meetings, as do representatives of the Medical Research Council.

The general, over-arching, aims of HESCCO outlined above were linked in turn to a number of specific objectives, in the form of key outcomes and deliverables. In turning to a review and assessment of these tasks the aim is to evaluate their success as models of good practice that may be of benefit in other national, or international, contexts.

HESCCO: key deliverables are:

- To establish semi-standardised protocols for sourcing eggs and embryos (including consistent procedures for approaching patients for fresh and frozen eggs and embryos from IVF and PGD programmes)
- To design, pilot, assess, and evaluate a consent form to produce a semi-standardised document for use nationally that conforms with HFEA and the Stem Cell Bank Steering committee requirements
- To collect data on patient perceptions of egg and embryo donation, informed consent procedures, and attitudes toward stem cell research (and to make this data available on a public website and through publication)
- To create two websites, one public and one restricted access, which would contain information about stem cell research and the national embryo supply (database, FAQ, publications, links, etc.)
- To establish a working national network facilitating feedback and evaluation of best practice in embryo procurement and hES cell derivation and to provide resources for professionals involved in this area
- To design, test, evaluate, and standardise the protocols for a national database through which all embryos donated to hES research are tracked
- To establish guidelines for feedback to patients, including standard policies on anonymity, commercialisation, traceability, testing, and the derivation of artificial gametes
- To achieve compliance with existing regulation, and to contribute to the development of evolving guidelines and protocols
- To contribute to the formation and improvement of regulatory guidelines within the UK and in Europe

In addition to patient information and consent protocols, and issues concerning compliance with evolving regulation, HESCCO meetings have provided an opportunity to address specific technical issues of quality control, in particular where these overlap with patients' concerns – such as the point at which embryos can no longer be withdrawn from research, and questions about the criteria for freezing embryos at each clinic and the availability of surplus embryos. The contribution by the MRC towards the costs of building/renovating of five GMP-standard derivation labs linked directly to ACUs, in Sheffield, Manchester, London (King's), Newcastle, and Birmingham in 2005, and the completion of the first of these in 2006 (in Sheffield), has intensified the need to develop, and to agree upon, a minimum set of nationally semi-standardised protocols, despite the difficulties of so-doing.

While debate about informed consent, patient screening and feedback, and paid donation, continue to indicate the lack of a uniform frame of reference on the question of 'ethically-sourced' eggs and embryos, the imperative to provide clear standards of best consenting practice is increasing. Thus, in addition to the fact that consent procedures for embryo donation to stem cell research must respond to novel technical issues such as immortalization and artificial stem cell-derived gametes, and are developing in a context of high-profile incidents of serious scientific misconduct, they are also *required to conform to a new, and in some senses higher, standard*. For these and other reasons, consent has become a priority issue for HESCCO, and occupied much of its attention.

While efforts in conjunction with the UK stem cell bank to develop semi-standardised national protocols for GMP and relevant aspects of derivation (such as appropriate standards of air quality, flow, and monitoring) are ongoing, the primary deliverable of the HESCCO initiative in the UK has been the establishment of practical, legal, up-to-date, and nationally-agreed-upon criteria for informed consent, as well as two new four-page consent forms (for fresh and frozen embryos). The forms are the outcome of an extensive, time-consuming, and laborious consultation process involving representatives from the Human Fertilisation and Embryology Authority, the Stem Cell Bank, the National Blood Service, the Department of Health, and the MRC.

5.4 Standardising Consent Criteria

Free and informed consent are key principles of the Human Tissue Act (2004), the HFE Act (1990), and the EU Tissue Directive (2004), as well as being standard principles of best practice in medical practice, random-controlled trials (RCTs), and any scientific research involving human subjects. For consent to be robust, legitimate, and ethically sound, comprehensive information must be given in a form that is readily accessible and allows a free and informed decision to be made by potential donors. In the UK, all written (patient) information provided and consent forms have to be approved by Local Ethics Committees (COREC[9]), and for research involving embryos also by the HFEA.

To meet some of the additional consent issues raised by egg and embryo donation to hES cell derivation, the Steering Committee of the UK Stem Cell Bank has, in collaboration with the HFEA, drawn up a list of 'minimum criteria' that must be addressed in patient information leaflets and consent forms. Before patients give consent to donation of their embryos for use in research projects to derive stem cell lines, they must be given oral information supported by relevant written material which confirms that:

- The research project is directed towards the creation of stem cell lines.
- Very few cell lines are derived from donated embryos at present.
- Any cells lines that are derived will be deposited in the UK Stem Cell bank, may be used for other projects within the UK and/or overseas, and may eventually be used for treatment, although the research will not lead to any direct medical benefit to the donor.
- The UKSCB is overseen by an independent Steering Committee which has the responsibility to ensure legal and ethical accountability for hES cell lines once they are accessioned by the bank.
- Donation to research will in no way affect the donors' treatment and the decision whether to donate is voluntary.

---

[9]The Central Office for Research Ethics Committees is run by the UK National Health Service, see www.corec.org.uk

- Donated embryos will be anonymised, although this anonymity must, under the terms of the EU Tissue Directive, be reversible under exceptional circumstances when grave matters of public health may be at stake.
- No information emerging from tests done on the cell lines will be fed back to donors (unless under the exceptional circumstances mentioned above).
- Donors can withdraw their consent until the point that the embryos are used for research.
- Cell lines, or discoveries made using, them may be patented and used for commercial purposes, but the donor will not benefit financially from any future profit generated by their cell line.

In addition to these minimum criteria (which must be provided in the patient information leaflet accompanying the consent form, and fully discussed by an independent party who is not part of either the clinical or the research team), each gamete provider must consent *in writing* to the following:

- *To the use of embryos created using their gametes in the research project for the derivation of stem cell lines.*
- *That they understand that a sample of any stem cell line will be deposited in the UK Stem Cell Bank and that the derived stem cell lines may be used in other research projects.*
- *That they are under no obligation to take part in the study and that a decision not to participate will not alter the treatment that they would normally receive.*
- *That they understand that they have a right to withdraw their consent without giving any reason, at any stage until the gametes and/or embryos have been used for research.*
- *That they understand that any cell line derived from their donated gametes/embryos may eventually be used for treatment purposes (including cell replacement therapies) in the future.*
- *That they understand that cell lines or discoveries made using them may be patented and used for commercial purposes, but that the donor will not benefit financially from this.*
- *That they agree to be contacted in the future in the unlikely event that the Stem Cell Steering Committee considers that they should be contacted in relation to confirmed test results performed on stem cell lines that are of direct relevance to their own, their family's or public health.*

5.5 Areas of Ongoing Concern

The collective national effort to draft, pilot, revise and gain ethical approval for a consent form that is now semi-standardised in the UK has proven a useful exercise on several fronts; for example by confirming minimum consent criteria, collecting information on patient perceptions, and improving patient information protocols. The consent form also offers a practical outcome in the form of a .pdf that is available as an easily down-loadable prototype or template. In many respects this process has also affirmed the value of working at a national scale in order to make efficiency gains and clarify areas of shared uncertainty.

However, inevitably this process has revealed areas of concern, which bear at least passing mention as they are likely to remain the subject of ongoing disagreement for the foreseeable future both in the UK and elsewhere. These areas are:

(a) Artificial gametes and 'fettered' or conditional consent
(b) Feedback to patients
(c) Payment of egg and embryo donors, or discounted treatment costs in exchange for donation
(d) Donor screening

5.5.1 'Fettered' Consent

That it may be possible for germ cell lines, or artificial gametes, to be derived from embryonic cell lines provides the first concern. If these were to be distributed outside the UK, potentially they could be used for reproductive purposes regardless of what UK law has to say on the subject (i.e. the ban on reproductive cloning). Patients may wish to allow their embryos to be used for research or for development of new forms of medical therapy but not for their cells to be used to develop a new individual in years to come. The possibility of introducing conditional consent, a decision which would respond to the legitimate concerns of potential donors, is one to which the HFEA has devoted serious consideration, but would set a new precedent for tissue donation which many regard as contrary to best practice in terms of either experimental research or tissue banking. Simply not allowing patients with these concerns to donate would reduce the number of embryos available for research into stem cell derivation and processing even if they were non-GMP and not intended for any therapeutic purpose. An additional objection to making an exception of embryo donation as a candidate for the unprecedented use of conditional consent is that adult cells might also be able to be made into gametes in the future, and that too little is known about the distinction between these two types of cells. To date, consent to donate embryos to stem cell research has remained unconditional in the UK.

5.5.2 Feedback to Patients

A second area of concern is feedback of information to patients. At one end of what might be called the 'feedback spectrum' is the voiced ethical concern that relevant information belongs to the donor and that no agency should hold secret identifiable information of relevance to the individual's health. However, information such as genetic predisposition to disease susceptibility may only come to light through research many years after donation, and might be clinically unreliable. How and by whom such information would be imparted also presents difficulty. On the other hand is the requirement that donors remain traceable through a system of reversible anonymity, so that they can be contacted in the event of a major threat to public health (the most commonly cited example being the detection of infective diseases such as BSE which could be spread through other contaminated tissue products). Beyond this requirement, current best-practice as recommended by the bank Code of Practice for the Use of Human Stem Cell Lines (2006) is for donors to be informed that 'no individual feedback will be given on tests performed by the UK Stem Cell Bank or research results of subsequent studies' although 'general information (...) including the results of research using embryonic stem cell lines will be published on the [Bank's] website'. Hence, while the option to be informed if their embryos develop into cell lines has largely been foreclosed, the possibility that patients would be contacted either in the event of a public health concern, or indeed should 'test results of direct relevance to the donor or the donor's family' be discovered remains open, and the more indirect feedback provided by the bank as research reports may become more detailed over time.

5.5.3 Payment to Egg and Embryo Donors

In July 2006, the HFEA granted its first permission to Newcastle University for 'egg sharing for research' – a practice designed to reduce the cost of IVF treatment in exchange for the donation of eggs and embryos to research – and announced a public consultation into the wider question of 'whether it is appropriate for women to donate their eggs for use in scientific research'. Egg-sharing for research, which is based on clinical egg-sharing programmes in which similarly-discounted IVF treatment is offered in exchange for donating eggs to other couples, has been the subject of a public consultation due to the potential for exploitation, and the spectre of commodification which attaches to any kind of remuneration for tissue donation (Fontaine 2002) and especially the perceived sale of tissue. In February 2007, the HFEA confirmed its decision to allow egg sharing for research and to amend their Code of Practice accordingly.

5.5.4 Donor Screening

While debate continues concerning the suitability of molecular assays to detect contamination in cell lines, thus eliminating the need for the kind of comprehensive donor screening that is standard practice in the National Blood Service for so-called master-stocks, a divergence of opinion attaches to the importance of screening, or even documentation, of donor characteristics. To some, this paramount question of public safety can breach no slackening of the reins, while, for other, perhaps more technologically-confident, observers, the costly and labour-intensive donor-selection processes characteristic of post-war blood-banking is no longer the paradigm of either best practice or fiscal prudence.

5.6 Conclusion

In sum, it is clear that the stem cell field continues to raise challenging ethical questions that will require an evolving process of innovation at the level of practice as well as principle. Like the complex questions of GMP arising at the IVF-Stem Cell interface, for which new definitions of quality control and monitoring are being devised, so too is this the site of ongoing innovation in definitions of 'best consenting practice'. Patients, who often share the more box ticking view of the myriad informed consent forms they complete during even a single IVF cycle, frequently ask challenging questions for which it is difficult to provide definite answers.

This is fortunate, because an active, critical dialogue is essential to robust informed consent procedures, and to the development of effective patient information materials. Especially at this early stage in the process of devising standards of best practice in the context of informed consent for embryo donation to stem cell research, difficult questions should be welcomed, openly debated, and valued because of the genuine uncertainties they reveal. Literally papering over the difficult practical questions raised by this unique context of tissue donation with overly reassuring patient information leaflets, or euphemistic informed consent forms, will significantly compromise the opportunity to learn more from patients directly about their hopes, fears, doubts, expectations, and direct experience of thinking through the many issues raised by this important and highly emotive field.

As has been evident in the wake of 'Hwang-gate', public scepticism is inevitably heightened by such dramatic episodes of scientific fraud, but underlying hopes for medical progress through ongoing scientific experimentation are robust and in many cases seemingly unquenchable (Braude et al. 2005). Though it may seem counterintuitive to claim that there are, indeed, grounds for welcoming public scepticism as a means of protecting the integrity and improving the quality of public debate over controversial areas of science, there is some evidence to support this view. Such evidence is also increasingly finding its way into public policy (Liddell and Wallace 2005). In the most recent recommendations to emerge from the Australian *Reports* on the *Prohibition of Human Cloning Act 2002* and the *Research Involving Human Embryos Act 2002*, for example, data collection on public perceptions of human embryonic stem cell research, and public outreach via a website are recommended as a priority.

In returning to the opening reference to a legacy of ambivalence toward IVF – including within the feminist literature – and the rapid expansion of technologically-assisted conception over the past half century, during which IVF and embryo research have continued to develop in tandem, and have become, if anything, even more closely interconnected, or symbiotic, as a result of the advent of hES derivation, we can reach a number of conclusions. For example, the case of the UK, especially in contrast to the US, demonstrates how significantly national contexts vary in terms of ethical oversight and regulation of hES derivation, and in particular the IVF-stem cell interface – with significant effects for basic science. Whereas in the US, research has been hampered both by government opposition to the use of human embryos in research (a position that is inconsistent with support for IVF), and by the lack of clear or coordinated national, or even regional, guidelines, the emphasis on national coordination in the UK has yielded a number of important benefits – from practical information sharing to the advantages of scale (and leadership) provided by the UK Stem cell bank. The pattern of widespread public support for basic scientific research that has the potential to yield significant medical benefits appears to be a more uncomplicated driver in the UK, where a generally pragmatic approach to health improvement is apparently motivated by an equally widespread belief in the moral value, or even imperative, of scientific progress in the name of improved human futures.

This brings us to an important concluding point about the benefits of national hES coordination demonstrated by the UK case, which is that the improvement of informed consent procedures and patient information are not only practical: they also assist in the creation of a dialogue about the rights and wrongs, the hopes and fears, and possibilities for improvement and the possibilities for exploitation, that in the long run (or even in the short term) are as likely to determine the eventual success or failure of the hES field (in which far more public than private funds currently invested) than the scientific knowledge or technology on which it is based.

We can thus summarise that to the extent the practical benefits of hES coordination are ethical as well as technical, we can also consider them sociological too. One of the interesting outcomes of the HESCCO initiative in its first three years has been the amount of dialogue and deliberation occasioning its formation, duties, and evolution into a national network. Like the bank, HESCCO now operates as an important public interface for hES research, and is likely to increase in importance in this role over time – locally, regionally, nationally and internationally. From this perspective, the importance of the time and attention devoted to negotiating consent procedures becomes more evident – and its value as a process, as well as a requirement, becomes more visible. Rather than an empty ritual of ticking boxes that induces cynicism and doubt, patient perceptions of the experience of being approached about the possibility of donating their embryos to hES derivation prove to offer a rich seam of opinion on this controversial topic. Far from unanimity, the diverse range of views recorded in even a brief survey of 166 patients provided extensive material for the FAQ section of the public HESCCO website, as well as suggesting several areas for further research.

In all of these respects, we see in the often confusing diffractions by which we commonly imagine social and technological change to be related – if not mirrored in one another, then still, somehow, a cohesive whole – that one important problem may be the assumption we can understand the ethical, the social and the technological as distinct domains. Ironically, but not surprisingly, this is one of the main lessons to be learned at the IVF-stem cell interface, which appears less and less to have its future pegged to principles and techniques of scientific intervention into cellular processes, and more and more to be itself the germ of a new understanding in which it is the principles and techniques of understanding social interaction that will play at least as large a role in its future, be it viable or otherwise.

Acknowledgements I am very grateful for the extensive input of research colleagues and collaborators in the research reported here, and in particular I would like to acknowledge the contributions of Peter Braude, Stephen Minger, Sue Pickering, Karen Throsby, Celia Roberts, Clare Williams, Glenda Cornwell, Emma Stephenson, Charles Hunt, Chris O'Toole and Sue Avery for helpful contributions, comments, and corrections.

References

Alberda, A. Th., Gan, R. A., & Vemer, H. M. (Eds.) (1995). *Pioneers in In Vitro Fertilization: The proceedings of a symposium held in Oss, The Netherlands, November 5, 1993*. London: Parthenon.

Banchoff, T. (2004). *Embryo politics: Debating life in a Global Era*. Unpublished manuscript, cited with permission from the author.

Bangsboll, S., Pinborg, A., Yding Andersen, C., & Nyboe Andersen, A. (2004). Patients' attitudes towards donation of surplus cryopreserved embryos for treatment or research. *Human Reproduction, 19*(10), 2415–2419.

Bjuresten, K., & Hovatta, O. (2003). Donation of embryos for stem cell research: How many couples consent? *Human Reproduction, 18*(6), 1353–1355.

Braude, P., Minger, S., & Warwick, R. (2005). Stem cell therapy: Hope or hype? *British Medical Journal, 330*, 1159–1160.

Brown, L., & Brown, J. (1998). *Our miracle called Louise: A parents' story*. London: Paddington Press.

Burton, P. J., & Sanders, K. (2004). Patient attitudes to donation of embryos for research in Western Australia. *The Medical Journal of Australia, 180*(11), 559–561.

Challoner, J. (1999). *The baby makers: The history of artificial conception*. London: Macmillan.

Choudhary, M., Haimes, E., Herbert, M., Stojkovic, M., & Murdoch, A. P. (2004). Demographic, medical and treatment characteristics associated with couples' decisions to donate fresh spare embryos for research. *Human Reproduction, 19*(9), 2091–2096.

Clarke, A. E. (1998). *Disciplining reproduction: American life sciences and "the problems of sex"*. Berkeley, CA: University of California Press.

Corea, G. (1985). *The mother machine: Reproductive technologies from artificial insemination to artificial wombs*. New York: Harper & Row.

Corrigan, O. (2003). Empty ethics: The problem with informed consent. *Sociology of Health and Illness, 25*(3), 768–792.

Doring, M., & Zinken, J. (2005). The cultural crafting of embryonic cells: The metaphorical schematisation of stem cell research in the Polish and French Press. *Metaphorik.de, 8*, 6–33. Retrieved November 20, 2007, from http://www.metaphorik.de/08/doeringzinken.htm

Edwards, R. G. (2001). The bumpy road to human in vitro fertilization. *Nature Medicine, 7*(10), 1091–1094.

Edwards, R. G. (2004). Stem cells today: A origin and potential of embryo stem cells. *RBMOnline, 8*(3), 275–306.

Edwards, R. G. (2005). Introduction: The beginnings of in-vitro fertilization and its derivatives. In R. G. Edwards & F. Risquez (Eds.), *Modern assisted conception* (pp. 1–7). Cambridge: Reproductive Healthcare.

Fishel, S., & Symonds, E. M. (Eds.) (1986). *In vitro fertilisation: Past, present, future*. Oxford: IRL Press.

Fontaine, P. (2002). Blood, politics, and social science: Richard Titmuss and the Institute for Economic Affairs, 1957–1973. *Isis, 93*, 401–434.

Franklin, S. (1999a). Dead embryos: Feminism in suspension. In M. Michaels & L. Morgan (Eds.), *Fetal subjects, feminist positions* (pp. 61–82). Philadelphia: University of Pennsylvania Press.

Franklin, S. (1999b). Making representations: The Parliamentary debate of the Human Fertilisation and Embryology Act. In J. Edwards, S. Franklin, E. Hirsch, F. Price, & M. Strathern (Eds.), *Technologies of procreation: Kinship in the age of assisted conception* (2nd ed., pp. 127–165). London: Routledge.

Franklin, S. (1999c). Orphaned embryos. In J. Edwards, S. Franklin, E. Hirsch, F. Price, & M. Strathern. *Technologies of procreation: Kinship in the age of assisted conception* (2nd ed., pp. 166–170). London: Routledge.

Franklin, S. (2001). Culturing biology: Cell lines for the Second Millennium. *Health, 5*(3), 335–354.

Franklin, S. (2003a). Rethinking nature/culture: Anthropology and the new genetics. *Anthropological Theory, 3*(1), 65–85.

Franklin, S. (2003b). Ethical biocapital: New strategies of stem cell culture. In S. Franklin & M. Lock (Eds.), *Remaking life and death: Towards an anthropology of biomedicine* (pp. 97–129). Santa Fe, NM: School of American Research Press.

Franklin, S. (2003c). *Definitions of genetic knowledge in the context of preimplantation genetic diagnosis: An ethnographic study*. End of Award Report (ESRC Award Reference Number L21825036)

Franklin, S., Roberts, C., Throsby, K., Braude, P., Shaw, J., Lashwood, A., & Pickering, S. (May 2005). *Factors affecting PGD patients' consent to donate embryos to stem cell research*. Paper presented at the Sixth International Symposium on Preimplantation Genetics, London.

Ganchoff, C. (2004). Regenerating movements: Embryonic stem cells and the politics of potentiality. *Sociology of Health and Illness, 26*(6), 757–774.

Gunning, J. (2000). *Assisted conception: Research, ethics and law*. Aldershot: Ashgate.

Henig, R. M. (2004). *Pandora's baby: How the first test tube babies sparked the reproductive revolution*. Boston: Houghton Mifflin.

Hogle, L. (2003). Life/time warranty: Rechargeable cells and extendable lives' culture. In S. Franklin & M. Lock (Eds.), *Remaking life and death: Towards an anthropology of biomedicine* (pp. 61–96). Santa Fe, NM: School of American Research Press.

Holm, S. (2002). Going to the roots of the stem cell controversy. *Bioethics, 16*(6), 493–507.

Kannegiesser, H. (1988). *Conception in the test-tube: The IVF story: How Australia leads the world*. Melbourne: Macmillan Australia.

Kerr, A., & Franklin, S. (2006). Ambivalence and the new genetics. In A. Webster(Ed.), *Innovative health technologies*. London: Palgrave.

Kitzinger, J., & Williams, C. (2005). Forecasting science futures: Legitimising hope and calming fears in the embryo stem cell debate. *Social Science and Medicine, 61*(3), 731–740.

Kovacs, G. T., Breheny, S. A., & Dear, M. J. (2003). Embryo donation at an Australian university in-vitro fertilisation clinic: Issues and outcomes. *Medical Journal of Australia, 178*(3), 127–129.

Laruelle, C., & Englert, Y. (1995). Psychological study of In Vitro Fertilization-embryo transfer participants' attitudes toward the destiny of their supernumerary embryos. *Fertility and Sterility, 63*(5), 1047–1050.

Liddell, K., & Wallace, S. (2005). Emerging regulatory issues for stem cell medicine. *Genomics, Society and Policy, 1*(1), 54–73.

Lock, M. (2002). *Twice dead: Organ transplants and the redefinition of death*. Berkeley, CA: University of California Press.

Lornage, J., Chorier, H., Boulieu, D., Mathieu, C., & Czyba, J. C. (1995). Six year follow-up of cryopreserved human embryos. *Human Reproduction, 10*(10), 2610–2616.

Magnus, D., & Cho, M. (2005). Issues in oocyte donation for stem cell research. *Science, 308*, 1747–1748.

McMahon, C. A., Gibson, F. L., Leslie, G. I., Saunders, D. M., Porter, K. A., & Tennant, C. C. (2003). Embryo donation for medical research: attitudes and concerns of potential donors. *Human Reproduction, 18*(4), 871–877.

Mulkay, M. (1997). *The embryo research debate: Science and the politics of reproduction*. Cambridge: Cambridge University Press.

Newton, C. R., McDermid, A., Tekpetey, F., & Tummon, I. S. (2003). Embryo donation: Attitudes toward donation procedures and factors predicting willingness to donate. *Human Reproduction, 18*(4), 878–884.

Parry, B. (2005). The new human tissue bill: Categorization and definitional issues and their implications. *Genomics, Society, and Policy, 1*(1), 74–85.

Parry, S. (2006). Reconstructing embryos in stem cell research: Exploring the meaning of embryos for people involved in fertility treatment. *Social Science and Medicine, 62*(10), 2349–2359.

Pfeffer, N. (1993). *The stork and the syringe: A political history of reproductive medicine*. Cambridge: Polity.

Pickering, S., Braude, P., Patel, M., Burns, C. J., Trussler, J., Bolton, V., & Minger, S. (2003). Preimplantation genetic diagnosis as a novel source of embryos for stem cell research. *Reproductive BioMedicine Online, 7*(3), 353–364.

Rapp, R. (2003). Cell life and death, child life and death: Genomic horizons, genetic disease, family stories' culture. In S. Franklin & M. Lock (Eds.), *Remaking life and death: Towards an anthropology of biomedicine* (pp. 129–164). Santa Fe, NM: School of American Research Press.

Robertson, J. A. (1995). Ethical and legal issues in human embryo donation. *Fertility and Sterility, 64*(55), 885–894.

Rose, N. (2001). The politics of life itself. *Theory, Culture and Society, 18*, 1–30.

Rose, N. (2006). *The politics of life itself.* Princeton, NJ: Princeton University Press.

Söderström-Anttila, V., Foudila, T., Ripatti, U. R., & Siegberg, R. (2001). Embryo donation: Outcome and attitudes among embryo donors and recipients. *Human Reproduction, 16*(6), 1120–1128.

Sperling, S. (2004). From crisis to potentiality: Managing potential selves: Stem cells, immigrants, and German identity. *Science and Public Policy, 31*(2), 139–149.

Svanberg, A. S., Boivin, J., Hjelmstedt, A., Bergh, L. A., Collins, A., & Bergh, T. (2001). The impact of frozen embryos on emotional reactions during In Vitro Fertilization. *Acta Obstetricia et Gynecologica Scandinavica, 80*(12), 1110–1114.

Thompson, C. (2005). *Making parents: The ontological choreography: reproductive technologies*. Cambridge: MIT Press.

Throsby, K. (2004). *When IVF fails*. London: Palgrave.

Tutton, R. (2002). Gift relationships in genetics research. *Science as Culture, 11*(4), 523–542.

Waldby, C. (2002). Stem cells, tissue cultures, and the production of biovalue. *Health, 6*(3), 305–323.

Waldby, C., & Mitchell, R. (2006). *Tissue economies: Blood, organs, and cell lines in Late Capitalism*. Durham: Duke University Press.

Waldby, C., & Squier, S. (2003). Ontogeny, ontology, and phylogeny: Embryonic life and stem cell technologies. *Configurations, 11*, 27–46.

Westlander, G., Janson, P. O., Tagnfors, U., & Bergh, C. (1998). Attitudes of different groups of women in Sweden to oocyte donation and oocyte research. *Acta Obstetricia Gynecologica Scandinavica, 77*(3), 317–321.

Williams, C., Kitzinger, J., & Henderson, L. (2003). Envisaging the embryo in stem cell research: Rhetorical strategies and media reporting of the ethical debates. *Sociology of Health and Illness, 25*(7), 793–814.

Chapter 6
Dividual Systems & Ultraneoteny

Elena Gagliasso Luoni(✉)

Abstract: The forms of collective thinking are based on the concept of identitarian individuality and on the relation between individuality and alterity as *external to the self* and seem to be unaffected by the ontological fact that some beings – women – can be singular *individuals* for a long part of their life, and yet they may become *dual* (containing and interacting with the other) and returning to an individual state by a separation/loss of duality (what may be called *dividuality*). We refer to embryonic development *or* to pregnancy, to the birth *or* to the parturition and there are no words for referring to the *unity of two* that makes pregnancy and development an *integrated process*, and makes birth and parturition the same event, observed from two different standpoints.

Is questioning this categorial frame useful?

If languages are fundamentally embodied, and in predicating world or time they project forms of kinaesthesia and corporeality, where can we find meaningful definitions of reality that may encompass this specific state of embodiment: the *embodiment of the two*?

In order to explore this subject, two fields are especially interesting: the contemporary *epigenetics* and the studies of *hominization*.

An overview of contemporary bio-evolutionary knowledges may indirectly corroborate the theme of an individuality being both dual and dividual, since the scientific focus now highlights (more clearly than in the past) the *interactive dynamics self/other in development as well in evolution*. Paleoanthropology underlines the dual bond between structural and behavioural changes in the maternal body in new anthropoid species, and the interesting consequences of neonatal prematurity: we are '*ultraneotenous*', with a complex celebralization, because we are also a particular dual system.

Keywords: Individual, feminism, hominization, ultraneoteny, epigenetic frame, language, nature/nurture

---

Elena Gagliasso Luoni
Facoltà di Filosofia, Università di Roma 'La Sapienza', Via Carlo Fea 2, 00189, Rome, Italy
elega@fastwebnet.it

6.1 The Unthought and Its Nonplace

> *Androids can't bear children," she said, then. "Is that a loss?"(...)"Is it a loss?" Rachael repeated. "I don't really know; I have no way to tell. How does it feel to have a child? How does it feel to be born, for that matter? We're not born; we don't grow up; instead of dying from illness or old age we wear out like ants (Philip K. Dick 1968).*

Nobody creates languages out of the blue, nobody makes grammars anew or invents non-existent dictionaries. Yet, what is significant from a theoretical, philosophical and epistemological point of view, is this: in any science dealing with generation, we work on the borders of an expression-lacking place, a symbolic fault. We shall thus be aware that in this sphere we always deal with an opaque zone, where in the course of time no chance was given for predicating a certain configuration of the states of being. This is due to manifold reasons, ranging from a difficulty in translating the proprioceptive experience of generation to the exclusion from the creation of discoursive practices of the very subjects that *enact* and *are acted* upon by the process of generation of another being.

We will keep this premise in mind, as a sort of conceptual background, since our mental categories feed on what exists and on what does not exist; and what is illuminated by languages has an unexpressed dark side.

Among the fundamental features of the living beings in our species, one is (up to now) experienced only by woman-subjects. It has not found a direct representation, but only a series of mutually excluding interpretations and thorough analysis mediated by someone else's sight. I am talking about the reproduction by intrauterine gestation, typical of all mammals: a phenomenon objectified by its observable manifestations and explained not in first person, but as a 'natural' phenomenon, *external to the living experience*, by means of a conceptual duplicity that *separates* the discourse universes of the two acting and interacting subjects (mother and child, container and content). As a matter of fact, we usually refer to embryonic development *or* to pregnancy, to the birth or to the parturition. There are no words for referring to the *unity of two* that makes pregnancy and development an integrated process, and makes birth and parturition the same event, observed from two different standpoints.

Similarly, the forms of collective thinking seem to be unaffected by the ontological fact that some beings – women – can be *singular individuals* for a long part of their life, and yet they may become dual (containing and interacting with the other) for limited and possibly repeated periods of time, eventually returning (with altered body and psyche) to an individual state by a separation/loss of duality (what may be called *dividuality*). The forms of collective thinking are actually based on the concept of *identitarian* individuality and on the relation between individuality and alterity as *external to the self.* The return to the in/dividual, that is, an indivisible unity, is quite an oxymoron, since it actually happens by means of a division. Separating from the Part of herself/Other, the human being experiences that event (commonly referred to as 'giving birth' to the new being) as a detachment, a division. She is not an individual anymore, and she was not an individual in the last nine months, when she was a dyad by dual sharing. And she is not an individual now, because she may be called a 'dividual': clearly a non-existent concept, since these modifications of the being (shared by the internally growing other), typical of one subject becoming two (pregnancy/embryonic development) and again becoming one for the loss of the other (birth/parturition), *do not exist* as categories.

In the vast, though from time to time circumscribed, space of the multifarious discoursive practices, this standpoint of observation and proprioception did not unfold as significant for the reflection about the real. There are actually some subjects that may have articulated it as moving from their own experiences. Yet, these subjects have been long restricted to the external margins of creative thinking about the world and themselves. When they finally gain access to it,[1] they find an apparatus of categorial realities already created by an alterity belonging to another gender: that is, by someone whose biological sexuality does not host within the body another growing being, nor interacts with it and eventually detaches from it.

Man's different biological and corporeal reality allows him to legitimately channel his thought towards an uninterrupted and solid living individuality. This uniqueness, even if fragmented in a plurality of intrapsychic states, is from time to time a complement or an antithesis to the multiplicity of other beings, and it is dealt with by means of the dialectic between the individual and society. As such, the relationship with the Other is necessarily external, mainly a 'meeting'.[2] This specific feature shows the deep genderness of human thinking, since it is substantially thought by *one* gender: the male one. With regard to reproduction and birth, in various contexts this thinking has scientifically stressed either the side of the

[1] A recent entry, intercepting the anthropological changes of the twentieth century due to feminist political practices. As it has been recognized, 'the ability not only to think but to think oneself, to be a thinking subject and not only the object of the thinking of the other, to be thinking instead of being thought, is the 'gain' of the feminine Self, allowed by the practice of selfconsciousness that favored the growth of a new 'inner life', born within the relation with the other woman' (Fraire 2002).

[2] The feature of the meeting with the Other as external and alien, is however typical to both men and women. Peculiar to women is the *intermediateness* between external/internal and alien/own. This feature is characteristic of a specific kind of 'meeting' with the newborn creature, just turned 'external' but marked by an extremely peculiar mode of 'alienness'. This creature is the same one that the woman had inside just an instant before: thus it is at the same time known and unknown, disturbing in some sense, when it appears outside of the self, exiting from the body it was inhabiting a moment before. This alienness still astonishes, even if the many forms of prenatal ecoscans have already 'showed' the image of the 'new' one we will meet at birth.

not-yet-born (embryology), or the side of the mother (gynecology), or the background genetics. Meanwhile, philosophical and mythopoietic reflection was allowed to talk about woman as 'nature', and about man as culture and transcendence. Observations and explanations proposed by physicians, gynecologists, geneticists, embryologists, psychologists, and later by psychoanalysts and bioethicists, have progressively deepened our observational, experimental and interpretative knowledge of the normal and pathological processes related to reproduction, gestation and natality. Yet, this deeper knowledge does not and cannot enter a categorial area that covers an ever-changing subject, shifting in the course of time from being an individual, to being *both* an individual *and* a dual entity, to being a 'dividual', and going back to individuality. This order of categories of being could not be expressed while the knowledges were being established; these categories could find a space, if any, only in a private lexicon.

The subjects – the ones experiencing these variations in physical and psychical form, as well as the prolonged and maybe repeated *internal* co-presences of the Other – have not been able to create a symbolic thought based on themselves. This in turn prevented the establishment of a shared cultural medium. As a consequence, they have *received* the theories, the 'individualizing' criteria and thought created by the subjects that are uniformly identitarian in the course of their life (though linearly modified by individual growth and decline processes). Though rich in projections, interpretations and explanations, the words of these 'un-dividual' subjects fall short in expressing a part of the perspective.

Actual languages are fundamentally embodied,[3] and in predicating world or time they project forms of kinaesthesia and corporeality. Yet, they don't have meaningful definitions of reality that may encompass this specific state of embodiment: the embodiment of the two. This prevented both the projection of those networks that intercept the real by metaphoric processes, and the creation of influential concepts or *sui generis* (actually, 'of a specific gender') categories based on the perspective of this embodying experience: a state of the being condemned to aphasia. As a matter of fact, what has never been thinkable is not spontaneously missed: also because, by contrast, the richness of the existing systems of interpretation and knowledge about reproduction has been and is constantly growing, fulfilling many demands in this regard.

[3]The projective basis of the three-dimensional corporeal dimension and disposition are revealed by a large part of our language. Many of the current words are embodied metaphors and refer to active perception, and at the same time are pertinent to the evaluation of the physical condition. 'Up', 'down', 'high', 'low', future (projected forward), past (unwound backwards), are all terms denoting not only time and space, but steeped (often unawarely) with positive or negative values, whose connotative sign is precisely related to the physical state they indirectly evoke. The study of embodied metaphors has been brought to the scientific attention by the semiologists and philosophers of language Lakoff and Johnson (1999).

6.2 Persistent Dichotomies

Today the very act of pointing at this categorial blind spot slips into an interesting knowledge melting pot, where philosophy of life sciences, phenomenology, semiotics, neurosciences and the evolutionary studies about hominization (that is, the protohistory of mankind), are mixed. This cross-fertilisation takes place in an interesting historical moment, because at the theoretical and political levels, women with feminist backgrounds redesign the meaning of cultural and scientific issues, precisely questioning the symbolic level they use in their own discipline discoursive fields.[4] This symbolic level proves to be not neutral, but fundamentally sexed by the specificity of one gender.

The typical categorial couples individual/population, or individual/other, or the issues of individual identity and plural identities, show – to a woman's gaze – an intermediate dark zone. If half of the human species may be dual, 'dividual' and individual, according to life stages; if alterity may be either external (the traditional Other) or internal and intermediate with respect to the world (the Other that develops inside us, to whom we provide a long lasting 'environment' in the neonatal stage); then the dichotomies of the being demand an integration, by means of a dynamic third dimension. This integration would involve most of today's worries connected to the status of birth-related states and to the various technologies for artificial reproduction.

A philosophically (scientifically and politically) relevant issue could be the chance to symbolize the *internal* experience of the duality self/other – even more, the *making of* this duality – and then the detachment from the actual duality (shared with the one that has inhabited us). Furthermore, we now may categorize the shift towards dividuality, that is, the deprivation of this inside-other as a consequence of its movement towards the outside of ourselves, though in strict proximity, parental care being almost an extension of the internal state.

As a matter of fact, scientific and ethical issues related to assisted reproduction and to bioethics of birth are today more than ever a target for the cravings of politics, religions and the market. As such, these themes tacitly imply the assumptions of a thought that, according to the ideologies and the interests at stake, automatically divides and distributes rights and priorities to one or the other component of the dual/dividual: it forces to choose between the mother individual and the embryo individual. However, the chance to mark the boundaries of the unthought not only

[4]In particular, Londa Schiebinger, an historian of science, has devoted her theoretical and critical research to these issues. *The mind has no sex* (1989) and *Has Feminism changed science?* (1999) are her most significant essays, analyzing how the state of the art changed in the course of ten years. The gender perspective has given new strength to new interpretations of reality, different from those traditionally considered as neutral. For example, primatology, anthropology, and medicine have been influenced by these new trends, and opened up new investigations. See particularly Hager (1997) and Hrdy (1981).

has to deal with a history of exclusion from a symbolic in which only one gender mattered, but also needs to step back from the present dichotomous feminist elaborations about motherhood issues.

Since early feminism, that questioned the 'role' rather than gender,[5] self-determination as a free choice was opposed to biological sex. The female biological sex was a burden, since it was a 'second class' sex: 'One is not born a woman, but becomes one', wrote Simone de Beauvoir (1949). In the French existentialist perspective held by this forerunner of feminism, nature was to woman only brute matter, visceral burden, to be jettisoned in the moment of the biographical 'invention' of themselves. Furthermore, in the nineteenth and twenty centuries, any scientific study of sex (sex as 'pure' nature and an exclusively biological data) was marked by a monosexed science (medical, of social hygiene, psychiatric, sociological), often knowingly or unknowingly misogynous. Science was also unaware of the extent to which its supposed objectivity was actually the result of culture, anthropological customs and of the earliest forms of biopolitics (Foucault 2004): the issue of 'female role' and the study of biological features of sex and procreation were directed according to the necessity of social control (Bleier 1984; Rose 1994).

In these philosophical and political ideas held by women thinkers in the 1960s and in others that later followed the thread of the autonomy from the biological,[6] there is a more or less manifest refusal of the passivity inherent to the process of maternity, favouring by contrast self-determination and transcendence from the flesh. In 30 years, with contemporary feminism, we have arrived at multifarious readings of artificial contaminations, conceived as a place where it is possible to go beyond the species' dual distinctions, and to move towards a post-human perspective in order to explore the cyborg's expanded freedom (Haraway 1988).

A shift then took place from accepting the predestination of gender to discussing the invention of the self, favouring a totipotentiality of the being expressed by means of a sexuation independent from the reproductive duty, and ultimately reaching the fragmented and manifold transgender identities (Butler 2004).

Therefore, on the one hand *gender* stands for everything dealing with the innovative construction of the self, freed from encumbrance of the 'role'; on the other

---

[5]In the 1970s, the focus was on the 'role', referring to a stereotype strictly related to classical and essentialist iconography of the encoded 'feminine' imaging. Today, the *gender*, mostly by the reflections of women philosophers, psychoanalysts, historians and anthropologists, has changed into a discourse on the *meaning of the dual categories* (Scott 1996), on their relations, on their intertwined genesis, on their indecidability borders. On one side, there is the philosophical categorial elaboration of the 'difference thought', and on the other the fragmentation of the very concept of 'identity', by the *transgender* studies, evidence of transformations, as well as endpoint and disintegration stage of the carnal materiality of sex (Butler 2004).

[6]From Firestone (1970) to the cyborg theme in Haraway (1988), to Butler's (2004) the deconstructionist and de-corporeal arguments (criticized in Duden 2002, cap.14). A *fil rouge* connects at a distance, over more than 30 years, the idea that the artificial in its various forms, be it technological or virtual, may be for women a powerful tool to emancipate themselves from the biological pole, marking the latter essentially as the immanence of an imposed and passive condition: the carnal corporeality and the biological sex received at birth.

hand, by contrast, we find dependence from natural constraints, the conditions of biological sexuation, the sex that 'binds' us in the reproduction of our species: 'natural' passivity is thus an antithesis to the activity of self determination by free subjects. The contrast between the cultural construction of gender and/or the 'acceptance' of received natural sex, is not limited to women, though the alternative nature/culture proves to be much sharper for them than it is for men. As a matter of fact, it was on the basis of the reproductive naturalness of the female being that the male's spirit or reason has equated female to nature, and attempted to dominate both.

The biological self means to the woman a reification of herself that has coupled for centuries with the role of brood-mare: a role to be freed from, gaining, together with other women, a free identity. The knot of female corporeality, binding self determined individual appropriation and the means of descent, is loosened, and only self determination is left.

Another branch of feminism of Anglo-Saxon descent has turned the conjunction of woman and nature into the crux of a positive reflection. The feminine dimension, in this case completely encased in the maternal – and, we add, in the values attributed to maternity by important currents of masculine thought –is acquired by eco-feminism (Merchant 1979, 1992) and by theories of the feminist ethics of care.

In the first case, a spontaneous closeness of women to Mother Earth is claimed, and their oblational integration with every living form. Yet, this view is not rooted in women's practice, rather in the Mitteleuropean romantic tradition: this is in turn related to the *Magna Mater* theme, present in many cultures and mythological traditions. Nature, moving from the term's very root *nascor*, 'procreates' all living beings, and any woman is akin to it because of similar features. A connection is thus drawn – especially within deep ecology – from the feminine to the defence of the integrity of the 'living planet', and to an extensive critique of the technological world, of manipulation of life, and of any form of artificiality. These are in fact thought of as aggressions by the male world (and particularly by industrial capitalism) to both the whole Earth and to woman, as its analog and champion as well.

By contrast, theories of the ethics of care hypostatize the mother's parental attitude as an ethical model. The fact of being responsibly in charge of offspring, permeated by a relational practice, may be the basis for a new ethic of responsibility, founded on shared relations (Gilligan 1982) and opposed to the normative ethics founded on the contract of the laws.[7]

A polarity is thus found: motherhood as a binding constraint imposed by the bios, or as a redeeming model of responsible care for life? The two positions are far apart and are part of a persistent dichotomy on the issue of being a woman, either willingly free from being a brood-mare, or resolved in and equal to this female feature. On the one hand, animality, to be transcended thanks to the opening of various elaborations and inventions of a different gender identity, on the other, an

[7]For a detailed and critical appraisal of the most recent developments of these arguments, assessing implications and comparisons with traditional bioethics, see Botti (2007).

essentialist archetype where an intrinsic positiveness of the feminine is expressed, to be extended to new forms and conventions of everybody's life.

The sharpness of this distinction does not fully encompass the multifarious nuances of the two options. Still, in both cases we remain within a field of values related to the reproductive dimension. That is, a partly pre-existing field, still lingering in the discourses about woman and motherhood built by a theoretical egemony that is historically given, and thus implicitly inherited: the antithesis active/passive, nature/culture, and I may add, organism/environment are very important from the 'dividual' viewpoint.

Leaving aside flights from reality, the male/female sexuation is our physical and biological basis, since we are part of a bisexually reproducing species. As such, fleeing from bios does not go a long way with regard to species survival. At the same time, motherhood as a process of embodiment of the two is an interesting challenge to our categories Self/Other. As a matter of fact, it goes through the various vicissitudes of the 'making and unmaking' (Fraire 2002) of the unborn as a field where acceptance is not the only choice, as shown by the wide diffusion of voluntary abortions in any time and place. By contrast, this choice has always been – and still is – feared by demographic policies and religions, that present woman as opposed to fetus for her own interests, thus justifying forced social, political and religious interventions (Botti 2007). Similarly, motherhood is not necessarily an added value of the person, a sign of an enhanced spontaneous oblativity: it may be experienced as such only by the offspring.

6.3 Epigenetics Between the Body of the Mother and Genes

I think that an exchange with our present bio-evolutionary knowledges may indirectly corroborate the theme of an individuality being as well dual and dividual, since the scientific focus is now highlighting (more clearly than in the past) the interactive dynamics self/other in development as well as in evolution.

Two fields are especially pregnant: contemporary epigenetics and the study of hominization. Since the development of molecular genetics, this approach slowly seeped into collective thought, and now in the mental dimension genes are the symbolic *par excellence* for any discourse that refers to an invariance in any kind of human feature (Duden 2002).

If we are, all of us, simple 'vectors of the genetic pool' of a population, or gene 'vehicles' (Dawkins 1976), ultimately there is no room in the biological world for real action by the subjects, and for the diverse forms of their constructive interactions in the internal and external world. Genocentrism, especially widespread in the United States, ultimately risks the anomy and the de-personalisation usually associated to an excess of reductionism (Gagliasso 2007).

Yet, from other corners of contemporary genetics, related to the most recent discoveries about embryonic development, new hints are coming for a possible recomposition of the individual action with the transgenerational species specific one. In the meantime, the split between innate, as genetic, and acquired, as environmental, is progressively blurring. Between the DNA predetermined invariants, and the 'freedom' of the mature organism in its life, lies a '*middle earth*'.

While the very word is unvaried on the lexical level, the cognitive contents of the theoretical term 'Genome' are transformed by the progress of knowledges, moving its level of meaningfulness. Our species, as any other, is the bearer of a genome; but to think of a self-replicating code as an invariant informational function, context-independent, and thus sharply separated from the *soma*, is radically different from discovering that DNA appears stable (albeit punctuated by casual mutation) only at a first approximated glance.

The folded macromolecule now appears to be a highly complex entity, continuously interacting with the intracellular environment; as such, it interacts – through many enzymatic levels – with the whole 'host/producer' body and its internal metabolic, hormonal and chemical modifications. Modifications that, in turn, are fine-tuned by the very presence of the 'developing guest'.

Our genes appear to be less and less embryo 'code strings' (or 'programs'), that is, locked carriers of the features of a future individual. They now are something different. We used to say that 'environment matters', referring to that second moment posterior to the genetic determinants and inaugurated, 'discovered', at birth: when we are cast in the world. That is, the meeting with the life forms that people have created from the world they belong to and their relations. In this view, environment was strongly opposed to the innatism of gene. Yet, today genomes are not anymore rigidly constrained programs, because environment and 'worlds' can be found inside the macromolecule itself. The genome interacts with its surrounding micro-environment, and its expression is regulated by other inclusions and reactions. Furthermore, maternal substances affect the unborn, modulating the cellular micro-environments in order to achieve its correct development (Jablonka and Lamb 2005). This is what is shown by most of the new findings about genetic regulation, cellular actions and feedbacks, and the enzymatic synthesis synergically acting. In many of these reactions in the early stages of development, the (maternal) medium has thus a key function. However, the medium itself is built up in cooperation: by the body of the mother and by the embryonic products, turning on a series of dual loops not yet explored in its details.[8] In fact, not only the mother intracellular environment enters directly in the embryonic development, but a whole lot of embryonic cellular products enter the mother's body too. These products linger in the body of the mother, and they may even become a constitutive part of her immune system for many years. It is for example the case of the fetal stem cells that may be traced for a long time among the mother's adult stem cells (Bianchi and Fisk 2007).

[8]The placenta is a typical example. About the structure of this 'no man's land' (or shared territory), already the subject of Buffon's reflection in the eighteenth century (Pancino and D'Yvoire 2006), there are stratifications (or negations) of sense and identity ranging from anthropology to the free market. It is not surprising that in many African peoples, placenta is highly regarded, pertaining to the child even after birth, and it is buried within the house, since it is thought to be a protection for the offspring. Fetus' or mother's placenta: due to this property indetermination the commercial use of placenta in the pharma-cosmetic industry has long been a black market.

In other words: what lies around and works interactively in the prenatal environment, contributes to modify the development of the uniqueness of the individual in hardly foreseeable ways. It is ultimately impossible to talk about 're/production' as the production of copies from a matrix, let alone in an extremely conventional way (Keller 2000).

These levels of theoretical and experimental investigation have undergone important developments, opening a new proper research field whose aim is to link individual development with species evolution dynamics. We may now talk of a paradigm of evolutionary epigenetics or evolutionary developmental biology (evo-devo) (Minelli 2007; Jablonka and Lamb 2005).

We shall even get, at some point, to question again (in a lay perspective) that area of knowledges, biases, doubts and folk beliefs about mother's 'influence' on the unborn. This area would then be considered not only as a cultural, anthropological and undoubtedly politically marked territory, but also as the place where approximate metaphoric and symbolic intuitions about connections between the mother and the fetus flourished in ancient times – often posing a threat for the women themselves.

These connections were evil or beneficial conditionings of the offspring's aspect or character by the 'active imagination'[9] of the expectant woman, or even by the women around them. Such beliefs have marked in many conjoint ways the history of knowledges about embryonic development, teratology, and the pregnancy process. As a matter of fact, they signalled a dramatic empowerment of the mother, inverse and complementary with respect to the deprivation of power that women experienced in the construction of the world.

In the world of epigenetics, a woman (as well as any other female in a sexed species) shifts from being a mere reproducer to the site where a constructive cooperation between the heredity of the species and the individual oscillations takes place.[10] Should this idea spread in popular culture, may it vary the self-perception of the subjects themselves? It could also modify accordingly the parameters of certain innatist prospections and the space of the contingent interactions of an environment, not anymore solely the space surrounding the organism.

We should then keep an eye on what is happening in studies about the genome, cellular environment and epigenetics, since there is striking evidence for a well-known mismatch. It reveals again, if needed, the gap between the predictive certitude flaunted

[9] See the richly detailed essay by Griffero (2003). In particular, chapter 4 is devoted to evil-eye and maternal cravings, analysing various beliefs about the 'transitive' feature of the reception of external images and projective fantasies, whose influence was thought to reach the unborn, generating, at best, similarities with the fancied person, or at worst, monstrosities. Maternal imagination, or transitive imagination as 'malefices' of the envious look of the others, like infertile women or with dead children. These beliefs persisted much beyond the Middle Age, reaching the eighteenth century.

[10] Different levels of interactions with the environment provide a new perspective in this 'space' that is traditionally intended as external to the bodies. Thus, there is a series of nested levels, reaching the cellular microenvironment that deals directly with the gene, fine tuning its expression. This internal environment, though, is connected to the macroenvironment through the general conditions of life of the mother. About how epigenetics affected the concept of environment, see Gilbert (2006).

by medicine, pharmacology and in the mass media, and what is being explored in less mediatically exhibited research fields that do not bear palingenetic promises.

These studies also affect indirectly the issues of categories of thought, since the pregnancy/development process is conceptualised differently from the past. Further, a new space is opened to the possibility of thinking an individuality that is also twofold and dividual (an oxymoron to our current logics), that is, the peculiar entity constituted by any future mother with her internal future offspring.

6.4 Ultra-Neotenic Combined Systems

It is by means of the structural and behavioural changes in reproductive sexuality, once shared with our cousins the primates (to whom we are strictly genetically related), that our evolutionary paths slowly diverged in deep times. There are many features affecting the hominizing organic and mental structures, slowly triggering the drift of our species, the most anomalous one. Among these features, the modalities of sexuation stand out, because of its role in the species specific evolutionary economy, its complex and often counterintuitive running.

A presumptive scientific practice, based on partial evidence left by ancestral anthropoids and their primordial habitats, is needed in order to understand how we have changed in the course of a million years in our hormonal, cerebral, physical and behavioural traits. In the range of deep time when the extremely slow hominization process started – about seven mya – significant genetic mutations correlated to shared living habits. When these new living modalities proved functional in the changing environments, they were selected and allowed the onset of new hominizing features.[11] Some revolutionary inventions, such as the gesture (allowed by the freeing of the upper limbs and the rotation of the thumb) that is the constructor of objects and relations; or the transformation of feeding

[11] The range of transformations primed by changing ecosystems in the life of ancestral primates is now analyzed not only by paleoanthropologists, but by neuro-paleoanthropologists and paleoanthro-physiologists as well. According to their specialist background, they provide detailed interpretative frameworks dealing with the appearance of the most significant features, those usually considered 'hominizing' traits. For an accurate analysis of the different ecosystemic and physiological systems of explanation of the networks of contingencies connecting the changing environment to the processes of organismic mutation, see Pievani (2002). From the loss of body hair, to the thinning of dentin and the lengthening and lightening of bones, from the appearance of long stages of mixed posture in the mixed vegetation zones (forest-savanna), preceding bipedality, to the consequences of the latter on the freeing of upper limbs and the transformations of the 'fine' prensility due to the opposable thumb; from the forward shift of the foramen magnum, with the reduction of the area of the lower insertion of maxillar muscles and the subsequent enlargement of the skullcap, particularly in the prefrontal areas; to the huge metabolic, enzymatic and structural modifications of the internal organs due to the changing feeding habits (Rotilio 2006): the wide and still expanding range of possible co-existing causes of the evolutionary process has originated a thick bush composed by many species, even coeval, of ancestral hominids. Among these coherently acting causes, one of the most significant ones is the peculiarity of a species that is anomalous with regards to the perinatal developmental rates: this is particularly important for the genesis of the human brain, cortex and psychism.

habits as well as the cooking of food, or the proto-symbolization processes –in coevolution with lowering of the larynx and the functional specialization of the Broca area in the brain – were 'invented' by small groups of hominids. The new traits conferred higher fitness to their bearer, and in a few hundreds of thousands of years, the novelties extended to the whole species, progressively involving other organic structures.

These transformations actually 'created' us, our bodies and nervous systems, and are placed in a spatio-temporal system that makes it now difficult to neatly distinguish the bios, the behavioral choices and the active adaptations.[12]

The biological vicissitudes of our species are thus closely entangled, since the inception of anthropogenesis, with the behaviors that choose and transform the environments of life, that is, in other words, the nature of the protoculture, or 'second nature'. By 'environments', we mean not only ecological niches, with their geographical and climatological landscapes, but also interactions with cospecific individuals, the Others, and especially parental relations. Among the latter, we find the mother/child relation, and the transformations connecting the fetus and later the newborn to the mind/body that generated it. Such transformations act within a feedback loop on future physical structure, increasing brain size, and the metabolisms typical of the peculiar hominid, later human, condition.

The new posture and relation with other members of the species create tight correlations between anatomical parts and their functions. Among these correlations, we are particularly interested in those that integrate the hominizing specificities of the female body and the developmental rates of the fetus and the newborn.

The upright walking and the life conditions implied by prolonged nomadism, affect a structural combination: the support basis of the feminine pelvis is enlarged and the vaginal labia move forward. These modifications couple with the disappearing of the visible estrus (typical of every primate) and the appearance of orgasm, two features that played a key role in shaping the relations between the two sexes and in defining their cultural and societal meaning. Mammary glands widen and are reduced in number (only two of them compared to the mammary chain along the ventral area), while they are placed in the upper area, within the reach of an offspring held in the arms, that are now freed by bipedalism.

---

[12] In most of contemporary biology, behavior is no longer considered as separated from fundamental biological equipment. At the beginning of the nineteenth century, Lamarck based his transformist theory on the inheritance of characters acquired by behavioral use and disuse of organs. While Darwin maintained that natural selection of individuals bearing the fittest casual hereditary variations was the major cause of species modification. Darwinian theory was later confirmed by biology, and particularly by genetics in the mid-twentieth century. Yet, epigenetics has put a new focus on the role of organisms' behavior in the 'canalisation' of selection (as Waddington maintained in the 1950s), or in promoting the 'genetic assimilation' of the most repeated behaviors, coupled a positive selection of the individuals bearing those behaviors. Today, genetic assimilation is considered as a milestone in explaining the establishment of the combined system where the coevolution of the lowered larynx (necessary to phonation), of the first forms of symbolic language, and of the Broca's area in the brain (involved in language processing) took place (Deacon 1997).

Many of these modifications may have been produced by genic mutations, while their own correlation and the integration with other independent organic and behavioral features have provided a stabilizing feedback.[13] Thus, the combined systems of the hominid evolutionary dynamics are not solely ecological. Among the many combined systems of relations internal to the species, there is an integrated correlation between the bodies of the generator and of the unborn. This correlation is increased by the changes in the tempo and mode of development and parturition.

As a matter of fact, the major difference between our species and the others, the one that bears the greatest lot of potential consequences with regards to our encephalization and thus for mentalization and human relationality, involves pregnancy, neonatality and the early years of the woman's offspring.

We are born with our cranial structures not yet joined, and birth is largely anticipated when compared to the physiological lengths of the other mammalian pregnancies. The open cranial structures (allowing for enlargement of the cortex and the extraordinary increase of synaptic connections in the early years) are due to a developmental incompleteness that makes us *premature* at birth: to bring a fetus to 'normal' maturity, and to give birth to an autonomous pup like the other primates do, it has been estimated that 20 months would be needed, instead of the human nine (Gould 1977).

Cerebral and physical immaturity leaves us dangerously exposed to the surrounding environment, with no innate resources to face it. However, if the open cranial structures in the 'fetal'[14] newborn are related to the bone structure of the maternal pelvis, in some respect immaturity makes us functionally analogous to marsupials. The latter in fact have to end their development in the maternal pouch, internally connected to the vaginal channel. Similarly we humans, subject to a sensory over-exposition unmatched among other mammalians, complete metaphorically our development outside the uterus. We finish the process in an intermediate area, that is, in the arms of the one that nourishes us; those arms now freed from the task of movement, and that have hands equipped with opposable thumb, capable of sophisticated manipulations and artefacts construction.[15]

---

[13] 'Exaptation' is the term coined by Stephen J. Gould and Elisabeth Vrba (1982) to define the presence and the dynamics of characters belonging to the structural heredity of the whole organismic form, and that, in the course of evolution, have been coopted in a given species in order to perform a new function, different from the one that caused it appearance before. A classic example of exaptation is the use – and the subsequent increase – of the flight structures in birds, previously significant as a thermoregulation features.

[14] Louis Bolk spoke about 'fetalization' of embryonic mammal in 1926 (Bolk 1926).

[15] The manufact typology long studied by paleoanthropology on the basis of the few available unaltered fossil remains (bones, graffitis, chipped flints), has been considered an interesting indicator for gender interpretations. A flint or a stone piercer were long interpreted as war or hunting tools. Only recently, they have been re-questioned in order to include in the hypothesis the possible existence of instruments made of perishable materials: cradles, baskets, dressed hides, woven textiles; the same flints have been reinterpreted as needles, cutting tools, scrapers and so on (Hrdy 1981; Hubbard 1990; Hager 1997).

There is a continuum between the fetus and the newborn. The cycle of pregnancy, parturition and parental care is uninterrupted. There is a dual bond between structural and behavioural changes in the maternal body in this new anthropoid species, and the consequences of the neonatal prematurity.

Prematurity at birth is matched by the slowing down of later growth, such that at sexual maturity we have juvenile physical and psychical traits, if compared to the our next of kin, the primates. This persisting dimension of juvenile traits, prolonged in the adult stage, is the so-called 'neoteny'. It exists in other living organisms, but in our anomalous and unique case, it is coupled to early birth, producing a peculiar mismatch in timing: acceleration at birth, and delay of growth. We therefore are not only premature at birth, not only neotenous but, as it has been said, *ultraneotenous*.[16]

Our origins are then coevolutionary, broadly and dually speaking, that is, internal to the two generating/generated bodies. Structural and functional correlation unfolds in the features that combine dimensions, form, and early birth, with the correspondent structural transformation in the body and in the behavioural attitudes of the one that generates it. So, we are entitled to speak about *external* coevolution (of species, between organisms and their habitats) and *internal* coevolution, such that some organisms – the females – are 'environment' for the development of the unborn, and persist in this 'environmental' role in order to ensure survival of the offspring during the process of autonomization: nutritional and microclimatic environment, a psychical and mental container, medium and defence with respect to the external macro-environment.

Therefore, in our species male and female relate to their environment by transforming it (or, in traditional words: intertwining nature and culture in their actions), whereas only the females may become organism *and* environment to another organism: containing the Other, as well as modified by it.

In this sense we may speak of a *combined dual system* and of a dividual with parturition that slowly goes back to the status of individual. The combined system is composed by these feminine bodies/minds bearing and feeding the offspring, on the one hand; and by the embryo, later fetus, and last the newborn, on the other. But in the course of the deep time of the evolution, the latter part of the dividual has contributed to shape the structural and functional traits of the one that gives life to it.

Our premature and ultraneotenous development completes in an intermediate context, placed between the womb and the external environment, where containment is gestural, relational, and encompasses maternal care and suckling. However, it extends progressively to the other parental figures, creating a series of important social ties, derived from the primary sexual and parental ties (Deacon 1997).

[16] In 1941 Adolf Portmann proposed a linguistic approximation in order to account for the controversial double-faced time of the ontogenesis of us humans, with acceleration of birth and late maturity: 'secondarily nidicolous' species. Today, his thesis, revisited by Gould within evolutionary thought, have been integrated in the phylogenetic dimension, so that we now speak of 'ultraneoteny' (Gould 1977).

This connection/sharing may be one of the causes of our becoming a symbolic species. Some ethologists and semiologists even suppose that most of the rough protolinguistic forms started in remote times as a significant form of emotional relations, as well as forms of social rules and practical instructions.[17]

Only in our species is this *sui generis* space-time so prolonged, spanning since the union of the gametes until the first years of life, and in a million years it stabilizes the physical and psychical system typical of a dual state. As such it is the dual dimension, as opposed to the individual, one of the pregnant driving forces of anthropogenesis.

Exactly this space-time, as an intermediate borderline, places itself as a joint between the biological and the historical: long repeated behaviours and psychic traits are assimilated in biology, then selection fixes their physical constraints.[18]

It is a pregnant time, a truly hinge condition since in the dual entity two individuals remain *united* and *distinct*: the offspring and the mother. The dual entity goes back to the individual state only gradually: parturition is a threshold and baby-carrying, we have seen, is very prolonged in our species, to the point that the threshold of birth seems much less sharp than it appears to an external observer.

Any human experience may become thought if it occurs not only physically, but if it also accesses some form of symbolization, such that it becomes constitutive of our mind-body set. It is now clear that to humans, when there is the chance for naming or representation of feelings and sensations, the consequent mentalization (abstraction, reasoning, symbolization, memory-project connection, categories creation and interpretation of reality) may have a feedback on global selfperception. This mentalization focuses the action of sensing or the sensing as an action (sens/actions), modifies self-knowledge of ourselves in the world, and of the potential interactions with our kin, including the sort of participative projections based also on the empathic understanding of someone else's mental and physical states. By the shared representation of perceptions, sensations and feelings, we act directly on the

[17]The moment of precocious communicative relations has been particularly studied by etologists, anthropologists and students of protolanguages. It is 'likely that on any human and hominid sound, before it became associated to a precise meaning, and even after that, a stratification of senses flourished, correlated to past and present individual and social situations, so that only on this basis meaningfulness may have risen' (Celentano 2000). Recognition of the continuity with the primates' communicative world does not establish a dialog with the studies carried on by philosophers of language: they emphasize the radical gap due to the abstract symbolization ability. In the middle, we may say that the analysis postulating the existence of archaic stratifications of human vocalizations (from the click sounds common to other anthropoids, to the lallations of fundamental phonemes common to nearly any culture, referring to primary figures and functions). The first phonemes are labial, and the semantic roots used to call the mother are similar across the various languages. Further, symbolization is deeply rooted in the highly emotional aspects that establish primary relations: the cognitive constituting of subsequent languages follows their emotional constitution. The problem of representation, of abstraction and of world models, immediately follows, so that by these criteria, private language precedes the social and conventional one. See the essays collected in Contessi et al. (2002).

[18]About the constrained role of selection and the relationship between selection, acquired structures and epigenesis, see the first part in Jablonka and Lamb (2005).

kinaesthetic and proprioceptive feedbacks of our being bodies and minds or, as it is now declined, 'embodied minds'. What is not representable, certainly exists: it passes through us, it enters in our vital processes. For example, all the realities under our perception threshold: from the light or magnetic waves outside our perceptive range because of our species structural limits, to the world of bacteria and symbionts that crowd our bodies and we don't 'feel', to atomic radiation that mutates our genes, but is obviously not 'known' to us. We are also permeated by a flux of multifarious automatisms, common to other 'knowing animals'. As a matter of fact, we practice a competence and an environment representation that is stratified in our neural processes (precortical and cortical), pertaining to survival: yet, we are unaware of these processes.

We are consciously part of the human world by the transformation of environments, by the creation of complex social bonds, and the awareness of ourselves and of our kin within natural and artificial systems, materially and symbolically produced, built by us and among us. So to speak, we are not only part of the *environment*, but we also belong to *worlds*,[19] that is, constructor and inhabitants of abstract and concrete worlds.

The typical *embodiment* of the most used linguistic metaphors is connected to the three-dimensionality of space, to the operational aspect of corporeality, either active or passive: we say 'up', 'down', 'behind', and any of these words are metaphors that move from the experience of our corporeality, of its movements, and from the internal states that these generate (Lakoff and Johnson 1999). Yet, the act of naming is obviously not only embodied, but it is necessarily inscribed within the implicit rules of discourses, that it, in that conventional field that allows – or hinders – the access to symbolic in the social: the so-called *discoursive practices* (Foucault 2004).

The sites of mentalization and consciousness – and thus indirectly, of perception – always cut only some portions of reality: not only because of the constitutional constraints of our species, but also by social and cultural conventions within organized groups. We can then speak only within the phenomenic fields that a given culture has thought as significant of the world, 'building' these fields by the shared access to practices and thought. As such – it is known – their significance does not proceed from the things or the functions in themselves, but is rather established by shared conventions that build a pregnant significance for some states of reality and not for others. Reality and the correspondent living experiences are articulated and disarticulated in time and space with the varied multifariousness of their symbolizations, according to the recognised and 'permitted' representational paths, and to

[19] The difference between world and environment is highly significant in philosophy, since according to phenomenology we inhabit 'worlds' because at birth we are cast in a reality produced by the symbolic and human activity, while the environment is the background made of the natural fixity, the bios, and the condition of the other animals. In life sciences, and particularly in ecology, our worlds are a highly sophisticatedly re-elaborated part of our habitats. These are in turn continuously constructed and de-constructed by their inhabitants, our relatives belonging to more or less close species, so that the border between natural and symbolic cultural environment is blurred, with a series of reciprocal acting constrains.

their logics; forms and logics that are long explored by anthropology, sociology and history of thought.

It is not therefore surprising that there is no metaphor moving from the linguistic *embodiment* of bodies varying through time, not solely self-moving and compact, not only vectors in space, but rather contaminated by the transformation of the hollow spaces and of their forms and functions in literally 'pregnant' stages of life.

Which cognitive *embodiment* on these states of being? Which linguistic metaphors for a corporeality speaking of remaining ourselves and at the same time of being transformed together with the Other growing within? That is, individual and dividual?

References

Bianchi, D. W., & Fisk, N. M. (2007). Fetomaternal cell trafficking and stem cell debate. *JAMA, 297*(13), 1489–1491.

Bleier, R. (1984). *Science and gender: a critique of biology and its theories on women*. Elmsford, NY: Pergamon Press.

Bolk, L. (1926). On the problem of anthropogenesis. *Verhandelingen Koninklyke Akademie van Wetenschappen, Amsterdam, 3*, 465–475.

Botti, C. (2007). *Madri cattive: una riflessione su bioetica e gravidanza*. Milano: Il Saggiatore.

Butler, J. (2004). *Undoing gender*. Cambridge, MA: Harvard University Press.

Celentano, M. (2000). *L'etologia della conoscenza* (p. 387). Napoli: La Città del Sole.

Contessi, R., Mazzeo, M., & Russo, T. (Ed.) (2002). *Linguaggio e percezione: le basi sensoriali della comunicazione linguistica* (pp. 64–73). Roma: Carocci.

Dawkins, R. (1976). *The selfish gene*. Oxford: Oxford University Press.

Deacon, T. (1997). *The symbolic species: The co-evolution of Language and Brain*. New York: W.W. Norton.

De Beauvoir, S. (1949). *Le Deuxième Sexe*. Paris: Gallimard.

Dick, P. K. (1968). *Do androids dream of electric sheep?* New York: Doubleday.

Duden, B. (2002). *Die Gene in Kopf – der Foetus in Bauch*. Berlin: Offizin Verlag.

Firestone, S. (1970). *The dialectic of sex: the case for feminist revolution*. New York: William Morrow.

Foucault, M. (2004). *Naissance de la biopolitique: Cours au collège de France (1978–1979)*. Paris: Seuil.

Fraire, M. (2002). Vecchie ragazze, donne nuove. In M. Fraire (Ed.). *Lessico politico delle donne: teorie del femminismo* (pp. 171–189). Milano: Franco Angeli (ed. or 1978).

Gagliasso, E. (2007). Doppia appartenenza e parzialità situate. In E. Gagliasso & F. Zucco (Ed.), *Il genere nel paesaggio scientifico* (pp. 65–88). Roma: Aracne.

Gilbert, S. F. (2006). *Developmental biology*. Cambridge, MA: Sinauer Associates.

Gilligan, C. (1982). *In a different voice: Psychological theory and women's development*. Cambridge, MA: Harvard University Press.

Gould, S. J. (1977). *Ontogeny and phylogeny*. Cambridge, MA: Harvard University Press.

Gould, S. J., & Vrba, E. (1982). Exaptation – A missing term in the science of form. *Paleobiology, 8*, 4–15.

Griffero, T. (2003). *Immagini attive: breve storia dell'immaginazione transitiva*. Firenze: Le Monnier Università.

Hager, L. (Ed.) (1997). *Women in human evolution*. London: Routledge.

Haraway, D. J. (1988). Situated knowledge. *Feminist Studies, 14*(3): 575–599.

Hrdy, S. (1981). *The woman that never evolved*. Cambridge, MA: Harvard University Press.

Hubbard, R. (1990). *The politics of women's biology*. New Brunswick, NJ: Rutgers University Press.

Jablonka, E. & Lamb, M. (2005). *Evolution in four dimensions: genetic, epigenetic, behavioral, and symbolic variation in the history of life*. Cambridge, MA: MIT Press.

Keller, E. (2000). *The century of the gene*. Cambridge, MA: Harvard University Press.

Lakoff, G., & Johnson, M. (1999). *Philosophy in the flesh: The embodied mind and its challenge to western thought*. New York: Basic Books.

Merchant, C. (1979). *The death of nature: women, ecology, and the scientific revolution*. London: Harper & Row.

Merchant, C. (1992). *Radical ecology: the search for a livable world*. London: Routledge.

Minelli, A. (2007). *Forme del divenire: evo-devo: la biologia evoluzionistica dello sviluppo*. Torino: Einaudi.

Pancino, C., & D'Yvoire, J. (2006). *Formato nel segreto: nascituri e feti tra immagini ed immaginario dal XVI al XXI sec*. Roma: Carocci.

Pievani, T. (2002). *Homo Sapiens ed altre catastrofi: per un'archeologia della globalizzazione*. Roma: Meltemi.

Rose, H. (1994). *Love, power, and knowledge: Towards a feminist transformation of the sciences*. Cambridge: Polity Press.

Rotilio, G. (2006). L'alimentazione degli ominidi fino alla rivoluzione agropastorale del neolitico. In G. Biondi, F. Martini, O. Rickards, & G. Rotilio (Eds.). *In carne ed ossa* (pp. 83–145). Bari: Laterza.

Schiebinger, L. (1989). *The mind has no sex? Women in the origins of modern science*. Cambridge, MA/London: Harvard University Press.

Schiebinger, L. (1999). *Has feminism changed science?* Cambridge, MA: Harvard University Press.

Scott, J. W. (1996). Il 'genere': un'utile categoria di analisi storica. In P. Di Cori (Ed.), *Altre storie: la critica femminista alla storia*. Bologna: Clueb.

Chapter 7
'Pop-Genes': The Symbolic Effects of the Release of 'Genes' into Ordinary Speech

Barbara Duden(✉) and Silja Samerski

Abstract: Within the last two decades the word 'gene' has migrated from science into ordinary conversation. Gene-talk has spread epidemically in political and professional arguments and ethical debates, but references to 'genes' have also entered personal deliberations. 'Genes' have now reshaped not only political, social, or medical concepts, but the very perception of the self. The intrusion of the term into common parlance and particularly the drastic encroachment of 'genes' into personal deliberation prompted our research project on 'genes' in ordinary prose: 'genes' have now even come to impose themselves as the ultimate answer to such primordial questions as, 'Where do I come from? Who am I? What is my future?' In the shadow of human genetics the first person singular or the personal pronoun, the 'I' of the speaker, is subtly, profoundly, and probably irreversibly affected because 'genes' in ordinary speech have the capacity to blend incompatible spheres of meaning. Outside the walls of laboratory science and DNA mapping the word has acquired an extraordinary alchemistic power: it refers to the most concrete, personal, and intimate – the soma of the speaker – but simultaneously also invokes statistical probabilities and aggregate risk profiles of populations. The alchemistic potency of the term makes it well-suited for exercising a crucial symbolic social function: references to 'my genes' and 'your genes' implant population statistics, probability calculations, and the demand for risk management in the corporeal makeup of the person using these terms or the person identified as a gene carrier. 'The gene' in ordinary prose imparts bodily substance to the nature of personhood in an era of dependence on professional guidance and the denigration of common sense perception and self-perception.

Keywords: Pop gene, risk management, genetic counseling, public understanding of genetics

---

Barbara Duden
Kreftingstrasse 16, 28203 Bremen, Germany
b.duden@ish.uni-hannover.de

Silja Samerski
Fesenfeld 104, 28203 Bremen, Germany
s.samerski@ish.uni-hannover.de

F. Molfino, F. Zucco (eds.) *Women in Biotechnology*,

7.1 Introduction

We started our research with an observation: in contrast to any other technical terms, there is an immense incongruity between the scientific concept and public perceptions of 'genes'. While molecular research has disproved the hypothesis of definable, distinct units of inheritance and development – that is 'genes' – outside of the laboratory the word has become the epitome of a 'building block of all life' and its technical manipulation. Geneticists admit that the concept of 'genes' in use a decade ago is no longer valid. In the preface to a history of genetics the authors state: 'somewhat detached from the gene as a public icon, but also unknown to many biologists (...) new findings have caused a watershed during the last decades. The more molecular biologists learn about genes, the less sure they seem to become of what a gene really is' (Beurton et al. 2000). Yet the public's belief in these genes seems firmer than ever. Within just a few years this one-syllable word has migrated from the laboratory into less rarefied terrain: today the term is being debated in classrooms, in pubs and around dinner tables. What are the semantic contours of the entity 'gene' that people all over are discussing? What do people say in this gene-talk? What kinds of certainties, emotions, fears, and hopes does the word elicit and what are the social consequences of this belief in 'genes'?

We, a historian of the senses and a social scientist and geneticist, therefore conducted a research project to investigate the meanings imparted to 'genes' in ordinary speech and to analyze the social symbolic function of the term. What does 'gene' say, demand, and command when it appears in political, familial, and professional speech? This question calls for clarification because the 'gene' – which has a weak standing as a scientific term – is so often invoked in assertions about what humans are and what is good for them. 'Genes' intrude into notions and perceptions of me, you, and the person beside you.

We believed it was crucial to investigate the appearance of the term in ordinary oral speech acts, rather than in texts, media discourse, or educational or scholarly articles. We concentrated on 'genes' viva voce. One speaks to another person with the expectation that he or she will listen and make an effort to understand. We insisted on an investigation of 'genes' viva voce because we were convinced that 'genes' in ordinary prose are invested with characteristics that had not been truly acknowledged: the word connotes something corporeal, something substantial in the person, and it has a reflexive *deixis*, which means that it points back to the speaker. When someone talks about his 'genes' or the 'genes' of the child to come, the speaker ascribes 'genes' to himself or herself. Whoever ascribes 'genes' to another person in fact also points back to himself or herself as a carrier of 'genes' (Duden 2002). We set out to explore these somatic reflexive 'genes'.

Our project had two separate but interdependent parts: in one part we scouted out the notions and perceptions of a range of people by conducting open-ended interviews to probe interviewees' ideas about 'genes', particularly as these related to their biographies and experiences of body, family, and illness. The other part of the project recorded sessions and observed what experts – in this case genetic counselors – were saying to enlighten their clients about their 'genes'; we analyzed what this instruction demanded of the listener. One part of the project listened for echoes of the word 'genes' in everyday notions, while the other observed its passage from the mouth of a geneticist to the ear of a listening client. We will briefly report on each project and then draw some conclusions.

7.2 The Double Face of Genes in Heudorf

7.2.1 The Research Project

During 2003 and 2004 we conducted interviews with two dozen people living in a southern German village, which we will call 'Heudorf'. We chose this village because its inhabitants come from varied backgrounds and continued to spend their daily lives in widely varied settings; for instance, we spoke not only with farm people who had lived in Heudorf for generations, but also professionals with roots elsewhere, who commuted to work in the nearby university town. The choice of a village situated next to a university town with a huge hospital and big departments for molecular biology led us to expect gene-talk in very different registers.

Our goal was to elicit ideas and deliberations on 'genes' that would give us clues to a number of questions: what types of gene-talk exists in the village? In which ways do 'genes' touch on the experiences and apperceptions of the speaker? Who incorporates the term into his or her familiar life world and who refuses to use the term? From where does the knowledge about 'genes' derive, and how confident do interviewees feel about their understanding of the term? What personal ideas, images, and metaphors do they attach to the word, and how do the interviewees feel about these? The aim of our interviews was to register the existing range of ideas about 'genes' among these people, and to encourage our interview partners to articulate their beliefs as personally, experientially, and as closely to their view of reality as possible. Pierre Bourdieu's practice in *La misère du monde* (1993) guided our method for conducting these interviews: making every effort possible to listen to the ideas voiced, to probe deeper by asking further questions, and to take the interviewee seriously. Jeanette Edwards' admonition – that we need 'a more sophisticated way of hearing' how people speak about genetics and need to take seriously what they say – served as our guide (Edwards 2002). The interviews were transcribed and together resulted in 800 pages of what we had recorded. We investigated this material with three central questions in mind: what do people say? What do people associate with 'genes' or genetics? What does this material on the oral pop-gene talk reveal about the effect of 'genes' on self-perceptions?

7.2.2 A Semantic Fissure

We can only offer a brief summary of our findings here. Our initial question in the interview produced a bundle of divergent ideas, with answers going in two divergent, in fact opposite and incompatible directions. The speakers either spoke of heredity or of 'manipulation'. The first led to talk about their families, relatives, and the past, the second evoked ideas about Petri dishes, genetically modified corn, 'cloning', and the 'making' of humans; one chain of ideas leads into the past, the other into a fictional future. Our respondents invariably spoke of the heredity of traits being passed down as a given, something indisputably factual, while 'manipulation' remained open-ended and conditional on the unlimited potential 'progress of science'. A remarkable semantic fissure runs through the entire material, a rupture separating heterogeneous spheres of meaning. We were also astonished by the amplitude of the word field: the word 'gene' elicits disparate and heterogeneous spheres and references that ordinarily cannot be compacted into just one term or be used as a metaphor. Furthermore, the mode of talking that was most characteristic in these interviews expressed a simultaneity of attitudes that ordinarily belong to incompatible spheres: certainty and uncertainty, doubt and assuredness come out in one breath.

7.2.3 An Inherited Entity

References to 'genes' in our compilation of voices from 'Heudorf' signaled our respondents' belief in some entity that is involved in heredity from one generation to the next. People use the term as a synonym for the proverbial wisdom that children embody inherited traits from their ancestors – 'yes, these are father's genes, the black hair and how she sits…' Yet our respondents also understood 'genes' as concretizing some invisible causal agency under the skin. In many explanations 'genes' were a synonym for the German *Erbanlagen* (inherited traits), a term naming those characteristics and habits which a person feels she received from her ancestors. Here the term stands in for one's whole being, all that one was endowed with in the cradle. Yet at the same time the word can name the assumed cause behind becoming, growth, and development. In this way genes either denote the very being of a person in her flesh and her habits, or a presumed agency that was its underlying, deep-seated cause.

As was expected all of the speakers struggled with the whereabouts of those genes: are they in the blood, in the brain, or all over? Popular phrases that describe genes as 'building blocks' or of DNA as a string on which 'genes' 'sit' like pearls on a thread reinforce the tendency to endow genes with a concrete shape and to make them into a kind of 'thing'. In our analysis we focused particularly on the verbs and the *composita* which the noun attracts in the imagination. In colloquial speech genes 'sit', they 'make' something, they 'prescribe' and 'determine'. This is particularly striking when the interviewees use a neologism in the German language: the 'gene-defect' which carries echoes in it of the term 'birth-defect'; they visualize a *compositum* that endows the 'gene' with the materiality of some thing that can be a 'defect'. A defect is a flaw, a mistake: 'a gene-defect is when something is not going right… when something is defective. This means that it is broken and in need of repair', explained one interviewee. We listened to many lively descriptions of how this 'defectiveness' is imagined. Our respondents likened a 'gene-defect' to crooked rods or sticks, something bumpy, a dent, or a chain that is broken. Their adjectives all point toward an entity that is crooked, bent, skewed, uneven, displaced, or in the wrong spot. The faulty 'genes' do something: 'a gene-defect makes things go wrong'. The attribute 'defect' gives substance and concrete shape to the referent of the noun. Thus 'genes' are hypostatized into a concrete, tangible thing, and this image in turn supports the idea of 'genes' as something that can be isolated and replaced.

7.2.4 Individual Destiny

When genes are used as a synonym for 'Erbanlagen', they belong to the sphere of that which is inevitable and beyond one's will or control. One of our respondents, a woman working in the bakery, explained: 'When I can influence things about my body with my will and thinking, then this for me isn't genes… I cannot influence what my height will be. I cannot influence what my eye color will be. I have no influence on the color of my hair. For me these are genes. But when I get fat, or ruin my body by not washing myself, or when I let myself go, that is not a gene.' In this sense 'genes' are associated with individual destiny or givens. That which simply 'is as it is' we may call nature or natural endowment. In the eyes of some Heudorf residents such as the bakery clerk, 'genes' not only appear to be a causal agent in development, but also to have the same capacity for determining such serious illnesses as cancer.

The local physician in the village explains how diseases are already sitting in the genes: 'When we look at it', she says, 'we are nothing but a vessel in which our genes are stirring and pushing'. She explains that there are certain diseases 'where you can tell exactly, this person will get sick if he has that gene – you have to look, does he have the gene? – the gene is fixed. You know on which gene the disease sits. Yes, and then, you can investigate that, and you can tell the affected person: "You have this gene, you will get sick from it, that is inevitable.'' The determining power of heredity, which was responsible for the person, is also transferred to 'genes' when they are reified as distinct causal agents for development. They will eventually and inevitably affect something in the future. In this way genes are conceived of as 'causes' in the past or as causative agents which will do something in the future. 'Genes' thus pertain to two temporal dimensions: they are shorthand for the common sense perception that the present is the outcome of the past, and they are taken as a determinative power 'for' the future. As in the figure of the 'gene-defect' and 'genes' as 'building blocks', the referent of the term invariably embodies the power of a unilinear causal agency under the skin. The term gives carnal plausibility to the idea that the future already exists inside one's body.

The statements we recorded in Heudorf are a vivid illustration for the belief that the connection between 'genes' and an illness is causal, yet the connection between DNA and an illness is – with the exception of a few cases – merely statistical. No one can – beyond those exceptions – predict what will happen to a person on the basis of a DNA sequence. Only probabilities can be calculated, that is, the frequency with which certain illnesses occur in the population of those individuals who carry a certain marker in their DNA. The expressions 'a gene for' or 'a gene-defect' are nothing but a shorthand for this statistical correlation. In colloquial speech, however, talk about 'flawed genes' or 'genetic defects' implies that some fault already exists in the body, that something is wrong and will ultimately lead to disease. The colloquial expressions in gene-talk obfuscate the gulf between flesh and statistics, and in this way give body to the concept of the 'risky self' which has become the object of biomedical procedures and techniques of prevention.

7.3 Gene Carriers as Faceless Risk Profiles

7.3.1 Genetic Counseling as a Research Topic

In the second part of our project we observed, recorded and transcribed 18 genetic counseling sessions which were undertaken by three different university institutions. Genetic counseling is an educational service training people to become so-called informed and responsible consumers of genetic tests and preparing them to cope with the test results. Thus, the goal of the one-to-two-hour crash course in biostatistics and genetics is not compliance with professionally prescribed conduct, but 'individual assistance in reaching a decision'. Geneticists encourage their clients to make their own decision – after being professionally prepared for it. In the case of pregnant women, who are the main clientele of genetic counselors, this decision is a historically unique one: the pregnant woman is being asked in the light of her current risk profile if she wants to keep her pregnancy or to make its continuation conditional upon further test results (Samerski 2002).

The reasons that had brought the clients to the sessions we observed varied broadly. Pregnant women and parents-to-be came because their gynecologist had classified their pregnancy as 'at risk'. An aunt with a severe lung condition, a disabled nephew, a deviant blood test, an abnormal sonogram, or simply her age (being older than 35 or younger than 18) – a welter of risk factors make sure that more and more women are diagnosed as being 'at risk'. The geneticist then lectures on the genetic causation of different diseases such as hay fever and Cystic Fibrosis, the probabilities of congenital heart defects, harelip, spina bifida and Down's syndrome, the pregnant woman's location in various risk diagrams and their ominous prognostications about the unborn's future, and lastly, a modern woman's obligation to feel responsible for decision making. A growing number of clients land in genetic counseling centers because they have heard of 'hereditary cancer' and the possibility of predictive testing and are worried because of incidents of breast or colon cancer in their families. Here, the geneticist explains the structure of the DNA, goes through the list of possible errors in the genetic make up, detailing their potential effects on a person's health; warns of the various increased cancer risks connected to a potential 'gene defect' and insists on regular medical check-ups; offers predictive testing and points to its limited meaning; and then leaves it up to the clients to decide what to do.

In order to analyze the professional redefinition of subjectivity in the shadow of genetics, there is hardly a better example then genetic counseling. During the session, the expert's lessons on 'genes' and 'risks' directly clash with the client's delicate condition. She is worried about the well-being of herself or her coming child – and the geneticist saddles her with potential genetic defects and risk management strategies. But in the end, it is the client who has to make a decision. Thus, the counselor has to try to bridge the chasm between his statistical constructs and her concrete situation. He tries to spell out his knowledge in words that sound meaningful to her. Technical terms mutate into popular scientific notions loaded with colloquial meaning. An abstract probability turns into a personal threat, and the correlation between genotype and phenotype jells into an illustrative 'gene for' (Samerski 2006). Thus, the educational session with a geneticist exemplifies how professional talk on the pop-gene reinterprets deliberation and autonomy.

7.3.2 The 'Gene' in Genetic Counseling Sessions

In genetic counseling the 'gene' is polymorphic: it appears as a 'thing', a material entity, a causal 'agent', something that can be 'defective', a building block for the organism, a carrier of information. It appears as something that has existed from the beginning and that is everywhere in the body. These different characteristics make it possible for 'genes' to smuggle a physically latent, predictable future into the person. What might happen tomorrow is, so it would seem, already present in one's genes. Should a disease at some point become manifest, this then seems to be only the last phase in a temporal, yet so far imperceptible, chain of becoming (Greco 1993). What had already gone wrong in the interior of the body appears at the surface. The statistical 'risk' thus seems to measure a pre-existing fault; it appears as a threat in one's own flesh.

7.3.3 Inconsistency in Talking

Characteristic in genetic counseling is a new inconsistency in talking that is typical for professional education about genetics. When client and counselor are talking about statistical chances and risks, the counselor addresses the person facing her or him as a concrete person, yet pontificates about statistical populations in the abstract. The counselee feels – rightly so – that she is being addressed by the geneticist: the counselor speaks to her, the woman, sitting opposite, and instructs her. More explicitly this expert speaks about 'her risk' for breast cancer and the 'gene-defect' she is meant to have. Inevitably in this geneticist's talk, the addressed person and the predication diverge, they pertain to incompatible spheres of meaning. A risk by definition pertains to frequencies in a statistical population and never to a concrete person. Adolphe Quételet, the astronomer and social scientist, stressed in the nineteenth century that laws about society or large numbers cannot be applied to an individual person. These laws have, depending on how they have been determined, nothing personal about them, which is why they can only be applied to individuals with certain restrictions. Applying them to an individual would be as erroneous as using a mortality table to determine the day when a certain person will die (Ewald 1993). Yet the counselor continually speculates about possibilities in the woman's future on the basis of risk curves and calculated probabilities.

The talk about 'genes' and genetics obscures this gulf between the individual person addressed and the referent of the predication. Genes endow clients with a conception of themselves that make them compatible with the statistical frequencies and populations of which the counselor speaks. In these counseling sessions about 'genes' it seems as though the statistical calculations offered mirror the physical constitution of the counselee, while in fact they only calculate and enumerate frequencies in statistical populations. The counselee is enjoined to understand herself as a construct whose very being is pieced together from the properties of statistical risk classes. The gene as a technogene abstraction, hypostatized correlation, and essence of being equates the person with a constructed subject that is the object of 'therapeutic' techniques of prevention. As Robert Castel has observed, 'What the new preventive policies primarily address is no longer individuals but factors, statistical correlations of heterogeneous elements. They deconstruct the concrete subject of intervention, and reconstruct a combination of factors liable to produce risk' (Castel 1991).

Professional education about 'genes' and 'genetic risks' transforms the person into a statistical profile, a faceless member of different risk classes. This fundamental transformation of self-perception is not an avoidable side effect of the counseling, but its declared goal. Only when the counseled understand themselves as a set of risk profiles and gene carrier are they prepared for the decisions they are meant to make: to weigh the chances and risks of different tests and treatment options, to be responsible consumers of genetic tests and risk-reducing preventive measures. When the counseled understand themselves as gene carriers and risk profiles they become, physically, compatible with the logic of cost-benefit analyses and economically oriented decision-making (Samerski 2007).

The goal of genetic counseling is to lead the clients to a new form of maturity, that is, to self-determined decisions in a world in which a sensuous reality that can be perceived is replaced by laboratory data and statistical calculations. Self-determination here is a contradiction in terms: to be able to make decisions in a technogene world presupposes that the 'self' – the individual person that the conversation refers back to – is replaced by a statistical construct, a risk profile. By convincing their clients that the probabilistic destinies that experts calculate and ascribe to them are already pre-programmed in their bodies, the clients are enjoined to refashion themselves into a resource for the management of statistical populations.

7.4 The Transformative Power of 'Genes'

Our project on the pop-gene set out to explore the symbolic and iconic characteristics of genes in ordinary speech. By investigating the *lemmagen* in two directions – in relation to the everyday understanding of people in a village and observing how 'genes' cross over from the laboratory into the everyday life world in a genetic counseling session – we analyzed the latent social function of 'gene' as a neologism in the German language. In both cases we focused on oral speech, occasions when speakers use the term to emphatically say something that is important to them. While speaking they expect that the listener will make an effort to understand their utterance. We came to the conclusion that this 'oral gene' possesses a unique alchemistic power: it connects heterogeneous spheres. It blurs that which people can say about themselves based on experience and sensual certainty with technogene abstractions from the laboratory. It identifies one's own physical constitution, the somatic being as one is, with units in a genetic program. The stubbornness of that daughter or the varicose veins one inherited from one's mother are reduced to mere indications of a hidden genotypic reality. And, finally, it transforms statistical risk calculations into personal, predetermined destiny. This confusion of speech that is based on experience and common sense with abstract technical terms based on models and algorithms can also be observed when other migrants from science infiltrate ordinary speech, for instance, in talk about the 'immune system'. But in gene-talk this confusion takes on a new dimension.

First, the gene, unlike any other word, is the object of ethical and political debates. The metaphors and the symbolism of the word 'gene' in ordinary speech determine to a large degree the public controversies about cloning, genetic testing and gene food. Because 'genes' in ordinary speech have no clear and distinct denotation, the word has the social and symbolic function of concatenating laboratory and everyday worlds, scientific objectivity and subjectivity, statistical calculations and individual being.

Second, our interviews show that gene-talk refers to the human being in a particular way. Invoking this word, one debates not what someone has, but his or her very being as it is. Inevitably, the diffusion of gene-words leads to a redefinition, even a re-coining of the human being. As our analysis of the genetic counseling sessions demonstrates, the diffusion of genetic terms transforms the human being into an object of modern bio-administration. When the genetic counselor addresses his clients as gene carriers, that is, as carriers of causal agencies and genetic programs, that expert ascribes a technogene body to them that makes them compatible with the risk calculation of modern health management. By being regularly coupled with the adjective 'genetic' and the idea of 'genes' as causal agencies risk projections (which by definition say nothing about a concrete person) appear as diagnoses of a flaw already present in one's own body. The gene, which fuses the concrete and the abstract, the personal and the statistical, thus transforms the person addressed as a carrier of genes into a faceless conglomeration of data. This new self-perception should, then, serve as the basis for the decision-making of the counselee, for genetic counseling aims not only to instruct clients, but also to enable them to do something (in the case we observed, to make a decision about whether or not to genetically test themselves or their coming child). This demonstrates the concrete social effects of the epistemic transformation of a person into a profile to be managed: one can request economic cost-benefit analyses and risk management as rational and responsible decisions about one's own person.

In ordinary speech 'gene' has a social symbolic function that transcends the power of a 'cultural icon'. Outside of the laboratory the word 'gene' has an extraordinary property: in one breath it refers to the most concrete and personal, one's somatic being, while it denotes statistical correlations and probabilities at the very same time. The pop-gene has an extraordinary, historically unprecedented transformative power: it mingles abstract and concrete, calculable and individual, soma and statistics. Thus gene-talk transforms the flesh and blood of a person, someone taking part in gene-talk into a product of genes; the person, her being, the embodied 'I' of the first person singular becomes a case without a face or a name. The pop-gene transports an image of the human being whose application of modern management strategies to herself or himself – risk evaluation and rational choice – seems not only plausible, but also reasonable.

References

Beurton, P. J., Rheinberger, H.-J., & Falk, R. (Eds.) (2000). *The concept of the gene in development and evolution. Historical and epistemological perspectives.* Cambridge: Cambridge University Press.

Bourdieu, P. et al. (1993). *La misère du monde.* Paris: Seuil.

Castel, R. (1991). From dangerousness to risk. In G. Burchell, C. Gordon, & P. Miller (Eds.), *The Foucault effect* (pp. 281–298). Chicago, IL: University of Chicago Press.

Duden, B. (2002). Meine Gene und ich. In *Die Gene im Kopf - der Fötus im Bauch* (pp. 253–264). Hannover: Offizin.

Edwards, J. (2002). Taking 'public understanding' seriously. *New Genetics and Society, 21*(3), 315–325.

Ewald, F. (1993). *Der Vorsorgestaat* (p. 196). Frankfurt a.M.: Suhrkamp.

Greco, M. (1993). Psychosomatic subjects and the 'duty to be well': Personal agency within medical rationality. *Economy and Society,* 22, 355–372.

Samerski, S. (2002). *Die verrechnete Hoffnung. Von der selbstbestimmten Entscheidung durch genetische Beratung.* Münster: Westfälisches Dampfboot.

Samerski, S. (2006). The unleashing of genetic terminology. How genetic counselling mobilizes for risk management. *New Genetics and Society,* 25, 197–208.

Samerski, S. (2007). The 'decision trap': How genetic counseling transforms pregnant women into managers of fetal risk profiles. In P. O'Malley & K. Hannah-Moffat (Eds.), *Gendered risks* (pp. 55–74). London: Routledge Cavendish.

Chapter 8
The Organizational Construction of the Body in Assisted Reproductive Technologies

Manuela Perrotta(✉)

Abstract: This chapter explores the relation between biotechnologies and the body through the case of assisted reproduction. On the basis of a case study realized in an artificial reproductive technologies centre in Italy, this paper explores how the organizational construction of the body takes place, providing some instances and illustrations of working and organizing practices as a locus where this construction is materially performed: how bodies are subjected to a series of organizational practices whereby the organization inscribes them in its spaces and produces 'organizational bodies'; how the passage from body to fluids (follicular and seminal) is done in the laboratory practices; how gametes (oocytes and spermatozoa) are manipulated and 'transformed' into embryos; and how finally they return to the body.

The aim of the paper is to reflect on the complex relationships between the construction of the body and biotechnology, organization and institution, in the process of assisted reproduction, in which the patient body, especially the female body, is fragmented and reassembled, and subtly and smoothly removed as the centre of the scene. The organization is inscribed in and performed through the body, which becomes liquid and is transformed into different things during accomplishment of the organizational practices.

Keywords: Body, organization, practices, technology, assisted reproduction

8.1 Introduction

Biotechnologies have been a matter of global discussion and the source of argument among different scientific communities, probably because of their particular epistemological status deriving from their position at the confluence of biological and

Manuela Perrotta
Department of Sociology and Social Research, University of Trento, Piazza Venezia, 41, 38100, Trento, Italy
manuela.perrotta@soc.unitn.it

F. Molfino, F. Zucco (eds.) *Women in Biotechnology,*

other areas of knowledge, and technologies. Although biotechnologies have produced innumerable discoveries, they raise not only ethical problems but also more general ones, especially in the cases of their application to human beings.

Particularly, I will explore the relation between biotechnologies and the body through the case of assisted reproduction. The choice of the empiric field is due to some distinctive characteristics of assisted reproductive technologies (ART). The first is the ethical and legislative context in which biotechnology is performed, especially in Italy, where the law regulating ART was debated for months and led to a public referendum (held in June 2005) on abrogation of certain of its provisions. The result of the referendum was declared void because voter turnout did not reach the necessary quorum, but the ethical debate is still heated. Secondly, ART working practices have the further interesting feature that two diversifying reproductive techniques are used, which are involved in a scientific and ethical controversy. Finally, the case study allows us to show how the organizational context interacts in the practices of reproduction, which are constantly redefined and negotiated in everyday work.

On the basis of a case study realized in an artificial reproductive technologies centre in Italy, in this paper I will explore how the organizational construction of the body takes place. Following Karen Barad's concept of posthumanist performativity (2003) I will examine how bodies are subjected to a series of organizational practices whereby the organization inscribes them in its spaces and produces 'organizational bodies'; how the passage from body to fluids (follicular and seminal) is done in laboratory practices; how gametes (oocytes and spermatozoa) are manipulated and 'transformed' into embryos; and how finally they return to the body.

The study, thus, allows us to reflect on the complex relationships between the construction of the body and biotechnology, organization and institution, providing some instances and illustrations of working and organizing practices as a locus where this construction is materially performed.

8.2 The Theoretical Background

According to Chris Shilling the body has historically been something of an 'absent presence' in sociology (Shilling 1993). It has been absent in the sense that sociology has rarely focused on the embodied human as an object of importance in its own right. Although the physical, fleshy body was rarely an object of explicit classical sociological concern, facets of human embodiment, such as language and consciousness, became central to the development of the discipline. In other worlds, the existence of human bodies tends to be taken for granted in analyses of social life. Studies of 'society' or 'institutions' assume but rarely examine how social practices are embodied and, in this sense, rely upon human embodiment for their enactment. The analysis of social practices has been abstracted from the embodied media through which these practices are accomplished.

According to Torkild Thanem (2003) after the complaints against the lack of embodiment and body research in the social science (Turner 1992; Shilling 1993), the last years have seen an exploding interest in the body, even in organization studies (Barry and Hazen 1996; Hassard et al. 2000; Dale 2001). As pointed out by John Hassard and colleagues (2000) the body is a largely neglected component of organizational practices. However, the majority of the work on this issue has emphasized ways in which bodies are organized, regulated and normalized, but much of this research remains disembodied, approaching the body as a research object like any other rather than critically reflecting on it in relation to their own embodiment (Thanem 2003).

The desire to consider the body in relation to practices of organization forms part of a wider, postmodernist and feminist movement in social and cultural analysis. Michel Foucault's writings (1973, 1977, 1978) have been influential in body theorization. However, in these studies the body is not really present in its brute corporeality (Hassard et al. 2000); it is, rather, a discursively produced entity that is subject to administrative operations and decisions. Feminist studies go over this interpretation: the live body is neither brute nor passive but is interwoven with and constitutive of systems of meaning, signification, and representation (Grosz 1994). Furthermore bodies are never simply human bodies or social bodies. According to Rosi Braidotti 'the body, or the embodiment, of the subject is to be understood as neither a biological not a sociological category but rather as a point of overlapping between physical, the symbolic, and the sociological' (Braidotti 1991).

Alexander Styhre (2004) calls for a (re)embodied view of the organization that does not rely solely on what Elisabeth Grosz defined as 'discursivization' (1995) of organizational activities, but acknowledges the human body as a key entity in organizational life. The Foucaultian approach to the body is characterized by a substantive preoccupation with the body and those institutions which govern the body, but also by an epistemological view of the body as produced by and existing in discourse. If discourse is the most important concept in Foucault's work and it is centrally concerned with, although irreducible with, language, Barad argues that 'language has been granted too much power' (Barad 2003). In her understanding, discourse is not a synonym for language:

> Discourse does not refer to linguistic or signifying systems, grammars, speech acts, or conversations. Discourse is not what is said; it is that which constrains and enables what can be said. Discursive practices define what counts as meaningful statements. Statements are not the mere utterances of the originating consciousness of a unified subject; rather, statements and subjects emerge from a field of possibilities. This field of possibilities is not static or singular but rather is a dynamic and contingent multiplicity (Barad 2003).

Accordingly, moreover, discursive practices are not speech acts, linguistic representations, or even linguistic performances, bearing some unspecified relationship to material practices. Indeed, they are not human-based practices. They are 'specific material (re)configurings of the world through which local determinations of boundaries, properties, and meanings are differentially enacted' (Barad 2003).

As Judith Butler (1993) pointed out, we should come back to the notion of matter, not as site or surface, but as a process of materialization that stabilizes over time to produce the effect of boundary, fixity and surface, that we call matter. Barad (2003) proposes a post-humanist account of discursive practices: matter does not refer to a fixed substance; rather, matter is substance in its becoming – not a thing, but a doing. Matter refers to the materiality/materialization of phenomena, not to an inherent fixed property of abstract independently existing objects. Materiality is discursive, just as discursive practices are always already material. Discursive practices and material phenomena do not stand in a relationship of externality to one another; nor are they reducible to one another. The relationship between the material and the discursive is one of mutual entailment.

In Barad's understanding, all bodies, not merely 'human' bodies, come to matter through the world's performativity. Bodies are not objects with inherent boundaries and properties; they are material-discursive phenomena. 'Human' bodies are not inherently different from 'nonhuman' ones. What constitutes the 'human' (and the 'nonhuman') is not a fixed or pregiven notion: 'material' is always already material-discursive – that is what it means to matter (Barad 2003).

Moreover, she proposes a specifically posthumanist notion of performativity that incorporates important material and discursive, social and scientific, human and nonhuman, and natural and cultural factors. A posthumanist account of the production of material bodies calls into question the givenness of the differential categories of 'human' and 'nonhuman', examining the practices through which these differential boundaries are stabilized and destabilized. According to this account meaning is not a property of individual words or groups of words; it is not ideational but rather specific material (re)configurings of the world.

8.3 Research Setting and a Note on Method

The process of assisted reproduction (once the preliminary examinations have been completed) begins with hormonal stimulation. The patients are monitored on alternate days, and checks are also made of extradiol levels, in order to determine the maturation stage of the follicles and if necessary change the dosages of the stimulation drugs. When the follicles are mature, extraction has to be performed within twenty-four hours, and this operation had to fit with the center's organizational requirements.[1]

In the meantime, the test tubes containing the follicular fluid are analysed to establish whether they contain oocytes. The oocytes are aspirated and deposited in the culture medium. At this point the procedures for IVF and ICSI diversify. In the case of IVF, three oocytes (the law allows production of just three embryos) are placed on a plate together with a certain quantity of sperms and the plate is put in

[1] In the centre I observed, for instance, the extractions could be made from Monday to Friday, except for Thursday, when the anesthetists were absent. Consequently, the maturation of the follicles had to coincide with one of those four days in the week, and it had to be carefully timed in advance.

the incubator. In the case of ICSI, the oocytes must first be washed. Each oocyte is injected with a single spermatozoon by means of the needle of a micromanipulator. Around three days after fertilisation, the embryos are transferred to the patient's uterus.

The embryos produced are classified according to their characteristics as belonging to class 1, 2 or 3. Embryos in class 1 are those most likely to implant themselves in the uterus, but those in classes 2 and 3 can be transferred to the uterus as well. The others are discarded.

The account presented in what follows[2] come from observation in a public centre of assisted reproduction. The centre manages around 300 treatments per year. It is part of a university hospital, in which the personnel work in close contact with the adjoining gynecology department. It has a staff of about 60 members: two chief consultants in reproductive medicine (each of whom is in charge of a team of doctors and trainee specialists); a head of the ART laboratory; a biologist; two anesthetists in the operating theatre (who also supervise two trainee specialists); 18 obstetricians[3] (they help the work, for instance managing the admissions office on alternating shifts); four nurses (providing nursing care during medical examinations and assisting in the operating theatre); and two ward attendants (whose main job was to accompany patients on trolleys when necessary, e.g., taking them to and from the operating theatre). There is no psychologist at the centre, so that couples are sent to the nearby psychiatric clinic for consultation and advice.

I used a case study research strategy (Eisenhardt 1989; Yin 1994) through qualitative data collection techniques and analysis.

Access to the field was negotiated with the head of ART laboratory, who allowed me to participate in the center's everyday activities, with the agreement of the gynecologists. The observation was overt to all the staff of the centers, while it was covert for the patients. I wore a white coat, which made me unrecognizable to the patients.

Data collection involved the use of participant observation, interviews, and documents review. Observation took the form of following the activities in each centre (from operating theatre to laboratory practices, monitoring of ovarian stimulations, patient reception, and so on) taking notes on work practices. Professionals were encouraged to talk about what they were doing, and I asked additional questions probing particular issues.

[2] The data presented are part of the international research project 'Practising Learning in Context. Dynamic Capability Development Across and Between Sectors', founded by the Advanced Institute of Management Research. The research was supported also as part of the Ministry of University and Research (MIUR) Project (2004–06): 'Apprendimento tecnologico e tecnologie di apprendimento: organizzazione e nuove tecnologie come sistemi di conoscenza distribuita' (D. M. n. 174 Uff. VIII 8.11.2004, prot. 2004145957), nationally coordinated by Silvia Gherardi.

[3] Usually obstetricians do not have a great deal to do with sterility given that their work is to assist with childbirth, but when the former (and larger) gynaecology clinic was closed they had been divided between the hospital department and the ART centre.

8.4 The Production of Embryos

ART centers enable infertile couples to achieve their goal by means of technology. But while the couples' objective is pregnancy ('the babe in arms'), the objective in the course of the process seems to become production of the embryo: or more precisely, the three embryos permitted by Italian law, if possible belonging to class 1, but at least ones which are implantable (classes 2 and 3). The economic metaphor of 'production' is commonplace in the language of the reproductive process (Gribaldo 2005): oocytes are 'produced' by stimulation; male partners 'produce' semen; and embryos are 'produced'.

During the observation the assisted reproduction process moved through various stages, from the medical sector (gynecology) to the biological sector, and vice versa. In the first stage, the couple were subjected to various tests in order to determine whether their problem was sterility or infertility,[4] and to decide whether a cycle of treatment should begin. At this stage, the doctor and the biologist worked side by side: the doctor conducted the first interview with the patient, prescribed tests (which are performed outside the centre), and performed echographs and hysteroscopies in order to exclude certain physical causes of sterility. The task of the biologist was instead to determine the male causes by analysing the seminal fluid in the ART laboratory.

The preliminary examination phase was followed by patient stimulation. Here the doctor became the leading actor. Stimulation protocols, monitoring, echographs, and so on, marked the progress of the activity. The pick-up was the moment when the medical domain gave way to the biological one: a transition symbolized by the 'ritual' transportation of the test tubes from the operating theatre to the laboratory. Those responsible for the task went back and forth, carrying the test tubes of follicular liquid to the laboratory, and then returning to the operating theatre to announce the number of oocytes found. This pattern was repeated until there were no more follicles to aspirate.

From the point of view of physical structure, while medical practices were dispersed around the ART centre and also took place externally to it (blood tests, psychological counselling, etc.), biological practices were concentrated in the laboratory as the 'embryo production site'. The biologist was the sole actor in this phase of the process, until he definitively left the scene on consigning the finished product, i.e., the embryo, to the patient (with the help of the doctor). Symbolically, the transfer – and in particular the object with which it was performed, the transfer catheter[5] – was the moment when the doctor's work was reunited with that of the biologist.

---

[4] Sterility is the absolute impossibility to procreate, while infertility is a reduction of reproductiveness.

[5] The transfer catheter consists of two thin tubes: one part is for the biologist, and the other one is for the gynaecologist. The gynaecologist inserts his part into the uterus with the help of a speculum. The biologist aspirates the embryos into his part of the tube using an insulin syringe. When the outer part of the catheter has been inserted, the biologist pushes his part of the tube through it and injects the embryos using the syringe.

The procedure prior to conclusion of the process was centered on bodies. First the organization inscribed itself in and on the bodies of the patients; and then the bodies were disassembled so that life could be produced. When the process had concluded (positively) and the embryos had been produced, they were transferred to the natural site of reproduction: the womb.

The trajectory just outlined can be divided into four stages:

1. The production of organizational bodies. When a couple was deemed suitable for a cycle of ART treatments, their bodies were subjected to a series of organizational practices whereby the organization inscribed them in its spaces and produced 'organizational bodies'. The couples were separated, and they followed different routes: the men were examined in the andrologist's office; the women attended various places according to the tests necessary, although the monitoring room (where ovarian checks were made during the stimulation period) was the main locus of their treatment.

The bodies of the patients who undergo assisted fertilisation are at the centre of a process whereby they are simultaneously subjects of reproduction and objects of medical treatment. But subjecthood is immediately lost when couples go to the centre to seek medical help, and the body on which attention is focused is an object in the reproductive chain. The bodies of the men and women that I encountered during my observation period had marginal value compared with organs and their products, and the male body was almost non-existent with respect to the female body. There was no access to the male body; it only existed in the background. Appointments at the andrologist's office consisted of an interview, the prescription of medical tests or drugs, and the handing to the patient of a sterile container for the 'collection' of seminal fluid (by which expression – together with the even more generic one of 'production' – was meant masturbation for the donation or examination of seminal fluid). Once the patient's personal details (name, surname, days of abstinence[6]) had been recorded, he was sent to one of the two bathrooms of the centre.

Female bodies were instead at the centre of attention. During my observation I saw numerous semi-naked female bodies in the lithotomic position (for ovary echographs, oocyte extraction, and transfers). Yet the centrality of the female body was only apparent: after an initial embarrassment, I lost interest in the body as a whole, and the biologists and the student practitioners showed me the monitors where I could see inside those bodies. The women's bodies thus merged into the background, becoming invisible and transparent, making way for images of ovaries, follicles and uteruses. On numerous occasions, the persons whom I was observing told me how to interpret the grey images on the black and white monitor. Bodies were sectioned, while technology singled out further key elements in the reproductive process: the female reproductive organs, the gametes, and finally the embryos.

[6]The patient's sexuality was only referred to in relation to abstinence from sexual activity: before examination of the seminal fluid he was asked to abstain for three to five days, because ejaulation affected the number and motility of the spermatozoa.

2. From bodies to liquids. When the pick-up was performed, the role of the ovaries became crucial, and attention focused on the follicles visible on the screen in the operating theatre. When the fluid had been aspirated from the follicles and taken to the laboratory, the central role passed to the oocytes. The latter were both the product of the women's reaction to stimulation and the material on which the biologist worked for fertilisation. These were the product of joint work by the doctor and patient's reproductive apparatus. It was impossible to know in advance how many oocytes had been produced because this was not strictly related to the number of follicles. Moreover, immature oocytes or ones unsuitable for fertilisation could be produced. The announcement of the number of oocytes was eagerly awaited: the nurse who had taken the follicular fluid from the operating theatre to the laboratory then returned to the operating theatre to call out the number of oocytes.

By contrast, the seminal fluid was produced almost secretively, as if the action was unimportant and taken for granted. The man was sent to one of the bathrooms on the second floor, alone with his sterile container, while his partner was prepared for the operating theatre. The only case in which the sperm donation attracted attention was when it did not happen or was severely delayed. When the oocytes had reached the laboratory, the availability of the seminal fluid became crucial, and given the ease with which it was produced, the center's personnel viewed any delay as inexcusable.

The male and female gametes were produced in the reverse conditions. The woman was taken to the operating theatre and underwent an invasive operation for extraction of the oocytes; while for men the process took place in one of the bathrooms. Indeed, the center's premises were gendered, and the gender distinction was perpetuated when fluids were separated from bodies.

Fluids (follicular and seminal) were brought to the laboratory and treated to isolate the gametes (oocytes and spermatozoa) and then 'transform' them into embryos. The patients and their bodies were excluded from the fertilisation process, and they rejoined their bodily products only at the moment of the oocyte transfer.

The gynecologist and the biologist had diametrically opposed relationships with the patients and their bodies: the doctor dealt with bodies, the biologist with fluids. From a logistical point of view as well, analysis of the center's physical structure shows that the fertilisation process took place in the only room that the patients were not allowed to enter: the laboratory. The sole moment of contact between the patient and the biologist occurred during the transfer, which was symbolically performed in the laboratory's anteroom.

3. Manipulation. At this point, as we have seen, the procedures for IVF and ICSI diversify. The explanation about how one technique is chosen rather than the other given to me by the head of the laboratory was that ICSI is used when there are serious problems with the seminal fluid: a scarcity of spermatozoa, scant mobility, or the absence of an acrosomal reaction (i.e., when the sperm does not produce the enzymes needed to penetrate the oocytes). Another reason for preferring ICSI is the age of the patient, because it is usually the case that older women produce few and often

poor-quality oocytes. With ICSI the exterior of the oocyte must be washed so that it can be seen more clearly and the biologist can decide whether it is at the right stage for fertilisation.

According to the head of ART laboratory:

> There is a limit on the number of sperms per microlitre beyond which you can't do IVF. It rather depends on experience; in fact you prefer not to go below 4–5 million per microlitre with a minimum motility of 40%, below that you opt for ICSI. But then everyone does what their experience tells them (...) there are some practitioners who do ICSI even with 4 million directly and others who may do IVF with 4 million, perhaps with a lower fertilisation rate, whatever (...) there are no rigid technical margins, it's mainly decided by experience.[7]

It was still unclear to me how it was established that seminal fluid was within the norm in the borderline cases, and how other factors (for instance the oocyte quality) are involved in choosing one technique rather than the other.

In IVF there is no direct control over the fertilisation of the oocytes: the work consists of assembling a certain number of gametes, placing them in the incubator (in the dark), and crossing one's fingers. In ICSI, however, the biologist has an active role, and he knows that if there have been no hitches during the operation, the chances of obtaining embryos are good. The likelihood of failure (i.e., the non-production of embryos) is relatively low and in any case manageable. For example, while performing an ICSI the head of laboratory had problems with a needle, and there was a risk that the oocyte might be damaged. To prevent this from happening, he replaced the needle, and the process concluded successfully. The expressions that he used to describe ICSI were emblematic: 'I like it more; it makes me feel more secure; it gives me satisfaction.' The head of laboratory also told me that a biologist with good manual skill and plenty of practice could achieve a 90% to 100% fertilisation rate.[8] Manual skill and practice were aspects on which the biologist could work, while other factors of vital importance for the process were beyond his control: he could not know how the patient would react to the stimulations, how many oocytes she would produce, or whether those oocytes would be mature and suitable for fertilisation.

---

[7] The italic represents direct discourse transcribed from the field notes.

[8] According to the head of the ART laboratory the egg fertilisation rate for IVF is around 70%, while for ICSI it depends more on the manual skill of the biologist who performs it, and it varies over a range of 50–95%. In spite of my requests during the period of observation the organization did not provide data on the exact number of treatments, and pregnancy rate for each technique. However, according to the head of the ART laboratory, there are no significant differences in the pregnancy rate between the two techniques. Moreover, referring to the national data reported by the Italian Ministry of Health Livia Turco to the Parliament on the enforcement of the law on assisted reproduction (retrieved October 20, 2007, from http://www.ministerosalute.it/imgs/C_17_pubblicazioni_662_allegato.pdf), during the 2005 the pregnancy rate for IVF was 26.4%, while for ICSI it was 23.9%. These data show that the reason for choosing ICSI cannot be due to a major efficiency of the technique but it is grounded on a complex ecology of elements.

4. The return to the body. The transfer was the moment when the body, after its disassembly, was reintegrated through the doctor/biologist merger. The embryos reintroduced into the patient's uterus marked completion of the process, which two weeks later might yield results completely different from the patient's point of view: a pregnancy or a failure.

This may explain the scant attention paid to the transfer, which was regarded as a practice too simple to be interesting. The embryo under the microscope was instead well worth seeing. The transfer is a key stage in the reproduction process in the sense that it is the moment when responsibility for a successful outcome passes from the biologist to the patient. If there are embryos (and almost always there are some), the biologist has done his duty, and the team must wait and see if the patient's body will let the embryos implant themselves. Responsibility is not explicitly attributed, but it is clear that everything 'technically' possible has been done, and now nature must take its course.

8.5 Conclusions: How the Body Becomes Liquid

It is impossible to think of an assisted reproduction process that does not involve a number of patients (and their bodies). However, in the everyday activities of these centers patient bodies just disappear.

After the first week of observation I recounted the story of my first experience in a centre of assisted reproduction to my colleague. When I finished my tale he just asked me: 'what about the body?' I instinctively answered: 'which body?' I could not believe it, but the body was completely deleted from my tale as it was erased from my experience in the centre.

In the process of assisted reproduction the patient body, especially the female body, is subtly and smoothly removed as the centre of the scene. During the 'embryo production' bodies are fragmented and reassembled. The organization is inscribed in, and performed through, the bodies. The attention is focused on what Alessandra Gribaldo (2005) called the 'microreproductive process', in which eggs, sperms and embryos absorb all the care. In this way the body becomes liquid and it is transformed in different things during the accomplishment of the organizational practices.

The fundamental role in the first stage of the reproductive process was performed by the ovaries and the follicles produced by them. The passage from bodies to liquids also marked a crucial transition from the doctor to the biologist. Manipulation was the phase in which it was no longer medical knowledge (central to all previous and subsequent phases of the assisted reproduction process) that was crucial, but biotechnological knowledge. The transformation of gametes into embryos took place in the laboratory under the biologist's direct responsibility: whence arose the need for control over the process. Finally, embryos are given back to the patient womb.

In the process the whole body is not the central element, and it is even unclear what can be called the body. Quoting Donna Haraway (1989), we can say that bodies are not born, they are made. What we can say following Barad's posthumanist account is that the body is the locus where discursive-material practices are enacted.

Finally, the particular epistemological status of assisted reproduction, deriving from its position at the confluence of different knowledges and technologies, ethics and institutions, shows how micro-practices and macro-processes are part and parcel of the same biopolitics.

Acknowledgements I begin with acknowledging the organization where I conducted the field-work which I cannot name for privacy reasons. I wish to thank those who gave me their time and help during the field research, and particularly the head of ART laboratory. I am grateful to Attila Bruni for our long, valuable and fruitful discussions, and for his support and advice in doing the fieldwork. Finally, I am particularly indebted to Silvia Gheradi for time and help, and I am grateful to her for support and valuable discussions in conducting the research.

References

Barad, K. (2003). Posthumanist performativity: Toward an understanding of how matter comes to matter. *Signs: Journal of Women in Culture and Society*, *28*(3), 801–831.

Barry, D., & Hazen, M.A. (1996). Do you take your body to work. In D. M. Boje, R. P. Gephart, & T. J. Thatchenkery (Eds.), *Postmodern management and organization theory*. London: Sage.

Braidotti, R. (1991). *Nomadic subjects: Embodiment and sexual difference in contemporary feminist theory* (p. 4). New York: Columbia University Press.

Butler, J. (1993). *Bodies that matter*. London: Routledge.

Dale, K. (2001). *Anatomising embodiment and organisation theory*. Basingstoke: Palgrave.

Eisenhardt, K. M. (1989). Building theories from case study research. *Academy of Management Review*, *14*(4), 532–550.

Foucault, M. (1973). *The birth of the clinic*. London: Routledge.

Foucault, M. (1977). *Discipline and punish*. New York: Pantheon.

Foucault, M. (1978). *The history of sexuality, vol. 1: The will to knowledge*. New York: Vintage.

Gribaldo, A. (2005). *La natura scomposta: riproduzione assistita, genere e parentela*. Roma: Luca Sossella Editore.

Grosz, E. (1994). *Volatile bodies: Toward a corporeal feminism* (p. 18). Bloomington/Indianapolis, IN: Indiana University Press.

Grosz, E. (1995). *Space, time, and perversion: Essays in the politics of bodies*. New York/London: Routledge.

Haraway, D. (1989). The biopolitics of postmodern bodies: Determinations of self in immune system discourse. *Differences: A Journal of Feminist Cultural Studies*, *1*(1), 3–43.

Hassard, J., Holliday, R., & Willmott, H. (Eds.) (2000). *Body and organization*. London: Sage.

Shilling, C. (1993). *The body and social theory* (p. 19). London: Sage.

Styhre, A. (2004). The (re)embodied organization: Four perspectives on the body in organizations. *Human Resource Development International*, *7*(1), 101–116.

Thanem, T. (2003). Contested and monstruos bodies. *Ephemera*, *3*(3), 250–259.

Turner, B. S. (1992). *Regulating bodies: Essays in medical sociology*. New York/London: Routledge.

Yin, R. K. (1994). *Case study reasearch: Design and methods*. London/New Delhi: Sage.

Chapter 9
What's in a Name? The Importance of Nomenclature in Biotechnology

Katherine Harrison(✉)

Abstract: In June 2006 the World Health Organisation published a review of the issues concerning International Nonproprietary Names (INN) for biotechnological products, reflecting a growing awareness of the challenges posed by novel therapies to the established scientific naming conventions. Like many aspects of the biotechnology industry, the process of naming and branding new drugs is complex and often shrouded in technical vocabulary. Examining fictional representations of biotechnology in parallel with analysis of the 'real-life' processes for product naming reveals some of the implications of nomenclature whilst rendering the scientific discourse more transparent. In this paper, I use Margaret Atwood's 2003 novel, *Oryx and Crake*, as an entry point for exploring these implications from a feminist perspective, examining biotech both as a space for critical analysis and its effects on the medical discourse surrounding women's bodies.

Keywords: Gender, fiction, biotechnology, nomenclature, Oryx and Crake

> *This point is the key to understanding the uniqueness of biotechnology today —capitalism's revolutions of production are able to reconceive of biology and of nature itself as a technology (Thacker 2006).*

This paper* marks the intersection of two worlds: my professional life as Communications Manager of a UK biotechnology company and my academic life as a research student in the Humanities. I have worked in biotech since January 2004, during which time I have been involved in a wide range of communications

Katherine Harrison
Birkbeck, University of London, UK and Linköping University, Sweden
c/o Tema Genus, hus T, Linköpings universitet, 581 83 LINKÖPING, Sweden
katherine.harrison@gmail.com

*Financial support for writing this paper was provided by the European Community under a Marie Curie Host Fellowship for Early Stage Researchers Training, and by a studentship from Birkbeck, University of London.

including presentation of clinical data at scientific congresses, investor relations and corporate branding. I started researching biotech in January 2007 as part of a wider research project focusing on intersections of gender, language and new technologies, which I have been undertaking whilst enrolled as a PhD student at the Department of English, Birkbeck, University of London, UK and as a Marie Curie Fellow at the Department of Gender Studies, Linköping University, Sweden. I have carefully selected a range of case studies that exemplify the intersection of gender, language and new technologies, and represent possible meeting-points for feminist theory and practice. Given the wealth of practical experience I have obtained, biotech presented itself as an obvious choice for inclusion in my project and has proved to be one of the most fruitful case studies to date.

The starting point for this paper was the publication in June 2006 of a review paper titled 'International Nonproprietary Names (INN) for Biological and Biotechnological Substances' produced by the World Health Organisation (WHO), which states: 'More and more new kinds of biotechnology-derived products arise, which need specific policies to steer how to deal with these products…' (WHO 2006). The primary aim of the WHO paper is to review the current guidelines for naming a new drug. However, its status as a 'living' document reflects a growing awareness of the challenges posed by novel therapies to established scientific (naming) conventions.[1] Like many aspects of the biotech industry, the process of naming and branding new drugs is complex and often shrouded in technical vocabulary. Examining fictional representations of biotech in parallel with analysis of the 'real life' processes for product naming reveals some of the implications of nomenclature whilst rendering the scientific discourse more transparent.

Biotech has been variously described as 'an evolution in basic science' (Goodfellow 2007), and 'the corporatisation of biology' (Haraway 1997). What these different definitions touch on, however, and what, for me, makes biotech so interesting is the merging of cutting-edge scientific academic research with the commercial sphere in a way that changes both originating cultures. It is this integration, this combination of fields which is perhaps the most distinctive feature of biotech, as Eugene Thacker notes in his introduction to *The Global Genome*:

> Bioscience research and the biotech industry are increasingly organized on a global level, bringing together novel, hybrid artifacts (such as genome databases and DNA chips), with new means of distribution and exchange (most notably, the use of the Internet in exchanging biological data) (Thacker 2006).

Thacker's definition of biotech as a product of biology and information technology foregrounds the theme of hybridity, a hybridity which has spawned a plethora of definitions of the industry, and which sees biotech drawing on many different existing practices and discourses. Many of these are specialist discourses, for example those associated specifically with clinical trials, and also regulatory, ethical, and financial discourses, together with the highly technical narratives related to the technologies used in the laboratory to discover new products. It is, therefore, no surprise

---

[1] In the section titled 'Current Challenges' of the WHO review paper the following point is made: 'The nomenclature of the biological medicinal products is an area of increasing complexity' (WHO 2006).

that this industry regularly invents vocabulary to name new bio-technologies, new drugs, new mechanisms of action for drugs, and new aspects of the development process. The process and consequences of naming within this industry is the focus of this paper. Before turning to the naming process itself, however, I would like to touch on some of the relevant discourses and issues that frame my investigation.

Unlike 'traditional' medicine, biotech harnesses the natural processes and mechanisms of the body to develop products that mimic or enhance these processes, and in so doing starts to disrupt the boundary between natural and unnatural, as Eugene Thacker notes:

> The very concept of a biotechnology is thus fraught with internal tensions. On the one hand, the products and techniques of biotech are more "tech" than "bio"; biology is harnessed from its "natural" state and utilized in a range of industrial and medical applications. On the other hand, there is no "tech", only "bio"; the unique character of the technology is that it is fully biological, composed of the workings of genes, proteins, cells, and tisues. (...) The advantage claimed for biotechnology is that it is more natural, a direct working with "life itself" (Thacker 2006).

This unstable, hybridised identity makes biotech a fruitful area for investigation; by neither falling into the category of technology, nor that of biology, biotech poses some interesting questions about the bodies on which biotechnological medicines are used, perhaps even redefining them as a kind of technology? This has broader implications for identity, at which Rosalyn Diprose hints in her article, 'A 'Genethics' That Makes Sense', 'biomedical science has a role in the constitution of our being as a discrete entity and is not just a mode of reparation of that being' (Diprose 1995). Diprose here points out that what biomedical science does, and how it conceptualises the body, affects the boundaries of the self [2] and by extension raises a host of questions pertaining to creation and maintenance of identity. Is biotech, with its new technologies and increased focus on working in harmony with the body's own processes, constituting our being in a different way to traditional medicine? How is it different? Does it somehow make this type of treatment more acceptable? And, what effect, if any, does this have on women's experiences of this new type of medicine?

I found Susan Sontag's comments on cancer in *Illness as Metaphor* (1979) to be a useful counterpoint to the above premise. Sontag describes how doctors call cancer cells 'nonself', the 'unnatural' growth of malignant cells challenging the idea of the limits of the physical body representing the boundary of self. Many cancer patients describe their cancer as 'alien' (*Ovacome* 2007), suggesting that this distinction is made by both health professionals and patients alike. By invading the healthy space of the body, a patient's sense of self appears to be challenged by her/his disease.

These discourses invoke a traditional rhetoric of natural/unnatural, but in this case the 'man-made' biotechnologies become more natural than the patients' own (diseased) bodies. There is a confusion here which is characteristic of much of the

[2] Here she references David Schenk's premise (following Merleau-Ponty) that the body 'is literally our selves expressed', from David Schenk, 'The Texture of Embodiment: Foundation for Medical Ethics', quoted in Diprose (1995).

writing on biotech, and demonstrates that current critical discourse is somehow insufficient when it comes to capturing or reflecting the new challenges posed by biotech. Inevitably with so many important issues involved, a simple Google search demonstrates that popular perception of biotech is also still very mixed (partly because 'biotech' includes so many things from GM food to stem cell research to radiolabelled antibodies). Opinions range from *Wired* magazine's hyping of all things technologically novel – the April 2007 issue includes their list of the 40 most innovative companies in the world (The Wired 40 2007), in which biotech company, Genentech, ranks at number 3, with the line: 'When you target specific biological mechanisms, your drug can side-step the one disease rut: Avastin has been OK'd for a growing list of cancers'. Meanwhile, British tabloid newspapers such as the Daily Mail still run headlines such as 'GM blunder contaminates Britain with mutant crops' (Poulter 2002).

A number of the concerns seen in the press are also reflected in science fiction texts about biotech,[3] and Margaret Atwood's novel, *Oryx and Crake*, is no exception to this (Atwood 2003). Her suspicion of the industry thus serves as a counterweight to my own (potentially positive) bias, developed during four years of working for a UK biotech company. *Oryx and Crake* depicts a world destroyed by a human-made biological virus distributed in vitamin tablets by a biotech/pharmaceutical company. Through a series of flashbacks narrated by one of the few survivors, the events leading up to the global disaster are revealed. This novel paints a dystopian picture of the effects of biotech when controlled by a powerful few, and incorporates many of the themes of fear and distrust seen in popular press coverage of the biotech industry.

Atwood's other novels often feature women as the main protagonist, and subtly question gender rules in society, as, for example, in the well-known *The Handmaid's Tale* (Atwood 1985). Atwood's work also often deals with history, including family histories. For example, *Alias Grace* (Atwood 1996) uses many different sources of information on the trial of the central character, Grace, but it remains unclear throughout which one is the 'real' representation of what happened, while *The Blind Assassin* (Atwood 2000) uses a story within a story to reveal different layers of narrative in the novel. Methods such as these undermine the authority of the narrative and leave it open to question. In *Oryx and Crake* Atwood uses these techniques to good effect in her focus on the biopharmaceutical industry. Her trademark focus on language and gender, as outlined above, is also clear in this novel.

Science fiction has long been interested in futuristic biological experiments, and it is no surprise to see mention of biotech appearing in science fiction from the 1990s onwards. Notable amongst these are Justina Robson, Tim Pears and Greg Egan. Lately, more mainstream writers, such as Margaret Atwood, have also turned their attention to biotech. Science fiction not only reflects popular hopes and fears,

---

[3] As Russell Blackford points out in his article, 'Beyond the Frankenscientist: Biotech and Biomedical Themes in recent Australian Science Fiction' (2002), science fiction stories about biotech are almost exclusively dystopian in outlook.

but it also has the luxury of imagining the future. Furthermore, whilst the popular press may be limited to reporting 'facts', science fiction is at liberty to exaggerate and experiment, testing the limits of these facts. In doing so, we learn more about the underlying concerns of the author and contemporary popular culture. However, this is not to say that science fiction has no weight outside the realm of fiction, as Russell Blackford comments:

> Gregory E. Pence, an American philosopher, has argued that science fiction (sf) has portrayed human cloning, in particular, in a negative light, contributing to the fear and repugnance that now prevail in the general and political communities. While some of Pence's specific criticisms and examples could be challenged, his general point about sf and cloning is persuasive. Works such as *Spares* and *The 6th day* show sf continuing to encourage and feed upon this repugnance. When future uses of biotechnology are in contemplation, much sf actually seems to be *anti*-science fiction (Blackford 2002).

I shall use Atwood's fiction as a way into examining the network of factors which have resulted in the current dominance of biotech. This approach also has parallels with the historicist framework used by media theorist, Friedrich Kittler,[4] and seeks to recognise that biotech is very much a product of its time. At the same time as ARPANET (the precursor of today's Internet) was first emerging, antibodies were being cloned for the first time and the techniques which were to enable closer study of the genome were being developed. Professor Julia Goodfellow, previously of the UK Biotechnology and Biological Sciences Research Council locates it thus:

> So, it's really about can you produce new things from biology. It came about through a revolution in basic science, or an evolution in basic science actually in the 1960s, 1970s, 1980s, typically with bacteria, and being able to manipulate bacteria so you could produce different products (Goodfellow 2007).

Ten years after the emergence of early bio-technologies and the Internet, we also see the start of cyberfeminism, with the publication of Donna Haraway's 'A Cyborg Manifesto' and William Gibson's classic cyberpunk text *Neuromancer* both drawing on new technologies and exploring possibilities opened up by biotechnologies and computing technologies. In this sense, biotech is a very postmodern kind of business.

Biotech is the high-profile convergence of biology and entrepreneurial skill. Nourished by academic discoveries, it is at the forefront of developing and testing novel therapies, which include not only new medicines based on existing concepts of treatment, but completely new ideas for drugs, with novel mechanisms of action. The novelty and fast-paced environment of the biotech industry has posed new challenges to existing discursive, legal and biological frameworks and also to existing

[4] In his foreword to the 1990 English-language edition of Kittler's *Discourse Networks 1800/1900*, David E. Wellbery describes Kittler's approach thus, 'Hermeneutic understanding is not at all what human beings always do with written or spoken texts, it is not a foundational condition for the processing of significant marks. Rather, it is a contingent phenomenon within the evolution of discursive practices in Europe; it rests on a host of preconditions such as alphabetization, the expansion of book production, the organization of the modern university…' (Kittler 1990).

conceptions of health. I will now examine the current process for naming a new product, prior to turning to Atwood's novel *Oryx and Crake* to show how she uses the fictional narrative to reveal the far-reaching implications of this process.

During development a new drug will have many different names. It will usually be assigned a code number during preclinical development when it is still being tested in the lab. When the drug enters clinical trials in humans, and becomes 'visible' to a new, wider audience including doctors and patients, it will be assigned a new name/number. If successful, the drug will be allocated a further two names during its development process – an International Nonproprietary Name (INN) and a brand name. This list of names may grow further if the product is developed in partnership with another company or the biotech company is acquired.

The names held by a drug during its development reflect the stage of development, the mode of action, the structure of the molecule, the company, or the family of medicines to which that particular product belongs. For example, 'DMXAA' was the preclinical name given to a drug developed at the University of Auckland[5]; this name is simply an abbreviated form of the chemical structure of the molecule: 5,6-dimethylxanthenone-4-acetic acid. During clinical development a drug's name is usually made up of a combination of letters and numbers, as in the case of ACAM 2000[6] where ACAM is an abbreviated form of the company's name (Acambis) and the numbers reflect the drug's place in the development pipeline. An INN contains a 'stem' suffix or prefix common to all drugs that share the same mechanism of action, for example all monoclonal antibodies share the INN suffix – mab. Finally, brand names often make reference to the company selling them, as in the case of Rocaltrol and Roaccutane, both developed and sold by Roche. None of these names, therefore, are arbitrary choices but, instead, unique identifiers laden with commercial and scientific information.

INN are proposed by the company or individual developing a product, submitted to the WHO for review and finally receive approval. These names are also referred to as the 'generic' name for a product, and as such always appear in lower case, for example erlotinib. An INN acts as a unique name which can be used worldwide for a product and is not owned by the company/individual who proposes it (unlike a brand name, for example). The concept of an INN was first devised in 1950 by the World Health Assembly with clear goals that benefited both patients and scientists:

> The existence of an international nomenclature for pharmaceutical substances, in the form of INN, is important for the clear identification, safe prescription and dispensing of medicines to patients, and for communication and exchange of information among health professionals and scientists worldwide (WHO 1997).

The WHO also imposes strict rules governing the creation of an INN: 'International Nonproprietary Names (INN) should be distinctive in sound and spelling. They should not be inconveniently long and should not be liable to confusion with names

---

[5] A vascular disrupting agent being developed by biotech company, Antisoma.

[6] A smallpox vaccine being developed by biotech company, Acambis.

in common use' (WHO 1997). A marketed product will have a brand name and an INN, both of which must appear on the packaging of a product although the INN often appears in much smaller type than a brand name.

Inevitably, as the drug moves towards marketing the commercial considerations become more apparent in the names; brand names are chosen to be catchy and easy for doctors to remember when prescribing, whereas INN (or generic) names are often longer or more complicated, and reflect the type of drug. When a drug is no longer protected by its patent, generic drug companies are free to produce this drug, and market it under its generic or INN name, undercutting the pricing of the original development company. Therefore, generic names are often designed to be difficult to say or spell, in the hope that doctors will remember the brand name rather than the generic name when prescribing. Compare, for example, the generic name bevacizumab, with its brand name, Avastin.

Following Latour, I'd like to suggest that something like a web of 'actants' surround this process of naming a new drug, whose combined input produces the 'knot' that is the name of a product. Actants in this case include scientists, doctors, patients, lawyers, marking consultants, drug development companies, mice, primates, clinical trials, regulatory agencies, WHO guidelines and the Chemical Abstracts Service, as well as the geographical location of the company, the wishes of collaborative partners and broader industry trends.[7] Every time a name is used publicly it is further validated and enters a complex system of peer review and cross-referencing by other scientists. The citing of publications about the drug causes its name to become better known within the scientific community and locates it in relation to other drugs. This is particularly important when the drug is still in trials and is therefore not a fully realised, and marketed, medicine. The power of this process should not be underestimated, as Latour highlights in *Science in Action*: 'The presence or absence of references, quotations and footnotes is so much a sign that a document is serious or not that you can transform a fact into fiction or fiction into fact just by adding or subtracting references' (Latour 1987). The process of raising awareness of a new drug is commercially important; it is essential to make doctors aware of a new drug as early as possible, so the more occasions on which they see the name, the better the chances of commercial success. Publications and presentations about the drug make it more concrete and formalise its existence. Therefore, the commercial strength of a drug lies very much in its links or associations through which it gains scientific validity, kinship and commercial value. This applies to both the name of the drug itself and also the narrative of its development.

[7] In *Science in Action*, Latour (1987) uses this strategy to show how different groups' interests converged in Pasteur's idea of the microbe: 'When Pasteur and the hygienists introduced the notion of a microbe as the essential cause of infectious disease, they did not take the society to be made up of rich and poor, but of a rather different list of groups: sick contagious people, healthy but dangerous carriers of the microbes, immunised people, vaccinated people, and so on. Indeed, they added a lot of *non-human actors* to the definition of the groups as well: mosquitos, parasites, rats, fleas, plus the millions of ferments, bacteria and other little bugs.'

Turning now to an examination of how Atwood's biotech fiction picks up on some of these same concerns about naming and validation, I'd like to suggest that Atwood uses a number of different techniques to critique the biotech industry. For example, she plays with grafting names, populating *Oryx and Crake* with companies called 'AnooYoo', 'RejoovenEsense', 'OrganInc Farms', which produce products such as 'Rockulators', 'pigoons' and 'ChickieNobs'. These names produce a welcome light-hearted moment in the middle of the dystopian world view, whilst re-emphasising the heavily commercialised aspect of the industry, as is clear in her description of Rockulators:

> Huge fake rocks, made from a combo-matrix of recycled plastic bottles and plant material from giant tree cacti and various lithops—the living-stone members of the Mesembryanthemaceae — were dotted here and there. It was a patented process, said Crake, originally developed at Watson-Crick and now a nice little money-spinner. The fake rocks looked like real rocks but weighed less; not only that, they absorbed water during periods of humidity and released it in times of drought, so they acted like natural lawn regulators. Rockulators, was the brand name. You had to avoid them during heavy rainfalls, though, as they'd been known to explode (Atwood 2003).

In this extract Atwood moves from the science behind the object (including its development and scientific name) to commercialisation (seen in the list of selling points and catchy brand name). This narrative is thus very much in keeping with the current process of naming a new biotech product. The last line of the extract, however, disrupts the expected narrative. Atwood instead undermines the integrity and authority of science, destroying the gloss of the marketing with the words 'they'd been known to explode'.

These 'grafted' names are also a nod at the hybrid nature of the biotech industry itself, its own name conjuring up both biological sciences and information technology, as Donna Haraway comments in her article 'Biopolitics of Postmodern Bodies':

> The words for the overlapping discourses and their objects of knowledge, and for the abstract corporate names for the concrete places where the discourse-building work is done, suggest both the blunt foreshortening of technicist approaches to communication and the uncontainable pressures and confusions at the boundaries of meaning within 'science' — biotechnology, biomedicine, psychoneuroimmunology, immunogenetics, immunoendocrinology, neuroendocrinology, monoclonal antibodies, hybridomas, interleukins, Genentech, Embrex, Immunetech, Biogen (Haraway 1991b).

Furthermore Atwood parodies the process of validation, narrated through the main protagonist, Jimmy, who prior to the global virus was a marketing executive who wrote copy supporting launch of new drugs:

> It was his task to describe and extol, to present the vision of what — oh, so easily! — could come to be. Hope and fear, desire and revulsion, these were his stocks-in-trade, on these he rang his changes. Once in a while he'd make up a word – *tensicity, fibracionous, pheromonimal* — but he never once got caught out. His proprietors liked those kinds of words in the small print on packages because they sounded scientific and had a convincing effect (Atwood 2003).

Through Jimmy, Atwood emphasises the power of language, while at the same time firmly disconnecting it from the truth.

Biotech is a rapidly evolving field, rich with new concepts of medicine. A focus on biotech allows us to see the process of naming in action, and, associated with that, the process of creating a history and a family for a particular medicine, from the very early development period, during which there is more scope for changing names and from which, in the space of months, the mechanism of action can easily morph from 'vascular targeting agent' to 'vascular disrupting agent'[8] driven by little more than a single article in a journal. During this period, the way in which the drug works in the body may not be fully understood so multiple, conflicting narratives may emerge, conflict and subsume one another. However, this indecision and multiplicity disappears later as mistakes are erased/forgotten to shore up the authority of the drug narrative when it is being marketed. Because of the fast-paced and informal culture of biotech companies these decisions are more likely to be made 'on the spot', rather than going through a slower, more formal and hierarchical process as experienced with a pharmaceutical company. This need for a different approach is recognised in recent guidelines from the WHO on INN for biotech products. The creation of this separate document solely for biotech reflects the difficulties inherent in naming these new bio-technologies, or at least the difficulties as perceived by a regulatory body.

In her fascinating article, 'Product, process, or programme: three cultures and the regulation of biotechnology', Sheila Jasanoff examines the regulatory framework surrounding biotech developed by the UK, the US and Germany during the 1980s. In particular, she focuses on how each country chose to control the release of new biotech products into the environment, and how they managed public concerns about these novel substances:

> Through the vehicle of regulation, states provide assurance that the risks of new technologies can be contained within manageable bounds. Procedures are devised to limit uncertainty, channel the flow of future public resistance, and define the permissible modalities of dissent. Regulation, in these respects, becomes integral to the shaping of technology (Jasanoff 1995).

In comparing the three distinct approaches taken, Jasanoff illustrates the different types of risks considered to be posed by biotech: physical, political and social. This comparison highlights flaws in the different approaches, and also suggests that these were developed in an almost arbitrary manner. As can be seen in the extract above, Jasanoff is keen to make clear the long-term impact of these decisions on the discourse of biotech in that particular country.

These official frameworks effectively act as the 'creation story' of biotech, one in which new and exciting products are positioned by regulatory authorities as being 'sufficiently close to their prior experience of technological risks to permit effective public control' (Jasanoff 1995). This paternalistic approach by the state is echoed in the scientific explanations of the pigoon project offered to Jimmy by his father in *Oryx and Crake*.

> The pigoon organs could be customized, using cells from individual human donors, and the organs were frozen until needed. It was much cheaper than getting yourself cloned for

---

[8] This example is taken from the discourse I witnessed surrounding the development of a drug called ASA404 (previously AS1404).

spare parts – a few wrinkles left to be ironed out there, as Jimmy's dad used to say – or keeping a for-harvest child or two stashed away in some illegal baby orchard. In the OrganInc brochures and promotional materials, glossy and discretely worded, stress was laid on the efficacy and comparative health benefits of the pigoon procedure (Atwood 2003).

Atwood immediately undermines the calm, authoritative tone of this paragraph by following it with one that starts 'to set the queasy at ease, it was claimed that none of the defunct pigoons ended up as bacon and sausages'. Throughout she uses humour to undermine the authority of narratives of power such as science or religion, as her parody of the creation myth aptly demonstrates. In the following extract, Jimmy (who is referred to as 'Snowman' following the virus) is telling the genetically-modified people, the Crakers, his version of the creation story after they have caught a fish for him:

A story is what they want in exchange for every slaughtered fish. Well, I owe them, Snowman thinks. God of Bullshit, fail me not.
'What part would you like to hear tonight?' he says.
'In the beginning', prompts a voice. They're fond of repetition, they learn things by heart.
'In the beginning, there was chaos', he says (Atwood 2003).

Jimmy's own disillusionment and the Crakers' naive belief in his creation stories call into question 'creation' stories of our own society, including those produced by science. The Crakers themselves further undermine this story – the reader knows that they are the product of a genetic experiment. *Oryx and Crake* is the story of Noah's Ark turned on its head – a world destroyed by science out of control, leaving only a handful of survivors including a new, improved human race (the Crakers), who are the product of genetic experimentation that has given them characteristics of other animals, from jellyfish to baboons.[9]

The act of naming, and stories about naming such as the creation myth, has particular resonance for feminist theory and Atwood is keen to stress the importance of this. The scientist Crake plays the role of the creator and it is through Crake's rules that Atwood makes plain the implications of naming:

It was one of Crake's rules that no name could be chosen for which a physical equivalent – even stuff, even skeletal – could not be demonstrated. No unicorns, no griffins, no manticores or basilisks. But those rules no longer apply, and it's given Snowman a bitter pleasure to adopt this dubious label. The Abominable Snowman – existing and not existing, flickering at the edges of blizzards, apelike man or manlike ape, stealthy, elusive, known only through rumours and through its backward-pointing footprints (Atwood 2003).

To name something is a means of claiming ownership, categorising or controlling. This is most clearly seen in the adoption of the moniker 'Crakers' for the genetically modified people developed by Crake. He is thus transformed into a father, or god-like figure, with a clear power dynamic between the two groups. Crake's bio-power

[9] For example, see *Oryx and Crake:* 'The people move closer, men and women both, gathering around, their green eyes luminescent in the semi-darkness, just like the rabbit: same jellyfish gene' (Atwood 2003).

over the Crakers' bodies also suggests parallels with the paternalistic attitude of the medical establishment and the discourse relating to patients' bodies and symptoms. Naming this group also marks them as a distinct category or type, and it is this drawing-up of new categories that is one of the broader consequences of new biotechnologies – and one which Atwood's novel certainly touches on.

I'd like to follow this line of thought by considering how new biotechnologies change not only our idea of what constitutes 'medicine' but also what constitutes 'disease' or 'health'. The move away from a generalised approach to one which involves identifying the cells responsible for certain diseases, establishing how they are different from healthy cells and developing medicines which can target these cells, has resulted in a very narrow focus on particular aspects of the body. This evolution, in parallel with a growing interest in the gene, has changed the location of disease from the whole body to specific areas or organs, but also has allowed doctors to identify those at risk earlier on in their illness and thus administer preventative medicine which improves long term chances. Examples of this include identification of biomarkers, use of antibodies as targeted therapies for cancer and the idea of genes that 'cause' certain diseases. I would like to suggest that this has two effects particularly relevant to feminist readings of biotech: a disassociation from the rest of the body and an earlier diagnosis of 'at risk' status. In principle, neither of these appears to be a problem. On first glance, they suggest that treatment of patients is improving thanks to advances in science and medicine. However, the difficulty arises when these two issues become gendered. In her recent article, Kelly Happe examines this possibility in relation to ovarian cancer. Mutations in the genes BRCA1 and BRCA2 have been strongly linked to increased risk of breast and ovarian cancers,[10] and in her article 'Heredity, gender and the discourse of ovarian cancer', Happe suggests that the new technologies used for identifying genes are 'incorporated into old frameworks for defining, identifying, and treating disease' (Happe 2006). Starting from a similar premise to the one advocated by Donna Haraway – that scientific discourse, whilst appearing neutral, actually naturalises social structures – Happe examines the discourse of ovarian cancer and the role that oophorectomy (removal of the ovaries) plays in determining risk of disease.

The treatment recommendations for healthy women at risk for ovarian cancer because of the presence of a BRCA mutation suggest that, in this discourse, risk *is* the disease, a slippage that is theoretically and rhetorically possible because of the implicit importance in our culture attributed to heredity (Happe 2006).

Happe's focus on the way in which language works in this situation makes her article particularly useful here. Later in the same article, she analyses an extract from an article published in *Gynecologic Oncology*, and demonstrates that the validated discourse of medicine actively encourages women to change their pattern of child-bearing to protect themselves from ovarian cancer. However, as Happe points out, implicit within this advice is the assumption that child-bearing represents the

[10] 'According to clinical data, women with BRCA mutations face anywhere between a 10% to 60% chance of developing ovarian cancer depending on which mutation they inherit' (Happe 2006).

fulfilment of a woman's existence, and thus more attention is given to this than the risk of cancer or the issue of experiencing very early menopause due to oopherectomy. Referring to Ehrenreich and English's 1978 text *For Her Own Good: 150 Years of the Experts' Advise to Women*, Happe suggests that:

> This bears a great similarity to turn-of-the-century theories about the impact of reproductive organs on a woman's well-being. Physicians believed that a woman's uterus – present exclusively for the propagation of the species – could in turn be the source of a range of illnesses if a woman did not procreate (Happe 2006).

Ethical guidelines require efficacy of treatment to be balanced with consideration of the quality of life of the patient undergoing treatment. However, as Happe's example shows, 'quality of life' is determined by doctors or medical institutions and may not reflect the authentic wishes of the patient. In the case above, effective treatment may be sacrificed to a perceived imperative to have children, which appears to equate to 'quality of life' for women of child-bearing age. In this way the medical establishment controls what constitutes 'life' for women in a way that reinforces traditional gender roles and which clearly demonstrates an underlying essentialist approach within the medical profession.

Similar concerns emerge in Bettina Leysen's article, 'Medicalisation of menopause: From "Feminine Forever" to "Healthy Forever"', in which she critiques the focus on reproductive ability as determining personhood for a women, and notes the way in which commercial imperatives behind treatment become naturalised and validated.

Recently, hormonal replacement therapy has been promoted for the prevention of cardiovascular diseases, the chief cause of death among women in all industrialized countries. A critical analysis of the evidence of the protective effect is unconvincing. Nevertheless, at least in Belgium, organizations have been founded with the goal of informing the general public about the many advantages of hormonal replacement therapy. Needless to say, they are financed by the pharmaceutical industry (Leysen 1996).

Published in the mid-1990s, Leysen's article captures an important historical moment where the extension of HRT to a wider group of women was being attempted, and her article lists the justifications given for this wider application. HRT, however, has proved to be a controversial topic. Recent studies have examined its effects in relation to both ovarian cancer and heart disease, and suggest that taking HRT causes a significant increase in the former (Boseley 2007), with the latter dependent on age at time of starting HRT (Hall 2006).

In her conclusion, Happe locates the discourse of identifying genetic markers for these cancers within the broader economic framework in a move that reminds us of the commercial drivers behind development of interventionist biotechnologies.

In the case of ovarian cancer, the potential link between environmental carcinogens exposure and incidence of the disease is virtually unexplored while biomedicine pursues genetic interventions that will very likely take several decades to come to fruition, if at all. This tension, and ultimate trade-off, between the genomics model of disease and the environmental health model is the result of the competing political perspectives they both represent and help articulate, if only indirectly (Happe 2006).

Suspicion of new biotechnologies on the basis of commercial gain such as this is reflected in both critical writings and the press. Atwood also integrates this suspicion into her novel *Oryx and Crake* in various ways, ranging from her ironic use of brand names to more open criticism voiced through the character of Jimmy.

Happe aptly demonstrates the connection between new bio-technologies and changing conceptions of health. However, both her own and Leysen's reading suggest that the underlying discourses and value systems remain unchanged by the new technologies. Instead, existing gender stereotypes are reinforced under the guise of scientific achievement, and, in the case of genetic markers, something very like essentialism re-emerges. The development narratives of biotech lead to new types of drugs, new ideas of disease and new categories of patient, but may not disrupt deep-rooted assumptions about gender – these are the broader implications of the discourse of biotech of which we should be aware.

9.1 Conclusion

In this paper, I have tried to demonstrate the importance of nomenclature in biotech. In particular, I have sought to do two things: firstly, to reveal the situated and contingent nature of the naming process by showing the range of vested interests; secondly, to suggest some of the broader implications of naming.

As Wendy Harcourt noted (see Chapter 1), although there are some success stories in terms of women's participation in science, overall there is still a serious disparity; 'across the EU as a whole only 29% of scientists are women'. In my paper, I have tried to point out other aspects of the biotech industry where women's interests remain under-represented, and unnoticed, namely in the *discourse* of biotech, an activity which often involves non-scientists. Problematising the practice of nomenclature in this way is my attempt to raise awareness of these blind spots.

My own position, with a foot in two camps (biotech employee and also feminist researcher), prevents me from reading biotech as either simply good or bad, as I believe in both the good that biotech can do, whilst worrying about (and working against) the bias of its practice. Difficult though this internal tension may be, I believe it may also be useful in actively preventing me from reaching a definitive conclusion, a partial and contradictory standpoint which I hope would be in line with Donna Haraway's views on situated knowledges:

> I am arguing for politics and epistemologies of location, positioning, and situating, where partiality and not universality is the condition of being heard to make rational knowledge claims. These are claims on people's lives; the view from a body, always a complex, contradictory, structuring and structured body, versus the view from above, from nowhere, from simplicity. Only the god-trick is forbidden (Haraway 1991a).

As such, my conclusion is not really a conclusion but instead an attempt to encourage others to explore some of the questions raised here. To return to my title: what *is* in a name? Well, in short, the issue of nomenclature in biotech (from product name to development narrative) has both implications for women in terms of the treatment they might receive as patients and also, from a feminist standpoint, as a neglected area in need of further critical analysis.

What's in a name? that which we call a rose
By any other name would smell as sweet
W. Shakespeare, Romeo and Juliet, Act II, Scene II

References

Atwood, M. (1985). *The handmaid's tale*. Toronto: McClelland & Stewart.

Atwood, M. (1996). *Alias grace*. New York: Talese Doubleday.

Atwood, M. (2000). *The blind assassin*. London: Bloomsbury.

Atwood, M. (2003). *Oryx and crake*. London: Bloomsbury.

Blackford, R. (2002). Beyond the frankenscientist: Biotech and biomedical themes in recent Australian science fiction. In D. Pastourmatzi (Ed.), *Biotechnological and medical themes in science fiction* (pp. 335–350). Thessaloniki, Greece: University Studio Press.

Boseley, S. (2007, 19 April). HRT linked to 1000 deaths from cancer. *Guardian online*. Retrieved April 28, 2007, from http://www.guardian.co.uk/frontpage/story/0,,2060508,00.html

Diprose, R. (1995). A 'Genethics' that makes sense. In V. Shiva & I. Moser (Eds.), *Biopolitics: A feminist and ecological reader on biotechnology* (pp. 162–173). London: Zed Books.

Goodfellow, J. (April 2007). *Interview with Katherine Harrison*. Unpublished.

Hall, S. (2006, 22 December). Study rejects link between HRT and heart disease. *Guardian online*. Retrieved April 28, 2007 from: http://www.guardian.co.uk/medicine/story/0,1977416,00.html

Happe, K. E. (2006). Heredity, gender and the discourse of ovarian cancer. *New Genetics and Society, 25*(2), 171–196.

Haraway, D. J. (1991a). Situated knowledges: The science question in feminism and the privilege of partial perspective. In *Simians, cyborgs and women: The reinvention of nature* (pp. 183–201). London: Free Association Books.

Haraway, D. J. (1991b). The biopolitics of postmodern bodies: Constitutions of self in immune system discourse. In *Simians, cyborgs and women: The reinvention of nature* (pp. 203–230). London: Free Association Books.

Haraway, D. J. (1997). *Modest_Witness@Second_Millenium: FemaleMan©_Meets_Oncomouse™: Feminism and Technoscience*. New York and London: Routledge.

Jasanoff, S. (1995). Product, process, or programme: Three cultures and the regulation of biotechnology. In M. Bauer (Ed.), *Resistance to new technology: Nuclear power, information technology and biotechnology* (pp. 311–331). Cambridge: Cambridge University Press.

Kittler, F. (1990). *Discourse networks 1800/1900*. (M. Metteer & C. Cullens (Trans.) and D. E. Wellbery (Ed.)). Stanford: Stanford University Press.

Latour, B. (1987). *Science in action: How to follow scientists and engineers through society*. Cambridge, MA: Harvard University Press.

Leysen, B. (1996). Medicalization of menopause: From 'Feminine Forever' to 'Healthy Forever'. In N. Lykke & R. Braidotti (Eds.), *Between monsters, goddesses and cyborgs: Feminist confrontations with science, medicine and cyberspace* (pp. 173–191). London: Zed Books.

Ovacome magazine, Spring 2007.

Poulter, S. (2002, 16 August). GM blunder contaminates Britain with mutant crops. *Daily Mail*. Retrieved April 9, 2007, from http://www.dailymail.co.uk/pages/live/articles/news/news.html?in_article_id=133672&in_page_id=1770

Sontag, S. (1979). *Illness as metaphor*. London: Allen Lane.
Thacker, E. (2006). *The global genome: Biotechnology, politics and culture*. Cambridge, MA and London: MIT Press.
WHO (World Health Organisation) (1997). Guidelines on the use of International Nonproprietary Names (INNs) for pharmaceutical substances. Retrieved July 9, 2007 from: http://whqlibdoc.who.int/hq/1997/WHO_PHARM_S_NOM_1570.pdf
WHO (2006). International Nonproprietary Names (INN) for biological and biotechnological substances. Retrieved July 13, 2007, from http://www.who.int/medicines/services/inn/BioRevforweb.pdf
The Wired 40: Our 10th Annual List of the most innovative companies in the world (2007). *Wired, 15*(4), 118–130.

Chapter 10
Egg Donation in the UK: Tracing Emergent Networks of Feminist Engagement in Relation to HFEA Policy Shifts in 2006

Alexandra Plows(✉)

Abstract: This paper provides a critical account of recent controversial UK policy (Human Fertilisation and Embryology Authority; HFEA) moves, and public responses, in relation to the sourcing and use of human eggs for biomedical research; specifically egg 'donation', for cell nuclear transfer (CNT). The paper will focus primarily on growing feminist criticism, in terms of (health) risks and other issues, such as concerns over the remit of public engagement and the policy process; the possible commodification of women's bodies and issues of globalised political economy; broader socio-cultural concerns such as the stigma of infertility; and ethical problems with the concepts of 'informed consent' and 'informed choice'.

As part of a broader research project,[1] 'embedded' qualitative ethnography has traced and engaged with developing networks of opposition as they have emerged in 'real time'. Some feminists have formed 'strange bedfellow' alliances with pro-life groups (Hands Off Our Ovaries!); others have seen this as counter-productive and aimed to catalyse 'organic networks' of opposition through information dissemination and networking. The egg donation issue is just one of several linked arenas where it appears that its on women's bodies that some of the most significant moves in relation to genetic and reproductive bioscience will be having most of an impact. It is also important to emphasise the existence of complexity and ambivalence, and to avoid polarised 'pro and anti' positions – to voice concerns about these issues is not to be 'anti-science' or 'anti-cures' or indeed even necessarily anti-(embryonic) stem cell research per se, but to be alert to potential issues and impacts. Identified risks and purported benefits need far more unpacking.

Keywords: HFEA, egg donation, feminism, risk

Alexandra Plows
Centre for the Economic and Social Aspects of Genomics (Cesagen), Cardiff University, 6 Museum Place, Cardiff CF10 3BG, UK
plowsa@cardiff.ac.uk

[1] http://www.cesagen.lancs.ac.uk/research/projects/newgentechs.htm

F. Molfino, F. Zucco (eds.) *Women in Biotechnology,*

10.1 Introduction

The data for this paper has been taken from a broader project studying emerging (UK) public engagement across a broader arena of human (medical) genetic science, which includes public engagement with biobanks, nanotechnology, for example. This research took ethnographic 'snapshots' at a range of sites of public engagement. This has been a challenging research field; the sheer complexity of public engagement has been a core finding as multiple actors move 'beyond pro and anti', mobilising over multiple frames, stakes, identity fields, in a multiplicity of sites and contexts (Evans et al. 2007; Plows and Boddington 2006; Welsh et al. 2007). The emergent nature of the research – as these technologies are only just starting to impact the public sphere in terms of specific applications – necessitated the identification of 'early risers' (Tarrow 1998) and 'prime movers' (McAdam 1986) who are still grappling with framing the key stakes of engagement in a complex field. Such mobilisation is extremely diverse. Many 'latent' (Melucci 1996) publics are civil society actors predisposed to mobilise, whose discursive frames are converging in this new arena (Nelkin 1995; Plows 2008). They represent just some of the many publics engaging with bioscience, co-constructing meanings, and so on, who also include scientists, policy makers, and academics. Human/medical genetics, genomics is thus a multiple issue site: in terms of emergent public engagement, this is triggering complexity and ambivalence, because there are many lines of engagement, many different groups of actors with competing and converging knowledge stakes (Harcourt 2007).

This research enabled an identification of shifts in relation to UK HFEA policy[2] on human egg (oocyte) sourcing for Cell Nuclear Transfer (CNT) or cloning, as an emergent catalyst or trigger moment for public mobilisation in the UK, as it developed in 'real time' in 2006/7. Public engagement, theories of mobilisation and social movements, Public Understanding of Science (PUS) and Science and Technology Studies (STS) issues and theories, are developed here especially in relation to the significance of the HFEA policy moves as a catalyst for public action, identifying emergent networking and mobilisation repertoires of multiple publics, feminists in particular. This paper is an ethnographic narrative, providing qualitative data analysis of emergent actor responses to specific policy shifts, which aims to contribute to a broad and well established debate on feminism, women, reproduction, gender and science (Spallone 1992[3]; Haraway 1991; Throsby 2004; Franklin and Roberts 2006; Nahman 2006; Sexton 2001, 2005; Dickenson 2002, 2007; Schneider 2007; Waldby 2002; Parry 2006, 2008).

The paper's core aim is to identify some of the ethical, social and political issues raised by key actors raising concerns or criticisms of egg sourcing. It is thus firstly a reportage on critiques (for example identified risks, defects in the public consultation

---

[2] See HFEA (2006).

[3] The work of 'early risers' (Tarrow 1998) such as Spallone, writing in 1992 has much contemporary relevance for today's debates even in a new scientific context.

process) as they are being articulated by a variety of ‘prime movers and early risers’, and secondly provides a network analysis of (feminist) public engagement and emergent mobilisation, commenting on the significance of this network analysis. Highly significantly, for many emergent actors, HFEA policy moves on egg sourcing enabled an ethical debate about embryonic stem cells which focused on risk in relation to women’s bodies, and broader questions of political economy, thus finally developing the ethical debating stakes beyond the status of the embryo in the public sphere, for example on UK radio and in UK newspapers. Methodologically I have taken an ‘action research’ approach (Stringer 1999) which can be defined as reflexively embedded ethnography, whereby the researcher is consciously positioned both as an actor in the research process and a reporter on it. Action research is part of a legacy of feminist academics developing and theorising feminist research methods (Roseneil 1993; Stanley 1991). Action research is an important methodology, here enabling an identification and early analysis of developing mobilisation amongst predisposed, latent networks – in this case, identifying a number of ‘prime movers’ including high-profile feminist academics. Through information dissemination and networking, my action research has thus been aimed at identifying, and engaging with, ‘organic loose networks’ of feminists and others, who have been framing concerns over risk and other issues as identified in this paper. Prime movers mobilising like this have ‘strong ties’ and ‘weak ties’ (Granovetter 1973) links to social justice and sustainable development networks, and to ‘watchdog’ campaign groups framing criticisms about techno-scientific development under globalised market structures (framed increasingly as ‘the bio-economy’). This loose (feminist) network is discussing multi-layered, highly complex, and often ambivalent arenas and identifying many potential sticking points between individuals. These can be seen as ‘Venn diagrams’ where issue, ethical frames cross over, but which also constitute converged areas of discursive conflict – for example over the framing of egg donation in terms of reproductive autonomy, which is sparking some difficult conversations over reproductive rights and choices within feminism, as will be discussed. Such social and ethical ambivalence and complexity has also resulted in some social convergences of extremely ‘strange bedfellows’ (Evans et al. 2007); such as the convergence of *some* internationally renowned feminists, and pro-life prime movers, in the formation of the campaign group ‘Hands Off Our Ovaries!’ (HOOO)[4]; the significance of this is discussed in due course.

‘Action research’ has thus involved participating in this ‘organic networking’; diffusing repertoires through ‘strong and weak ties’ networks (Diani 1992; Carroll and Ratner 1996). This is being done through the dissemination of information and catalysing discussion in civil society, within academia, and in other spaces. The dissemination of information and the provision of supportive debating space as a resource for civil society is seen as an end in itself: to build social capacity and facilitate ‘knowledge transfer’. There are many spaces where information is being diffused, including within journals and more academic spaces such as seminars. There are examples of academic/civil

[4]http://handsoffourovaries.com/

society interaction over genetics and reproduction in many European contexts. In Germany, for example, the campaign group Reprokult,[5] comprised of (amongst others) politicians, academics, and health policy experts and workers, has long been developing a critique which articulates many of the central risks and grievances which have been re-framed in relation to HFEA policy moves, referred to in later sections of the paper. The network 'strong ties' and 'weak ties' between Reprokult and other UK, European and international – perhaps especially American – feminists have ensured a building of capacity amongst loose networks of predisposed 'prime movers' for several years, through organised events, briefing papers, and 'organic networking'.

In other 'discursive turns' (Dryzek 2000) completely outside of the academic and policy spheres,[6] UK and European feminist activists and 'anti-globalisation' activists are developing their own convergence spaces (Routledge 2003), for example through the World Social Forum (WSF) and the European Social Forum (ESF) (Welsh et al. 2007), and developing e- networks; the European e-list femACT is an example. Highly significantly, in certain culturally and nationally context-dependent spaces, there are also some clearly defined examples of public and policy critical engagement by more mainstream and academic prime movers, for example on the risk/benefit analysis issue, and specifically over the implications and remit of HFEA public consultation and policy process in relation to the egg sourcing policy issue, as discussed in this paper. Before providing a timeline of HFEA egg donation policy and emergent (UK, EU, International) critical feminist responses to it, I will firstly provides a short summary of the scientific research and related technologies and some potential future applications.

10.2 Cell Nuclear Transfer (CNT) or Cloning, Embryonic Stem Cells, and Stem Cell Lines

Stem cell research is at a very early stage of R&D. Stem cell 'lines' can be 'grown', in the right conditions, from a variety of sources including from adult stem cells. Human embryonic stem cell lines are grown from an embryo at a certain, extremely early and highly regulated, stage of development, sometimes referred to as a 'blastocyst'.

Human embryonic stem cell lines are thus sourced[7,8]:

(a) From 'discarded' embryos, often from embryos discarded after PGD (Pre-implantation Genetic Diagnosis). For example: a woman who is a carrier for the

---

[5] http://www.reprokult.de

[6] It should be noted that there are many social network and campaign ties between these more 'radical' grassroots organisations and more formal policy and academic-orientated actors.

[7] 'Hybrid' embryos which could become stem cell lines, using an animal egg and a human cell, are now also potentially allowable under very recent UK policy moves, as of September 2007; see HFEA statement, September 5, 2007 at http://www.hfea.gov.uk/en/1581.html and a typical UK news reportage at http://news.bbc.co.uk/1/hi/health/6978384.stm

[8] A usefully clear diagram explaining this procedure is available at http://www.isscr.org/public/nuclear_transfer.htm

disease cystic fibrosis (CF) (one of only very few diseases termed a 'single gene disorder') could have her eggs fertilised via In Vitro Fertilisation (IVF), thus enabling her to have PGD to determine which, if any, of the fertilised embryos carry the CF gene; if they do, and if she chooses not to implant these embryos, she can donate the discarded embryos which carry the CF gene for stem cell research and the creation of stem cell lines.[9]

(b) From human eggs or oocytes, from which a cloned embryo can be created via cell nuclear transfer (CNT). Unfertilised eggs or oocytes can thus be 'fertilised' not by a sperm, but by CNT (cell nuclear transfer) – cloning – to develop stem cell lines which are a genetic match for a specific donor. The research is at a very early, experimental stage. It is the sourcing of these eggs, specifically for their use in CNT research, which is the subject of this paper in terms of summarising the ethical and other issues raised by critical (feminist) prime movers.

10.3 Overview of 'Future Promise'

Scientific viability and efficacy in terms of the framing of projected outcomes, has an impact on policy and governance, i.e., assessing risks and benefits. As already indicated, this is early research in extremely experimental stages where, for example, even the storage conditions of any resulting stem cell lines in various 'banks' represent as yet unknown variables in terms of impacts on any cell line produced. CNT routes to embryonic stem cell line creation represent an extremely new mode of R&D in this field.

To summarise briefly, the long-term aims and outcomes of such research are as follows:

- Treatments, cures, of diseases and conditions: diseases being researched specifically in relation to R&D on CNT include: diabetes, neurological conditions such as Huntington's, Parkinson's, Alzheimer's.
- Short overview of potential applications (varies from short- to long-term – deriving from uncertainty of long-term cell viability):

• Basic R&D re-understanding/developing CNT process; stem cell lines, autoimmune response (CNT delivers potential genetic matches for the donor of the cell and/or the egg)
• Understanding disease expression
• Testing drugs (pharmaco-genetics, genomics – 'personalised' pharmaco)
• Gene therapy
• Organ and tissue replacement

[9] It should be noted that PGD is also a contested ethical issue with implications for feminist debates on reproduction and the body as women negotiating the difficult decisions over the use of PGD encounter Disability Rights perspectives, for example, and as such has been the subject of a public consultation in the UK; see Human Genetics Commission (2004). For a brief overview of the issues relating to feminism, see Plows (2006).

Such core frames of treatment, cure, understanding disease expression and so on are the dominant discourse within (health) policy, governance and relevant scientific domains.

Stem cell research offers enormous potential for major advances in clinical therapy. Stem cells could be used to replace missing or damaged cells in important diseases, such as diabetes and Parkinson's, and in the treatment of traumatic injury including paralysis. The establishment of the UK Stem Cell Bank is an important step along the way to realising this potential.[10]

Such frames have, not surprisingly, much discursive legitimacy amongst many public groups, and have strong political allies (policy, science, patient groups, UK and EU governance, biotech and pharmaceutical industry, and so on). It is important to reiterate that it is possible both to be hopeful that science may indeed find cures and develop better treatments for serious disease, and also be concerned about other risks and implications; this emphasises the significance of social ambivalence – beyond pro and anti – (Evans et al. 2007) in relation to the Ethical Legal Social Aspects (ELSA) of bioscience more generally, and specifically in relation to embryonic stem cell research. It is important to highlight, and discuss, the science goals, and the debate within and outside the scientific community about the actual scientific viability of CNT and of potential applications (also to note that scientists' own framings of what has scientific legitimacy and value, is not immune to political motivation (Eriksson 2004a, b)). However the scientific value of stem cell research is not the main focus of this paper; although (hopes for) potential outcomes in terms of treatments for disease, are very linked to the bigger picture themes discussed here in terms of, for example, what might be termed the political economy of health; public expenditure on stem cell R&D, health policy priorities for example, and the fact that the science is undeniably already having an impact on the public sphere and the female body. Realistic assessments of how viable the science is in terms of delivery of 'future promise' (Brown and Michael 2003) – for example, in the media, see early coverage of the Korean stem cell breakthrough[11] (Haran et al. 2008) – also has implications in terms of the impacts on female patient (potential donor) expectations. For example, it is possible that potential female donors could have formed overly optimistic expectations of the outcome of their 'altruistic donation'.[12]

---

[10]From UK stem cell bank website, retrieved July 26, 2007, from http://www.ukstemcellbank.org.uk/overview_new.html

[11]Arguably, the geopolitical backdrop of the Korean cloning scandal 2005 enabled a space for ethical concerns on egg sourcing to be heard more clearly in the public sphere.

[12]See Economic and Social Research Council (ESRC) Centre for the Economic and Social Aspects of Genomics (Cesagen) seminar statement 2006, p. 2. This feminist seminar, held in November 2006 by several CESAGen staff including the author, resulted in a statement produced by workshop participants (Cesagen feminist seminar participant statement 2006).

10.4 UK HFEA Policy

The broad terms of UK HFEA policy now are: that eggs can now be sourced for stem cell research, subject to application to the HFEA, via the following routes:

(1) 'Altruistic donation' – from women not undergoing IVF. These women will not be paid for their donation, though they will get basic expenses of £250.
(2) 'Egg sharing' – women 'donate' or 'share' or 'trade'[13] some of their eggs in return for free or reduced cost IVF cycles.

HFEA timeline is:

- July 2006: HFEA gives a 'temporary' licence to the North East England Stem Cell Institute (NESCI) allowing 'altruistic' egg donation, via IVF 'egg sharing', for CNT research.
- July 2006: Public consultation on egg sourcing initiated by HFEA entitled *Donating eggs for research: safeguarding donors.*[14]
- 21 December 2006: The terms of the temporary licence given to NESCI are extended by HFEA: women not already undergoing IVF could donate eggs for CNT research.
- 21 February 2007: Egg sourcing from both routes (altruistic donation, and egg sharing), both routes being the subject of public consultation, has now been officially allowed by the Human Fertilisation and Embryology Authority (HFEA).[15]

The HFEA policy moves were a line in the sand moment, catalysing mobilisation. As a sociologist embedded in the field it would have been hard to miss the seismic movement in loose, latent, predisposed social networks, including UK (feminist and other) academics and NGO actors, who can be identified as 'prime movers'. Responding to the issues raised in this specific policy consultation represented a clear line of engagement, a mobilisation opportunity in a very complex and contested field. Feminist and other prime movers identified a clear imbalance in terms of assessing a risk/benefit ratio, it was for many a final straw and thus a catalyst for engagement on several counts.[16] Some of these emergent issue frames are now summarised. It should be

---

[13] Terminology in this specific context is highly significant and reflects radically differing perspectives. Donna Dickenson (2007), for example, refers to the relationship as a 'trade', locating the arrangement within a broader political economy framework which she strongly critiques; see also Schneider (2007). The terms used by the HFEA itself in its consultation document, on the other hand, tend to refer to altruistic donors who 'share' eggs, implying an equal power relationship which can be 'safeguarded'. These terms as they appeared in the HFEA consultation were the subject of focused criticism in the aforementioned CESAGen feminist seminar.

[14] http://www.hfea.gov.uk/cps/rde/xchg/SID-F57D79BA850329D/hfea/hs.xsl/1417.html

[15] http://www.hfea.gov.uk/en/1496.html

[16] See also the extremely significant event organised by Edinburgh University academics as part of a policy feedback mechanism to the HFEA consultation: http://www.talkingstemcells.ed.ac.uk/index.php?action = ShowArticle&id = 72

stressed that this account provides a specific snapshot of issue framing amongst certain networks and is not intended in any way to be a definitive account of the way the issue is being engaged with by multiple public groups and policy makers. I also comment further on some of these ELSA critiques being framed by (UK, International) prime movers in relation to this specific issue.

10.5 Public Engagement and Regulatory Implications

As the timeline makes clear, on 21 December 2006 the HFEA extended the temporary license to NESCI *whilst the public consultation process on the topic was ongoing*. It should also be noted, that the initial temporary license was granted by the HFEA pre-dating the launch of the public consultation. The Cesagen feminist seminar participant statement (2006) which addressed the HFEA public consultation in detail, referred to in a previous footnote, identified a fait accompli embedded in the document's language and remit. Essentially, the HFEA consultation document asks the question – should 'altruistic donation' be allowed? But the sub-header of the document title: *safeguarding donors*, shows that such a framing means the debate has already moved to the ethics of 'safeguarding' the donors. Questions inside the document relate to a mitigation process in terms of donor risk management and 'informed consent' procedures. Prime movers during the Cesagen seminar, and in other settings too, have raised critical points (issue frames) relating to the viability of the public consultation process, which can be summarised in the following bullet points:

- How well the public was informed; in fact the 'public consultation' consisted of a confusing, highly complex, web-mounted document; this was not a proactive effort to meaningfully engage 'the general public' and falls short of identified 'best practice'.
- How the framing of the debate by the HFEA – should women be allowed to donate? – in the public sphere affected the outcome of the debate; it was too narrowly framed and presented a certain outcome as a fait accompli. The framing in the HFEA document further implies that to block women from choosing to donate is to deny them individual [reproductive] 'choice'; a difficult conundrum for feminism, and a counter-productive framing which closes down routes for constructive debate between stakeholders, perhaps especially among feminists.[17] The title of the Cesagen seminar statement which critiqued the

[17]'The questionnaire frames the issue of egg donation for research in terms of 'allowing' 'non-patients' or 'people' (that is, women) to 'choose' to donate eggs to research: the issue should be framed the other way around in terms of whether researchers should be allowed to approach women (…) The narrow and specific framing of the consultation questions prevent us from talking about, for example, the wider social context within which women will be expected to make decisions and the inequalities the schemes will exacerbate and produce' (Cesagen feminist seminar participant statement 2006).

> HFEA policy consultation addressed this issue by turning the HFEA's consultation question on its head, asking: 'Should scientific researchers be allowed to ask women to provide their eggs for disease research?' The Cesagen seminar document title thus emphasises the political situated-ness of the issue; even if women are in some cases proactively wishing to donate eggs, this is a situation which has been catalysed by the needs of the scientists and, as other prime movers have argued (Dickenson 2007) the demands of the 'bio-economy' (see later section).

The HFEA consultation process and document thus raises classic Science and Technology Studies (STS) and public understanding of science (PUS) issues (Wynne 1996, 2006; Jasanoff 1990) which can be summarised in terms of a contested legitimacy of policy practice. In this arena of public engagement with science and the policy process, as in many others, the following questions have continued to have relevance in an emergent critical discourse: is governance acting as 'midwife' to the technology? What is public engagement *for*? What impact does public engagement have, or should have, on policy outcome; public engagement and agenda setting; and who defines and enrols relevant expertise in policy processes? The pre-emptive nature of the 'temporary license extension' by the HFEA in December 2006 clearly foregrounded such problems – a crisis of confidence in (UK) public engagement and consultation mechanisms. The regulatory issues raised by the use of eggs/oocytes in biomedical research more generally, is discussed in detail in Schneider (2007).

10.6 Overview of Framing of Risks and Concerns: Health, Ethics, Political Economy

As discussed in the introduction, the HFEA policy moves have been a highly significant breakthrough moment in ethical debates on stem cells – because it is clearly ethical issues and identified risks which relate to *women's bodies*, not the status of the embryo, which are the key frame for these emergent loose networks of early risers and prime movers. Even in relatively secular countries such as the UK, with a liberal attitude to abortion, it is highly significant that in relation to stem cell research, it has been almost impossible to hear any other framing of ethical risk issues in the public sphere, outside of the ethics of the status of the embryo (see for example, House of Lords debates on stem cells, Select Committee on Stem Cell Research 2002). The black and white framing of 'science vs pro life' is still the usual suspect mode of debate, for much media discourse in relation to anything to do with stem cell research (Haran et al. 2008).

Women can be said to be in the front line of bioscience; they are negotiating new territory; they are 'moral pioneers' (Rapp 1999). Concerns about the impacts on/risks to women's bodies, and associated ethical and socio-political concerns, being framed by prime movers, relate in general terms, to how and why the eggs are sourced. The following points, emergent and not definitive,[18] represent some key issues being raised by prime movers:

- Health risks: the drugs given to women to stimulate multiple egg release can result in a condition known as ovarian hyperstimulation; this affects one in ten women; and the condition can be fatal in rare cases.

This results in an important risk/benefit analysis: this was the catalyst moment for many (UK, European) 'prime movers'.

> (...) We believe that the risks of hormonal hyperstimulation of the ovaries cannot be justified in basic research, in which the benefits are very uncertain: the risk/benefit ratio is far too high (...)[19];

- Health uncertainty: long term impacts of the drugs: 'health effects of egg donation may take decades to emerge' (Pearson 2006).

Such uncertainty over long term health outcomes has implications for the viability of 'informed consent', as the *Nature* article cited above also emphasised:

- Ethical concerns around informed consent: the HFEA ruling, to reiterate, means that egg sharing – getting free or reduced cost IVF, in return for trading eggs for CNT – is now legitimate practise in the UK (subject to a successful application for a license). In this context, there are ethical problems/concerns around the concept of informed consent, and also around the framing of this arrangement as informed choice and informed consent (Parry 2006, 2008).

For example, a clear ethical (emotional) problem with regard to informed consent given by a woman in an egg sharing or trading arrangement, relates to the woman having to see egg A as a potential baby, and egg B (the egg destined for CNT which she shares, i.e., trades for free or reduced cost IVF) as just a sac of cells which can be experimented on. Especially in the light of 80% failure rates of IVF; if egg A fails to develop into a viable embryo, how does the woman feel about her 'shared' or traded egg B? Such ethical concerns are not even mentioned in the HFEA consultation document *Donating eggs for research* (HFEA 2006).

During her presentation at the WONBIT conference, Sarah Franklin (see Chapter 5) emphasised an important ethnographic finding: that some women – women she has spoken to in the IVF clinic – will want to donate eggs, 'to give something back'. This is significant; there are strong arguments around altruism and autonomy here, and one should be careful not to frame women as passive victims of science and technology; of

---

[18]For a highly contextualised analysis of these issues as they relate specifically to the terms of the HFEA public consultation document, see the CESAGen feminist seminar statement. For specific discussion on these issues by key academics engaging with this topic, see Sexton (2005), Dickenson (2007), Parry (2006), Throsby (2004), Franklin and Roberts (2006), Schneider (2007).

[19]This is an extract from a jointly signed open letter in the *Guardian*, Tuesday May 9, 2006 comprising several UK and European academics and NGO actors including the author. See http://www.guardian.co.uk/letters/story/0,1770438,00.html

course, a long-standing debate within feminism. There exists a strong patient perspective: for example, the mother of a child with Cystic Fibrosis (CF) might want to donate her eggs for CNT stem cell research so that basic R&D can be done on the disease, for long term 'public good' outcomes. This is a difficult ethical arena, to deny a usually well informed (lay expert), campaigning blood relative the chance to donate, when they likely know better than anyone how experimental the research is; though also serious concerns about emotive exploitation remain. There are also highly difficult tensions over bodily autonomy and state control, over the 'right' to do what one wants with one's own body.

However where to draw the line in relation to health risks, and the policy framing of egg trade as informed choice and informed consent, remain significantly problematised ethical issues which cannot be brushed under the carpet. There are issues to bear in mind in relation to the framing of egg 'donation' in terms of informed choice, informed consent, altruism and the 'public good', and they are being articulated by a variety of prime movers. They include the following:

- Noting the 80% failure for IVF, viable eggs/oocytes are never 'spare'; from the point of view of the donor involved in an egg sharing scheme, eggs are always potentially a viable future baby.
- The possibility of misguided altruism in terms of the future promise viability of the science, even into the long term, as discussed earlier.
- It is essential also to embed the discussion in relation to cultural attitudes to infertility, the rise in the use of IVF (in the UK, elsewhere) and to be alert to the (re)framing of IVF as a reproductive right.

Thus informed consent and informed choice with specific reference to egg sourcing for CNT as discussed in the HFEA policy document are contested concepts in the clinic setting. Karen Throsby (2004), Michal Nahman (2006, 2008), Sarah Parry (2006, 2008) have all undertaken ethnography of women in IVF clinic settings in relation to a critical analysis of informed consent criteria. Parry, for example, identifies extreme confusion over informed consent criteria and how decisions are being taken by women at a difficult emotional time, which in several instances they later have no memory of taking. These are major ethical problems (which again, are not mentioned in the HFEA consultation). Sarah Parry, and also Franklin, Roberts, Throsby and Nahman all emphasise the situated nature of the conditions under which women, the significantly increasing number of them having IVF treatment it should again be noted, are supposed to make informed choices, and unpacking the political nature of these constructed choices.[20]

Ambivalence and complexity are highly significant social states. Franklin and Parry's ethnographic observations differ in terms of what and how they report about women's attitudes to egg donation; but both are extremely important and

[20] See also Schneider and Schumann (2002), Schneider (2007), Sexton (2001, 2005), Dickenson (2002, 2007).

both represent a social 'truth' in the context of their methodological approach and ethnographic setting. It is highly necessary, at this moment in time, to have a thorough debate on these issues (reproductive choice in particular) within feminism; but we must be careful not to allow the debate to be reified and polarised, as I would argue, the framing of the HFEA policy document inevitably does. This has produced an unnecessarily difficult clash between liberal feminists supportive of reproductive choice and those who, whilst equally supportive of female reproductive autonomy, are identifying broader issues in the context of such choices accruing to the fallout of global political [bio]economy, which is arguably the core driver of science innovation and science policy (Dickenson 2007; Sexton 2005; Birch 2006).

10.6.1 The Ethics of Bio-Citizenship[21]

Further in relation to a discourse of altruistic (female) egg donation embedded within the HFEA policy document, concepts of 'duty' and 'citizenship' (public good, altruism) are increasingly being invoked in relation to genetics, reproduction, and fertility in a number of settings. As stated earlier, Sarah Franklin notes altruism as a discursive frame amongst potential donors in her recent ethnography of the clinic. Very significantly, the bioethicist John Harris has stated that 'Biomedical research is so important that there is a positive moral obligation to pursue it and to participate in it' (Harris 2005).

What will being a good citizen mean in the realms of health, genetics and fertility? Who will decide, make policy and set standards, including acceptable levels of risk? (Plows and Boddington 2006). There are of course parallels with blood donation; people generally want to contribute to society (such framings are also relevant to calls for public donations of DNA to the UK Wellcome biobank). There are however major differences in terms of the potential risks accruing to such 'altruism', between different types of public donation of bio-material. The health risks and health uncertainties (and the emotional uncertainties) accruing to female citizens wishing to altruistically donate eggs or oocytes, are high, as outlined earlier. Further, issues of political economy are not well developed in many accounts of [bio] citizenship stakes (Plows and Boddington 2006), although significant work by several authors is starting to address this elephant in the room (Birch 2006; Raman and Tutton 2008) and it is a well articulated frame in the more underground, radical networks of young(er) European feminist activists from the 'alter globalisation' generation.

[21] The term bio-citizenship has been coined by Rose and Novas (2004) and is critiqued in Plows and Boddington (2006) and in Raman and Tutton (2008).

10.6.2 The Bio-Economy

The bio-economy can be defined as 'the discourses and practices of economic competitiveness that pervade biotechnology policy-making in the UK, Europe and the USA' (Birch 2006).

Within Europe, this means the prioritising of 'bio-competitiveness' and the marketing of bioscience as innovation (knowledge stakes) for economic growth, identified as 'the European Union's Lisbon Strategy to become the 'most dynamic competitive knowledge-based-led science'. It is essential to highlight the crucial relevance of market discourse in relation to patient/consumer/individual/donor choice in the context of the bio-economy. Several key academics have made explicit, critical reference to the impacts of 'the bio-economy' in relation to highly predictable global flows of human resources; specifically in this context, the gendered economic implications of egg sourcing for research purposes. Core concepts include a critique of commodification (Dickenson 2002, 2007; Schneider 2007) and bio-value (Waldby 2002). Sexton (2005) charts the route of eggs from Romania (see also Nahman 2008), and highlights concerns about exploitation of economically marginalised actors. One notes a further ethical ambivalence over fees, expenses for donation. Bharadwaj (2005) traces the routes of eggs from India, where IVF has been provided in exchange for eggs for stem cell research. This is, not to put too fine a point on it, where the infertility issue meets globalised Neoliberal bioscience; since the HFEA ruling, this also applies to the UK.

The dominant argument on the issue of bio-competitiveness is that it is good for scientific innovation and for Western economies. This is not the place to unpack the viability of this argument; what is important in this context, is to be alert to the fact that this is the dominant model, and it is causing globalised ethical fall out: eggs as a resource to be commodified. An egg which comes from a specific woman's body in a specific place and time, with a specific narrative attached, could become a globalised, perhaps even immortal, stem cell line with numerous patent applications accruing to it.

10.7 Conclusion

The above subsections have set out some key issues being framed by emergent prime movers in relation to HFEA egg sourcing policy as they have emerged in 'real time'. They represent examples of framings of risk and other critiques; the above summary is not intended to be a definitive analysis of discursive (frame) capacity but it aims to indicate some emergent trends in certain social and discursive spaces. The paper's final remarks return to the actor-network analysis of engaging prime movers and the academic and social significance of emergent mobilisation in a highly ambivalent, complex and contested arena. The paper has primarily focused on network analysis amongst emergent loose networks of feminist and other actors, including UK academics, who are developing a critique. To summarise, HFEA policy moves were a line-drawing moment for many, catalysing mobilisation amongst 'organic', latent loose networks of academics, feminists, social justice activists and others. This catalyst moment has in fact been brewing within 'latent' networks for some time, as theories on social movements and mobilisation would lead one to expect (Melucci 1996). As discussed in the introduction, they are not the only social actors developing a critique[22] – other social actors include convergences of 'strange bedfellows' (Evans et al. 2007). Strange bedfellows can be summarised as a counter-intuitive cluster of social actors from different lifeworld (Habermas 1984) positions, who in highly context-dependent circumstances, share a position over a specific issue, perhaps but not necessarily, for the same reasons. There are two examples of strange bedfellows in relation to prime mover critical responses to egg donation.

Firstly, HOOO's 'Hands Off Our Ovaries!' is a feminist and pro-lifer alliance formed in response to the sourcing and use of women's eggs for biomedical research. Highly significantly, the key frame visible of the HOOO website is a political economy critique of egg sourcing which emphasises the gendered power and economic issues referred to in earlier sections of this paper. The pro-life campaign group Comment on Reproductive Ethics (CORE) were discussing areas of shared critiques with American feminists at a conference in London, in April 2003. A month into my new job, this was when I first realised that something very important was happening in terms of understanding and theorising mobilisation, and the framing of the stakes of engagement. This is academically highly significant, as such a strange bedfellow cluster signals what is in fact a growing phenomenon in the public engagement/bioscience arena; where stakes are multiple and complex. Social complexity and ambivalence is thus an identifiable and highly significant trend, impacting with multiple publics, multiple issues. This has triggered multiple social assemblages (Irwin and Michael 2003) or clusters. There are complex 'identity politics' here (Plows and Boddington 2006) which account for these clusters of usual suspects and also of strange bedfellows cutting across lifeworld (Habermas 1984) positions in a conceptual 'Temporary Autonomous Zone' (Bey 1991); the Hands Off Our Ovaries feminists and pro-lifers group formation is an important signifier of social complexity in relation to engagement with bioscience.

Whilst the significance of ambivalence and complexity is encapsulated in HOOO, this strange bedfellow cluster is also of concern from a personal point of view as a feminist with an action research interest in tactics; as this has been one of the few sites where its been possible to clearly uncouple the ethics from debates solely accruing to the status of the human embryo, and instead discuss risks to women's bodies, as stated earlier. Whilst the HOOO website focuses on risks to

[22] Again it should be stressed that this autobiographical ethnographic snapshot has not covered the networks of prime movers mobilising in support of egg sourcing for stem cell research in any detail.

women's bodies triggered by the bio-economy, this frame is, I would argue, undermined by the pro-life connection. This is a difficult conundrum. A constructive, and supportive, debate within (UK, EU, International) feminism about the efficacy of HOOO as a political tactic is highly important. National cultural, political and legal differences are also extremely relevant in this context, particularly on embryo status and reproductive rights (Bender et al. 2005). It is also important to note, that *some* HOOO pro life prime movers (such as Josephine Quintavelle, the director of the pro-life campaign group CORE) have broadened their ethical frame of reference away from the status of the embryo, to unpack the framing of female choice in the context of egg donation; hence Quintavelle's involvement with HOOO, and, one can surmise, the rationale behind those prime mover feminists who have decided to mobilise with such well-known pro-life actors, over the egg sourcing issue, through the campaign group HOOO. However because Quintavelle is well known as a pro-life advocate, this will inevitably frame how HOOO's aims and agendas are viewed by others; in other words, it is likely that the pro-life connection will undermine attempts to raise broader concerns of political economy and risks to women's bodies.

A second (potential) strange bedfellow is stem cell scientist Stephen Minger, who has publicly made highly critical comments about health risk and ethics, and the scientific viability of projected outcomes and applications, in relation to the HFEA egg sourcing policy discussed throughout. Minger also critiqued the HFEA policy process (Plows 2007). This is highly significant in that these are some of the key critical frames also being articulated by other groups and prime movers as outlined in this paper; this is thus a *discursive* strange bedfellow cluster, rather than one where people are actually mobilising together, like HOOO. In other words, Minger has articulated critiques in the public sphere that correlate with key concerns raised by mobilising feminists in this paper; but there is no actual mobilisation, no alliance-building, no equivalent of HOOO. Further, there is also ring fencing going on here in terms of Minger's political positioning of his own research. Since December 2006, Minger has publicly[23] framed his own research, on CNT hybrid embryos[24] (using an egg sourced from an animal and a human cell) as more ethical than CNT via human egg sourcing, because of the health risks accruing to human egg sourcing. Whatever his motivations for doing so, what is perhaps most significant about Minger's articulations of (ethical, health) risk in the public sphere, is the fact that as a scientist actually involved with stem cell research, his comments got taken seriously. In other words, it would have been impossible to cast Minger as 'anti-science', as often happens to actors who frame risk or other criticisms in

[23] For example on the news programmes of UK radio 4, during late 2006 – Summer 2007.

[24] It is interesting that the HFEA's recent decision to allow hybrid embryos in 2007 has triggered much (UK) public and media debate on the bioethics of the issue (see footnote 7) in a way which failed to materialise in relation to the issue of completely human cloned CNT embryos for biomedical research, the topic of this paper. The animal/human boundary can perhaps be read as a more significant mobilising frame for UK publics.

relation to bioscience; and thus his interventions, perhaps inadvertently, have lent legitimacy to others framing risk critiques, contributing to a more open discursive space which can be productively utilised.

Such constructive and well-resourced discursive space is highly necessary for the public good and regulatory 'best practise' (see also Schneider 2007), whether in the UK or further afield. The HFEA consultation issue has meant that an important policy shift has been made without properly informed public input, which is perhaps especially significant in the context of rising public demand for IVF. These issues are having an impact in the here and now and thus need to be the subject of discussion and debate more thoroughly in the public sphere. The prime movers (McAdam 1986) discussed throughout this paper are developing discursive resources and are thus organically building civil society capacity through the dissemination of issue frames through weak ties networks (Carroll and Ratner 1996; Granovetter 1973). This needs further facilitation; perhaps in part by developing a programme of constructive debates on this and related issues by feminist academics and other key stakeholders and experts such as patient groups and NGOs. These could be resourced by policy as part of EU strategy on developing 'deliberative democracy'.

References

Bender, W., Hauskeller, C., & Manzei, A. (Eds.) (2005). *Crossing borders: Cultural, religious and political differences concerning stem cell research*. Münster: Agenda Verlag.

Bey, H. (1991). *T.A.Z.: The temporary autonomous zone, ontological anarchy, poetic terrorism*. Brooklyn, NY: Autonomedia.

Bharadwaj, A. (2005). Cultures of embryonic stem cell research in India. In W. Bender, C. Hauskeller, & A. Manzei (Eds.), *Crossing borders: Cultural, religious and political differences concerning stem cell research*. Münster: Agenda Verlag.

Birch, K. (2006). The neoliberal underpinnings of the bioeconomy: The ideological discourses and practices of economic competitiveness. *Genomics, Society and Policy*, *2*, 3. Retrieved October 31, 2007, from http://www.gspjournal.com

Brown, N., & Michael, M. (2003). A sociology of expectations: Retrospecting prospects and prospecting retro-spects. *Technology Analysis and Strategic Development, 15*(1), 3–18.

Carroll, M. K., & Ratner, R. S. (1996). Master framing and cross-movement networking in contemporary social movements. *Sociological Quarterly, 37*, 601–625.

Cesagen feminist seminar participant statement (2006). Should scientific researchers be allowed to ask women to provide their eggs for disease research? A statement of concerns in response to the current HFEA public consultation: Donating eggs for research: safeguarding donors (p. 6). Retrieved October 31, 2007, from http://www.cesagen.lancs.ac.uk/events/eventsdocs/HFEA_sourcing_eggs.pdf

Diani, M. (1992). Analysing social movement networks. In M. Diani & R. Eyerman (Eds.), *Studying collective action*. London: Sage.

Dickenson, D. (2002). Commodification of human tissue: Implications for feminist and development ethics. *Developing World Bioethics, 2*(1), 62.

Dickenson, D. (2007). *Property in the body: Feminist perspectives*. Cambridge: Cambridge University Press.

Dryzek, J. S. (2000). *Deliberative democracy & beyond: Liberals, critics, contestations*. Oxford: Oxford University Press.

Eriksson, L. (2004a). Maverick: Made in UK. Construction of dissent in Britain and Sweden. Submitted to *Social Studies of Science*.

Eriksson, L. (2004b). When scientists fight. *Science & Public Affairs*, June, 25. Retrieved October 31, 2007, from http://www.the-ba.net/NR/rdonlyres/1F300683-61D2-45F7-9F10-04506B7FDDA6/0/SPAJune04.pdf

Evans, R., Welsh, I., & Plows, A. (2007). Towards an Anatomy of public engagement with medical genetics: Strange bedfellows and usual suspects. In P. Atkinson, P. Glasner., & H. Greenslade (Eds.), *New genetics, new identities*. London: Routledge.

Franklin, S., & Roberts, C. (2006). *Born and made: An ethnography of preimplantation genetic diagnosis*. Princeton, NJ: Princeton University Press.

Granovetter, M. S. (1973). The strength of weak ties. *American Journal of Sociology, 78*, 6.

Habermas, J. (1984). *The theory of communicative action*. London: Heinemann.

Haran, J., Kitzinger, J., McNeil, M., & O'Riordan, K. (2008). *Human cloning in the media: From science fiction to science practice*. London: Routledge.

Haraway, D. (1991). A Cyborg Manifesto: Science, technology, and socialist-feminism in the late twentieth century. In *Simians, cyborgs and women: The reinvention of nature*. New York: Routledge.

Harcourt, W. (2007). Heading Blithely down the garden path? Some entry points into current debates on women and biotechnologies. Overview Paper in conference booklet produced for the Women in Biotechnology conference, Rome 2007. Retrieved from www.wonbit.net

Harris, J. (2005). Scientific research is a moral duty. *Journal of Medical Ethics, 31*, 242–248.

HFEA (Human Fertilisation Eggs Authority) (2006). Donating eggs for research: safeguarding donors. Retrieved October 31, 2007, from http://www.hfea.gov.uk/cps/rde/xchg/SID-F57D79BA850329D/hfea/hs.xsl/1417.html

Human Genetic Commission (2004). Choosing the future: Genetics and reproductive decision-making. Retrieved November 20, 2007, from: http://www.hgc.gov.uk/Client/news_item.asp?Newsid = 40

Irwin, A., & Michael, M. (2003). *Science, Social Theory and Public Knowledge*. Maidenhead-Philadelphia: Open University Press.

Jasanoff, S. (1990). *The fifth branch. Science advisers as policymakers*. Cambridge, MA: Harvard University Press.

McAdam, D. (1986). Recruitment to high-risk activism: The case of freedom summer. *American Journal of Sociology, 92*, 64–90.

Melucci, A. (1996). *Challenging codes: Collective action in the information age*. Cambridge: Cambridge University Press.

Nahman, M. (2006). Materializing Israeliness: Difference and mixture in transnational ova donation. *Science as Culture, 15*(3), 199–213.

Nahman, M. (2008). Nodes of desire: Romanian egg sellers, 'Dignity', and Feminist alliances in transnational ova exchanges. *European Journal of Womens' Studies, 15,* 65–82.

Nelkin, D. (1995). Forms of intrusion: Comparing resistance to information technology and biotechnology in the USA. In M. Bauer (Ed.), *Resistance to new technology: Nuclear power, Information Technology and Biotechnology* (pp. 379–390). Cambridge: Cambridge University Press.

Parry, S. (2006). (Re)constructing embryos in stem cell research: Exploring the meaning of embryos for people involved in fertility treatments. *Social Science & Medicine, 62*, 2349–2359.

Parry, S. (2008). (Re)Constructing embryos in stem cell research: Exploring the meaning of embryos for people involved in fertility treatments. *Social Science & Medicine* (in press).

Pearson, H. (2006). Health effects of egg donation may take decades to emerge. *Nature, 442*, 607–608.

Plows, A. (2006). PGD, PND and the challenge of "Informed Choice" for feminism. *PropEur Newsletter, 2*, 4–5. Retrieved November 20, 2007, from: http://www.propeur.bham.ac.uk/NewsletterVol2.pdf

Plows, A. (2007). Regulating egg donation for research: Putting the cart before the horse? Retrieved February 27, 2007, from http://www.bionews.org.uk/commentary.lasso?storyid = 3315

Plows, A. (2008). Convergence: Nanobiotech and the politics of technology. In F. Jotterand (Ed.), *Nanotechnology: Framing the field* (in press).
Plows, A., & Boddington, B. (2006). Troubles with biocitizenship? *Genetics, Society and Policy, 2*, 3. Retrieved October 20, 2007, from http://www.hss.ed.ac.uk/genomics/vol2no3/documents/APGSPVol2No32006.pdf
Raman, S., & Tutton, R. (2008) Life, science and biopower. Submitted to *Social Studies of Science* (in press).
Rapp, R. (1999). *Testing women, testing the fetus: The social impact of amniocentesis in America*. New York: Routledge.
Rose, N., & Novas, C. (2004). Biological citizenship. In A. Ong & S. Collier (Eds.), *Global assemblages: Technology, politics, and ethics as Anthropological problems* (pp. 439–463). Oxford: Blackwell.
Roseneil, S. (1993). Greenham revisited: Researching myself and my sisters. In D. Hobbs & T. May (Eds.), *Interpreting the field: Accounts of ethnography*. Oxford: Clarendon Press.
Routledge, P. (2003). Convergence space: Process geographies of grassroots globalization networks. *Transactions of the Institute of British Geographers, 28*(3), 333–349.
Schneider, I. (2007). Indirect commodification of ova donation for assisted reproduction and for human cloning research - proposals for supranational regulation. In M. Steinmann, P. Sykora, & U. Wiesing (Eds.), *Altruism reconsidered: Exploring new approaches to property in human tissue* (in press).
Schneider, I., & Schumann, C. (2002). Stem cells, therapeutic cloning, embryo research: Women as raw material suppliers for science and industry. In *ReproKult (Women's Forum for Reproductive Medicine), Reproductive Medicine and Genetic Engineering: Women between Self-determination and Societal Standardisation: Proceedings of the Conference held in Berlin from 15 to 17 November 2001P* (pp. 70–76). Retrieved October 31, 2007, from http://www.reprokult.de/e_forum_3.pdf
Select Committee on Stem Cell Research (2002). Stem cell research. Retrieved October 31, 2007, from http://www.parliament.the-stationery-office.co.uk/pa/ld200102/ldselect/ldstem/83/8302.htm
Sexton, S. (2001). If cloning is the answer, what was the question? Power and decision-making in the geneticization of health. *International Journal of Sustainable Development, 4*(4), 407–433.
Sexton, S. (2005). Transforming "Waste" into "Resource" from women's eggs to economics for women. Retrieved October 31, 2007, from http://www.thecornerhouse.org.uk/pdf/document/eggs.pdf
Spallone, P. (1992). *Generation games: Genetic Engineering and the future for our lives*. London: Women's Press.
Stanley, L. (1991). Feminist auto/biography and feminist epistemology? In J. Aaron & S. Walby (Eds.), *Out of the margins: Womens studies in the nineties*. London: Palmer.
Stringer, E. T. (1999) *Action research: A handbook for practitioners*. London: Sage.
Tarrow, S. (1998). *Power in movement: Social movements, collective action, and politics* (2nd ed.). New York: Cambridge University Press.
Throsby, K. (2004). *When IVF fails: Feminism, infertility and the negotiation of normality* (1-403-93554-8). Houndsmills: Palgrave.
Waldby, C. (2002). Stem cells, tissue cultures and the production of biovalue. *Journal for the Social Study of Health, Illness and Medicine, 6*(3), 305–323.
Welsh, I., Plows, A., & Evans, E. (2007). Human rights and genomics: Science, genomics and social movements at the 2004 London Social Forum. *New Genetics and Society, 26*, 2.
Wynne, B. (1996). May the sheep safely graze? A reflexive view of the expert-lay knowledge divide. In S. Lash, B. Szerszynski, & B. Wynne (Eds.), *Risk, environment & modernity: Towards a new ecology* (pp. 27–83). London: Sage.
Wynne, B. (2006). Public engagement as a means of restoring public trust in science: Hitting the notes, but missing the music? *Community Genetics, 9*, 211–220.

Part III
Environmental Effects of Biotechnology

Chapter 11
Seeking the New Biotechnological Fix? Public Health Genetics and Environmental Justice Policy in the United States[1]

Giovanna Di Chiro(✉)

Abstract: This paper analyzes the Environmental Genome Project (EGP), a new public health research initiative of the US National Institutes of Health that focuses on the role that gene-environment interactions play in disease causation. The public health goal of the EGP is to identify populations that may be genetically predisposed to contracting environmental illnesses such as cancer, asthma, and heart disease. This information about potential genetic susceptibility to disease is intended to be used to design public health policies to protect 'vulnerable' populations, i.e., low-income and poor communities of color who display disproportionately high rates of environmental illnesses. In this paper, I use a feminist-environmental justice lens to analyze the extent to which the EGP signifies the practice of biotechnology in 'the public interest' and whether it supports its stated goals of developing a progressive genetics/genomics tool to serve the aims of environmental justice. Based on interviews with scientists and activists, I discuss what many critics have identified as the genetic reductionism underlying the project. This critique by public interest scientists and women environmental justice activists argues that the EGP focuses more attention on discovering 'susceptibility genes' in 'vulnerable subpopulations' and less attention on the social, political, economic, and environmental factors that are productive of environmental health and illness.

Keywords: Environmental Genome Project, environmental justice, science in the public interest, public health genetics, community participation in science

Giovanna Di Chiro
Environmental Studies Department, Mount Holyoke College, 50 College St., South Hadley, MA 01075, USA
gdichiro@mtholyoke.edu

[1] This article incorporates work previously published in an earlier essay (Di Chiro 2004b).

11.1 Introduction

With the recent flurry of research into gene-environment interactions and their role in disease causation and environmental illness, many feminist scientists, scholars, and activists from around the world have called for a broad critical assessment of these new biotechnologies and their promises to 'revolutionize' the conventional approaches for protecting public health. As Wendy Harcourt argues, a feminist analysis of the potential benefits and risks that may arise from these new developments in biomedical research must incorporate a 'social justice, gender, and ethical perspective', starting with the 'assumption that we need to find and encourage biotechnologies that promote and improve the quality of life for all' (Chapter 1). Moreover, she continues, interdisciplinary feminist conversations lay the foundation for developing a 'big picture vision that gauges the implications of biotechnologies for communities, environments, bodies, food, work, and safety' (ibid.). An inclusive feminist analysis, therefore, adopts a critical stance towards the *presumption* of scientific progress often associated with the discourse on biotechnology, and supports the *democratization* of the decision-making processes that establish the goals and priorities of scientific research. The promotion of a genuine 'science in the public interest' in the field of biotechnology would start with questions such as: Who defines which research questions are most important for improving people's health and solving environmental problems? What constitutes meaningful participation in the future directions of science and technology?

In this paper, I examine the responses and analyses of women activists in the US environmental justice movement (EJM) to the new wave of public health research initiatives introduced by the National Institutes of Health (NIH) that focus on the genetic bases of environmental illnesses. Women environmental justice activists (who are predominantly from low-income communities and communities of color in the US and who live in neighborhoods that are overburdened with polluting facilities such as chemical plants, oil refineries, and toxic waste dumps) have for many years called upon the diverse fields in the environmental sciences to be more responsive to the needs of their communities, many of which suffer disproportionately high rates of environmental illnesses such as asthma, cancer, and heart disease (Di Chiro 1996; Hofrichter 2000; Bullard 2005). A growing number of women activists has joined forces with toxicologists, medical doctors, epidemiologists, agroecologists, foresters, and GIS experts to shape a new 'public interest science' that re-appropriates the tools and resources of science and technology in the service of human and environmental rights (e.g., Heiman 2004; Di Chiro 2004a; Corburn 2005).[2] In so doing, these activists are creating the conditions – epistemological *and* political – for the development of a new 'science of

[2]For example, the EJ organizations West Harlem Environmental Action (www.weact.org), The Louisiana Bucket Brigade (www.labucketbrigade.org), Communities for a Better Environment (cbecal.org), and the Center for Health, Environment, and Justice (www.chej.org), among others, mobilize scientific resources in their organizing strategies while retaining a critical perspective.

environmental justice'. But, as Native American activist Debra Harry asks, 'on whose terms' are such alliances being built?[3] Whose sciences and environmental knowledge systems will comprise the emergent 'science of environmental justice'? Furthermore, are the new developments in biotechnology and public health genetics promoted by the US government serving the goals of environmental justice and moving us toward a healthier, more peaceful, and just future?

11.2 The Environmental Genome Project in Focus

For over a decade, the developments in the sciences receiving the most attention and arguably generating the most controversy involve the field of genetics and the new advances in biotechnology. This paper discusses one of these new developments, a research initiative funded by the NIH that focuses on the use of biotechnology to identify and ultimately 'fix' gene variants that are thought to increase the risk of developing environmental illnesses such as cancer, asthma, diabetes, and heart disease. This government-run project, called the Environmental Genome Project (EGP), is touted as being on the cutting edge of the new 'genetics revolution' and it caught my attention because it was presented as a *progressive* biotechnological 'tool' that would advance environmental justice and *democratize* modern biomedical knowledge – i.e., genetic or molecular knowledge – about disease causation and prevention.[4] The NIH has declared a 'biomedical imperative' to understand disease at the molecular level, an assertion that raises questions about the role of high-tech science – in this case biotechnology – in bringing about new forms of governance in the domain of environmental regulation and public health policy (Wilson and Olden 2004). How does the development and promotion of this new 'molecular knowledge' about disease risk signal or authorize a re-focusing of the 'public health consensus' (Duster 2003) from its presumed emphasis on regulating industrial activities to ensure a clean and healthy environment, to a *new* emphasis on regulating (or, expecting *self*-regulation by) the 'susceptible' individual?

It is clear that the EGP (and other related genetic research initiatives) represents the emergence of the geneticization or molecularization of the environmental health sciences and of public health policy, and many women EJ activists are concerned about what this means for their communities, the so-called 'environmental justice populations' in the US, which are the target populations of the EGP. What the US Environmental Protection Agency (EPA) labels 'environmental justice populations' are the predominantly low-income and poor communities of color who are five times more likely to live next to a polluting facility such as a waste incinerator or an oil refinery and who exhibit disproportionately higher incidence and mortality rates of environmental illnesses such

---

[3] For more information on the work of Indigenous activist, Debra Harry, see Di Chiro (2007).

[4] NIEHS Conference on Human Genetics, Environment, and Communities of Color, Columbia University, New York, February 4–5, 2002.

as cancer, respiratory disease, neurodegenerative disorders, and birth defects.[5] The low-income African American, Latino, and Native American communities who comprise the EJ populations in the US have long argued that their health problems are the result not of their racial or class differences or their genetic make-up but, rather, the result of living in contaminated neighborhoods, breathing polluted air, drinking impure water, living in substandard housing, and not having access to affordable healthcare (Bullard 1994; 2005). Activists from these communities have raised concerns about the US government's support of the EGP and the overall geneticization of the environmental health sciences, which they fear will direct *more* attention to the molecular level of disease causation (that is, to the individual body and its alleged genetic flaws), and *less* attention to the macro-level factors such as racial and gender inequalities, socioeconomic status, political disenfranchisement, and disparities in environmental exposures (reflecting the current neoliberal trend in the US toward environmental de-regulation). The macro-level issues, activists maintain, represent the most significant forces creating 'susceptibilities' to environmental illnesses and, moreover, since they are the historical outcomes of social, political, and economic decisions and policies, they can be changed. To learn more about the diverse perspectives on the forward motion of this 'genetic revolution' in environmental health, I interviewed several scientists involved with the EGP, a group of public-interest scientists who have raised critical questions about the project, and a group of women leaders in the environmental justice movement who live in the communities targeted for genetic research.

11.3 An Overview of the Environmental Genome Project

The Environmental Genome Project is a joint venture by the National Institute of Environmental Health Sciences (NIEHS) and the Human Genome Institute at the NIH and was launched in 1997 as one of the first 'functional genomic studies' emerging from the technological advances on human genetic variation generated by the Human Genome Project and is said to be 'the genomics of the future' (Carrano 1998; Schmidt 2000). In recent years, the consensus in modern biomedical science has concluded that illnesses are the outcome of a 'complex array of factors' including genetic susceptibility, environmental exposures, and aging. Until now, according to government scientists, the largest information gaps in knowledge needed to accurately assess human disease risk have been located in the arena of human genetics, and consequently, the EGP aims to close those gaps. To increase the precision of our understanding of human variation in sensitivity or resistance to disease, EGP researchers are developing a catalogue of all the genetic variances, or single nucleotide polymorphisms (SNPs), that may render certain populations 'more susceptible to, or more resistant to, substances they may encounter at work, at home, or more

[5] See the EPA environmental justice website, retrieved October 20, 2007, from http://www.epa.gov/ebtpages/envienvironmentaljustice.html

generally, in the environment'.[6] Project researchers intend to collect blood samples from 'approximately 1000 Americans from five ethnic groups – Asian American, African American, Hispanic, Caucasian, and Native American – in order to learn what the allelic variations and frequencies of these "environmental disease" genes are in the US population overall and within ethnic groups' (Albers 1997). Former NIEHS director, Dr. Kenneth Olden asserts that this pathbreaking research will help his office identify 'susceptible subgroups' and will 'provide more precise information for regulators, such as the Food and Drug Administration and the Environmental Protection Agency, who now arbitrarily apply large safety margins to cover what is not known about the variations in human susceptibility. More precise information would permit the best protection at the least cost'.[7]

During its initial phase, the EGP is cataloguing relatively common sequence variations, or SNPs, in genetic samples taken from different populations. At present, scientists in the EGP and the Genome Institute have assembled a 'DNA Polymorphism Discovery Resource', which is a set of 450 immortalized cell lines[8] taken from 450 US citizens whose ancestry represents all the major geographic regions of the world – Europe, Africa, Asia, and the Americas.[9] Although the popular discourse about genetic variation research (for example, the Human Genome Diversity Project), appeals to the multi-cultural rhetoric of race/ethnicity, that is, 'equal scientific resources and attention for all racial/ethnic groups', NIH scientists, invoking the contemporary scientific retreat from the idea of race as a legitimate object of biological investigation, assert that race or ethnicity is not a valid basis for genetics research.[10] Rather, they adopt the more rigorous and biologically acceptable concept of 'ancestral lineage' and assert that the Polymorphism Discovery Database contains most of the genetic variation, or genetic archetypes, that exist in human beings. From these 450 cell lines, all obtained with strict institutional review board approval, scientists aim to discover the *normal* range of SNP variations for the so-called environmental response genes that exist in the human species.[11] Today, scientists have 'estimated that there are approximately 11

[6] National Institutes of Health News Advisory (October 10, 1997).

[7] Ibid.

[8] These are cell samples, usually taken from blood, that have been subjected to a series of chemical, bacterial, or genetic manipulations that promote continual replication, thereby providing the researcher a steady supply of the particular cell line.

[9] Author's interview with Dr. Lisa Brooks, National Human Genome Research Institute (NHGRI), Bethesda, MD, January 25, 2001.

[10] Recent work by sociologist, Troy Duster, suggests that genomics research in law and medicine is at the same time rejecting the study of 'race' and developing new 'proxies for race', which he likens to 'putting old wine into new bottles'. For example, see Duster (2002).

[11] Author's interview with Dr. Lisa Brooks, NHGRI, January 25, 2001. The project scientists acknowledge, however, that given the melting pot or mixed-blood composition of the US population, the notion of a pure line of genetic archetypes is unrealistic. This means, of course, that the Polymorphism Discovery Resource will be unable to 'cover all the sources of polymorphisms in the US and thus be a comprehensive survey of the range of sensitivities in the US' since its limited samples could not possibly account for the genomic diversity inherent in a nation of immigrants with immeasurable mixed racial/ethnic backgrounds.

million SNPs in the human population', with only a small percentage of those being associated with higher disease risk (Wilson and Olden 2004).

Project scientists have targeted over 500 environmental response genes, or 'susceptibility genes', that code for proteins critical in the body's response to environmental agents and that appear to be associated with increased risk for disease. The environmental response genes of interest to EGP scientists include, for example, DNA repair genes, cell cycle control genes, cell signaling and receptor genes, and the cytochrome P450 family of genes, which produce enzymes that metabolize or detoxify chemical toxins. Polymorphisms in these environmental response genes are thought to play a role in causing disease or in susceptibility to disease and 'might determine how a person responds to and metabolizes drugs or carcinogenic compounds after exposure' (Olden and Wilson 2000). Kenneth Olden and Samuel Wilson explain the EGP's focus on the 'environmental response genes' and their relation to the environment:

> Every organism is exposed to hazardous agents in its environment on a continual basis. As a result, organisms have evolved sophisticated pathways that can minimize the biological consequences of hazardous environmental agents. These pathways constitute "the environmental response machinery". All human genes, including those that encode components of the environmental response machinery, are subject to genetic variability, which can be associated with the altered efficiency of a biological pathway. So, a person's risk for developing an illness as a result of an environmental exposure might be dependent on the efficiency of their own unique set of environmental response genes (Olden and Wilson 2000).

To date, the EGP scientists have completed the re-sequencing of 371 of the 550 candidate 'susceptibility genes' and have uncovered 50,000 new SNPs within these genes. These susceptibility gene SNP assays are now publicly available for use in genetic analysis in a GeneSNP database, which is housed and maintained at the University of Utah's Genome Center (Wilson and Olden 2004; Nickerson et al. 2005). After accomplishing the resequencing and cataloguing of all the possible polymorphisms contained in the 550 selected environmental response genes, phase 2 of the EGP will determine the functional significance of those allelic variants.[12] In other words, phase 2 will study how, if at all, one or more polymorphisms in these candidate environmental response genes predispose an individual to contracting a disease whether or not the individual is exposed to an environmental 'trigger'. This phase of the project will study the association of particular SNPs with particular diseases by conducting population studies examining the 'frequency with which given alleles occur in a population of affected people, compared with the frequency of the same alleles in a population of unaffected people' (Olden and Wilson 2000). For example, if a group of people has a sequence variation, or a combination of sequence variations, in one or more of the histocompatibility genes that code for proteins that resist damage from the toxic substance *beryllium*, then if those individuals are exposed to beryllium, will they be more susceptible than others to developing berylliosis, a deadly lung disease?[13]

---

[12] Author's interview with Dr. José Velasquez, NIEHS, Research Triangle Park, North Carolina, January 30, 2001.

[13] A group of scientists has recently determined that approximately 97% of workers suffering from berylliosis carry a genetic susceptibility to the disease; see Wang et al. (1999).

Finally, phase 3 of the EGP will develop animal models to increase our understanding of human susceptibility to disease and the precise mechanisms leading to disease. After identifying and isolating a particular polymorphism that appears to increase an individual's susceptibility to environmental illness, scientists will then investigate the exact function of that SNP variation by conducting so-called knockout and knock-in experiments on laboratory animals. These studies will help scientists to determine the biological outcomes, or the phenotypic expression that surfaces, when a specific gene is removed from or added to the organism under experimental conditions (in Olden and Wilson 2000).[14]

An allied project of the EGP, also sponsored by the NIEHS, is the National Center for Toxicogenomics (NCT). The promise of toxicogenomics lies in determining at a molecular level exactly how a particular toxic substance damages the DNA itself. Using new genetic methodologies and technologies, scientists can expose experimental samples of certain target genes or gene fragments to a carcinogenic chemical like trichloroethylene (TCE), for example, in order to determine exactly how those target environmental response genes are impaired or altered as a result of exposure. Toxicogenomic research aims to develop computer-assisted genetic assays that might identify protein biomarkers of exposure and effect, enabling scientists to detect specific genetic 'signatures' for a given pathway of toxicity. These signatures might help 'identify the agent and dose to which individuals or populations have been exposed', which will, according to the NCT, 'dramatically change our understanding of human disease risk...provid[ing] new opportunities for the nation (and individuals) to protect human health and prevent disease'.[15]

11.4 Environmental Health Policy Implications

By collecting information on the genetic basis of environmentally-associated diseases and increasing our knowledge of gene-environment interactions, the EGP's (and NCT's) environmental health goals are to help with more accurate assessments of disease risks, thereby increasing multi-fold the explanatory and predictive power

[14] Martin Teitel, former director of the Council on Responsible Genetics, explained to me that 'knockout' experiments regularly result in what is known as genome redundancy. That is, a gene is knocked out (i.e., chemically removed from the genetic material), and the associated trait or function that the scientist was expecting to *disappear*, appears phenotypically anyway, demonstrating that several genes may perform the same function. The far-reaching goals articulated by promoters of human genome research, including the EGP, assert a greater level of confidence in science's putative mastery of the workings of the human genome than the experimental data provide. Author's interview with Martin Teitel, Council on Responsible Genetics, Cambridge, MA, January 26, 2001.

[15] National Center for Toxicogenomics website, retrieved October 20, 2007, from http:// www.niehs.nih.gov/nct/impact.htm. For an analysis of the intersections of biological and cultural meanings in toxicogenomics, see Shostak (2003).

of environmental health risk assessment and epidemiology. The EGP scientists argue that due to our incomplete knowledge about disease causation, standard epidemiological studies are largely based on educated guesses, an imperfect decision-making process that forces government agencies, such as the EPA or the FDA, to either *underregulate* or *overregulate*. By linking the power of molecular studies to epidemiology, scientists argue, regulators can increase the scientific validity of the government's environmental health policies.[16] Furthermore, the EGP's SNPs catalogue will provide information to the pharmacogenetics industry, which, when armed with this new knowledge of susceptible subpopulations, can design action-specific drugs. A drug may be designed, for example, to supply lung tissue protection from the damaging effects of beryllium exposure that the aforementioned hypothetical subpopulation's particular genome doesn't provide.

The EGP also hopes to aid in earlier diagnosis of disease or to screen populations at higher risk of disease, thereby 'taking the guesswork out of risk assessment' (Olden and Guthrie 2001). Armed with this new monitoring tool, regulators would have access to the potential range of sensitivities to a particular environmental agent and would be able to control it with greater accuracy, or to develop better preventative measures. According to Kenneth Olden and Janet Guthrie, the precision afforded by this new information will allow regulators and medical professionals to 'reduce risk…by several mechanisms: (i) eliminating or reducing exposure, (ii) pharmacological intervention, and (iii) gene therapy' (Olden and Guthrie 2001). Indeed, this list represents the primary risk-preventative measures that were enumerated during my interviews with EGP geneticists. The first line of defense consists of cautioning at-risk individuals that they should keep themselves out of harm's way, an admonition that might proceed as follows: 'If you have the SNP that makes you genetically sensitive to, say, beryllium, you are advised not to work in the Brush-Wellman defense contracting plant in Tucson, Arizona, because there you may be exposed to beryllium, an element that is vital to the manufacture of nuclear weapons and to the aerospace industry'. Other second-order 'preventative' measures that were mentioned during my interviews included manufacturing and prescribing genome-specific drug and gene therapies, and DNA screening for 'susceptibility genes' to use in genetic counseling on heritability and reproduction.[17] Labeling these post-exposure measures as 'preventative' is what's known in environmental regulation parlance as an 'end of pipe' solution – *not* in fact what most environmentalists would consider *prevention*. Prevention of environmental health problems usually means not producing the pollution, or the environmental 'trigger' in the first place.

More in line with what is generally understood as prevention, the EGP scientists pointed to the significance of the role of genomic research on susceptibility genes for the development of more effective environmental regulations designed to protect 'susceptible' individuals. Of course, this begs the question of what in fact does it mean to be 'susceptible'? That is, can or should someone be labeled 'susceptible',

---

[16] Author's interview with Dr. José Velasquez, NIEHS, January 30, 2001.

[17] Author's interview with Dr. Lisa Brooks, NHGRI, January 25, 2001.

or genetically-disadvantaged, *before* they are exposed to a toxin that most would argue they should not be exposed to in the first place? If an individual carrying a specific 'faulty' genetic variation or variations is *not* exposed to a potential environmental 'trigger', they most likely would not become ill. The discourse of susceptibility assumes that we *will* live with environmental toxins – it *naturalizes* environmental toxicity and at the same time *pathologizes* some genomic subsets of the human population and constructs a new geneticized subjectivity – the genetically at-risk person (or group) (Novas and Rose 2000). The idea of inherent genetic susceptibility to environmental illness and the identification of the genetically at-risk individual or subpopulation makes sense only if it is accepted that environmental pollution is a given or a 'natural' and inevitable component of modern society. Ultimately, this approach to environmental health policy is about placing the responsibility for ameliorating risk squarely on the shoulders of the *individual* and shifts the focus away from government responsibility and corporate accountability. Furthermore, this approach to environmental health policy can be understood as an example of Foucault's (1978) concept of *biopower*; being identified as 'genetically at risk' is a *productive* identity formulation creating the conditions for individuals to enact the specific practices of what Novas and Rose call 'responsible genetic subjects' (Novas and Rose 2000).[18] Some of the 'responsible' practices of the genetically at-risk individual citizen might include choosing to undergo genetic counseling, Pre-implantation Genetic Diagnosis (PGD), or choosing *not* to live or work in a particular place that may expose you to those toxins to which you know you are 'susceptible'. As I learned more about the EGP, it was becoming clear to me that the proclaimed research focus on gene-environment *interactions* that supposedly undergirded the project seemed to display much more curiosity about an individual's genetic make-up than about the toxic substances that may be contaminating the individual's environment.

11.5 Environment, Genes, and Justice: Democratizing Human Genome Research

To gain other perspectives on the promises and limitations of human genome research I also interviewed two geneticists allied with environmental justice organizations who are interested in genetic research on gene-environment interactions that may be of use to communities suffering from high rates of environmental diseases.

[18] See Plows and Boddington (2006) for a critical analysis of Novas and Rose's concept of biocitizenship, which they argue is insufficiently attentive to the social inequalities that construct who gets to count as a 'citizen' and which citizens are able to exercise their civil rights. Furthermore, they examine how the discourse of biocitizenship, with its attendant concepts of rights and responsibilities, verges on the *presumption* of the existence of a democratic system governing the biosciences, which has created the 'space for open debates about the directions and benefits of biotechnology' (Plows and Boddington 2006).

Trained in genetics and molecular biology, both Dr. Paul Billings (GeneSage and Council for Responsible Genetics) and Dr. José Morales (Biotechnology in the Public Interest) support the growth of scientific research initiatives, such as the EGP and the NCT, that study how allelic polymorphisms confer resistance or susceptibility to a particular environmental agent. For Morales, such advances in biotechnology help us to understand how the diverse mechanisms creating disease pathways are encoded in the genomes of different populations and, more important, he argues, this new knowledge will provide powerful 'informational weapons' to arm impacted communities with the scientific tools necessary to fight for environmental justice.[19] While both scientists insist that genetics research greatly adds to our understanding of human health, they argue strongly against the genetic reductionism embedded in the promotion of technologies such as genetic testing and screening as the *key* to preventing environmental diseases. Billings emphasizes this point:

> One of the things that annoys me about [the HGP] is that people are talking about the causation of disease as being present in the genome. That's incorrect. The *mechanism* of disease is in the genome, in the sense that much of what diseases is an aberrant physiological response to an environmental agent. So the bacteria gets into your lungs, you pour out all these white cells...all that stuff is in genetic control. But, that doesn't mean the causation of the illness was the genes. The causation of the illness was the bacteria. In the same sense, [in the EGP] the environmental agent will be the causation. There may be genes that alter our responses and there may be damage that the environmental agent actually produces to the gene – all possible. But it's still the environmental agent that causes the problem.[20]

Genetic reductionism oversimplifies what Morales and Billings, and many of the EGP scientists themselves, identify as the real causes of disease. Although there *are* some diseases that are highly associated with particular genes, such as Tays-Sachs disease or cystic fibrosis, *most* diseases have their roots in multiple causes, whether it be the involvement of multiple genes or exposures to multiple environmental toxins. But, increased risk for disease is also linked to other factors including inadequate nutrition, poverty, and lack of quality health care. Furthermore, as Morales and Billings explain, although some diseases, like cystic fibrosis, are caused by a single, recessive, highly penetrant allelic variation, most environmental diseases, like cancer, are 'polygenic', that is, they are the result of the complex interactions of several genes that will modify an individual's health risks only in the event of exposure to specific environmental agents. Therefore, as these public interest scientists imply, the most rational and simplest measure to improve the health of low-income populations would be to prevent the dumping, incineration, or release of environmental contaminants into their or anyone else's communities, one of the central goals of the environmental justice movement.

---

[19] Author's interview with Dr. José Morales, Biotechnology in the Public Interest, New York, December 12, 2001.

[20] Author's interview with Dr. Paul Billings, GeneSage, San Francisco, February 16, 2001.

11.6 Whose Science? Whose Environmental Health Research Agenda?

As discussed in this paper's introduction, a feminist analysis of biotechnology would start with the issue of who determines which research questions are most important for improving people's health and solving environmental problems (Harding 1991). In my conversations with the NIH scientists, it would appear that there was considerable public interest in the toxicity of *beryllium* since the example of beryllium exposure was consistently put to use in helping to explain to me the goals of the EGP. But, are communities in fact calling for more research into the genetic origins of berylliosis? I had first heard about beryllium during a research trip to Tucson, Arizona, to interview women environmental justice activists working on the US-Mexico border. One of these activists, Rose Augustine, invited me on a 'toxic tour' of the 85% Latino neighborhood of South Tucson where we visited the Brush Wellman factory, a defense contracting plant that uses beryllium to manufacture the trigger mechanisms for nuclear missiles such as the Tomahawk and Patriot missiles deployed in the first Gulf War and the new generation of air-to-surface missiles manufactured in the US today.

Beryllium is a metallic element extracted from bertrandite and beryl mineral ores and is chemically converted into a hard, lightweight metal that is three times lighter than aluminum and six times harder than steel (Roe 1999a). Because of these properties, and because it enables a more efficient chain reaction in the nuclear fission process, beryllium has become a 'strategic metal' critical to the US government's military policy and to the expansion of its nuclear arsenal. Beryllium was first produced for use in the Manhattan Project in factories in Cleveland and Lorain, Ohio, and by the mid-1940s local doctors were treating dozens of beryllium workers and their families (including nuclear weapons scientists) who had become seriously ill with berylliosis, a respiratory disease that destroys the lung's delicate alveoli air sacs and is usually fatal. Alarmed at the rising incidences of berylliosis in factory workers and nuclear scientists, the newly formed Atomic Energy Commission realized it had both a public health and a public relations problem on its hands. Although the federal government knew that the metal was toxic at infinitesimally small doses, in the name of national security it continued to support production, to overlook Occupational, Safety and Health Administration (OSHA) inspections, and to kill plans to strengthen beryllium regulations. Despite considerable epidemiological evidence and a mounting death toll, only once in the last fifty years has OSHA tried to impose stricter exposure limits for beryllium. That was in 1975 and the then-head of the Department of Energy, James Schlesinger, warned that the new regulations would 'seriously limit our ability to develop and produce weapons for the nuclear stockpile' (Roe 1999b). Thus, with the endorsement of President Jimmy Carter, the OSHA plan to more tightly regulate beryllium was abandoned.[21]

---

[21] See the reprint of the six-day series on the history of beryllium in the United States (Roe 1999a; b).

As Rose Augustine and I stood outside the Brush Wellman factory in South Tucson, we saw the plant's stacks towering over an elementary school and a child-care center. Augustine commented: 'Many people in the neighborhood are sick with lung problems and the [Pima County Health Department] tells us it's our "genealogy" or the chilies and beans that we eat'.[22] Augustine expressed concern about how genetics arguments are used to explain the serious illnesses suffered by the low-income Latino communities who work in and live next to the plant. It became more clear to me that the scientific research issues of interest to the NIH geneticists were quite different from the environmental health concerns of the Latino residents and workers of South Tucson.

Today, in the name of defending the 'free world' and the war against terrorism, the US government is still in the business of nuclear weapons manufacturing and therefore requires a steady supply of beryllium and its alloys. It is in the interests of national security and military supremacy that the Department of Energy, the Department of Defense, *and* the NIH should be eager to determine how to prevent 'susceptible' subpopulations, who might comprise the labor force in nuclear weapons production factories, from getting sick from exposure to beryllium.

11.7 Dilemmas of Participation in Human Genetics/Genomics Research

A feminist approach to research in the new biosciences would be attentive to concerns about what constitutes meaningful participation in the future directions of biotechnological research. Particular concern is voiced by many Native American communities regarding the potential of the new genetics to improve the health of Indigenous peoples and other communities of color by collecting genetic variation data on the different 'races'. Native American women activists, such as Debra Harry, a Northern Paiute from the Pyramid Lake Reservation in Nevada, have consistently questioned the US government's plans to catalogue Indigenous Peoples' DNA and call for greater scrutiny of the nature and quality of public participation in this new scientific research. Harry argues that participation in projects like the (now suspended) Human Genome Diversity Project (HGDP) and the EGP, which for Native Americans most often has been limited to donating blood or other body parts, must be predicated on the 'return of the benefits of the genetic research to the donor community'.[23] A number of Indigenous groups in North America and worldwide, Harry explains, are offered an exchange value arrangement to remunerate them for their involvement in the research – either they are given a cash payment

[22] Author's interview with Rose Marie Augustine, Tucsonans for a Clean Environment, Tucson, Ariz., June 13, 2001.

[23] Author's interview with Debra Harry, Indigenous People's Council on Biocolonialism, Wadsworth, Nev., January 23, 2001.

for their body material, provided with limited health care for conditions like diabetes, or promised royalty payments for future products that may be manufactured from their genetic materials.[24]

For Harry, human genome diversity research like the EGP represents the newest genre of race-based science couched in the benign and objective language of either 'geographic lineage' or 'environmental disease prevention'. She argues that contemporary forms of eugenics and racialized science arise when researchers insist on 'collecting and maintaining racial group or tribally-identified labels on samples', a cataloguing practice that is paramount to 'operating with a zookeeper's mentality' and that inherently invites 'racialized outcomes' (Prakash et al. 2003). Furthermore, Harry contends, the health problems of most Indigenous peoples in North America and elsewhere are the outcome of the interconnected histories of colonialism, economic poverty, deteriorating infrastructure, and contaminated environments. The single-minded focus on genes obscures these fundamental causal factors and diverts attention away from the necessary actions to change them – actions that would call for political, cultural, and economic solutions to these *preventable* problems. Indigenous peoples around the globe, says Harry, must be fully informed about the goals and objectives of the most recent army of 'bio-prospectors' to invade Native peoples' lands and communities in order to decide for themselves whether they should participate in research projects like the EGP.[25]

With these caveats in mind, exactly what would meaningful participation in genetics research consistent with the goals of environmental justice look like? To respond to this question the NIH's program on the ethical, legal, and social implications (ELSI) of human genetic/genomic research (Sharp and Barrett 2000) offers grants for community-based initiatives that 'promote public understanding of the social, ethical, and legal implications of conducting environmental health research involving human subjects in areas such as gene-environment interactions'.[26] One of these initiatives, supported by a $1.2 million ELSI grant, involved a consortium of three universities: the University of Michigan, Michigan State University, and the Tuskegee Institute. The central aim of the 'Communities of Color and Genetics Policy Project' was to 'develop a process to engage communities of color, representing a range of socio-economic backgrounds, in dialogues concerning genetic research and its resulting technologies'.[27] The project convened 'dialogue groups' of ten to fifteen people per group with African American and Latino communities in Michigan and Alabama and assembled their responses in a report to be developed into recommendations for laws, professional standards, and institutional policies

[24] Ibid.

[25] Ibid.

[26] Request for Applications, RFA ES-02-005, Environmental Justice: Partnerships to Address Ethical Challenges in Environmental Health. Program Administrator, Shobha Srinivasan, Ph.D., NIEHS, Research Triangle Park, North Carolina.

[27] Interview with Vence Bonham published on March 2001 in The Alliance Alert, monthly publication of the Alliance of Genetic Support Groups, Washington, DC.

regarding the use and application of genome research and technology.[28] The most oft-repeated concerns raised by participants in the dialogue groups were about privacy and control of access to one's genetic information, fears of stigmatization and discrimination by employers or insurance companies, and distrust of government and scientific agendas in terms of who will be the primary beneficiaries of genetics research (Schulz et al. 2000).

The apprehensions voiced by members of the Michigan and Alabama communities are situated within a broader social paradigm of 'scientific inevitability', in this case, the inevitability of the forward progression of human genetics research. Given that the dialogue groups are funded with ELSI monies that are disbursed by the NIH, the very institution that is conducting the genetics variation research, there exists little opportunity for participants to raise questions that are outside the rhetorical universe of the forward advancement of this research. The participants in the ELSI community-based dialogue groups were limited to raising concerns about their taking part in genetic research projects exclusively within the context of how to minimize the potential dangers of the research process and its potential outcomes; it is assumed that the 'human genome research ship' has already sailed out of the harbor (or, as Wendy Harcourt titled her paper 'Headed blithely down the garden path', see Chapter 1). The predominantly African American and Latino respondents were not invited to question the research agendas themselves, nor to recommend other research agendas. This notion of 'participation' is about reassuring the donors of genetic material (that is, members of minority communities) that their privacy will be protected, their culture will be respected, and that they will benefit somewhere down the line – all worthy goals – but it is not about strengthening the tools of participatory democracy that would enable community residents to contribute meaningfully to the discussion on how to solve environmental health problems. Debra Harry argues that these community-based ELSI studies are essentially designed to get communities of color to rubber-stamp the research and provide the informed consent required to proceed legally with the genetic studies. As the head of the NIEHS, while speaking at the plenary session of an ELSI-funded conference in New York in 2002, succinctly put it: 'the research must not be stopped' (Olden 2002).

What would constitute appropriate guidelines for meaningful participation in community-based genetic research projects? Meaningful participation does not mean either donating one's blood or other body parts and then being remunerated for these parts, what I call 'exchange-value' participation, where bodily material is a commodity that is exchanged for cash or other compensation. Nor does it mean, as in the case of ELSI-sponsored dialogue groups, first assuming the inevitability of the implementation of the scientific project and the risks that go along with it, and then laying out a set of limited, predetermined options among which an at-risk community may choose. Genuine participation that supports environmental health

[28] See the project website, retrieved October 20, 2007, from http://www.sph.umich.edu/genpolicy/index.html

democracy means establishing at the outset exactly what are the community's most urgent health problems. Moreover, democratizing the research process would necessitate a commitment to the centrality of the community's role in setting research goals, in formulating and prioritizing research questions and protocols, and in contributing to decisions about use and dissemination of the research outcomes. The recent turn in government-sponsored studies toward 'community-based research' may, according to some activists, offer some promise in efforts to build collaborative research partnerships among environmental health scientists and members of communities of color. An assessment of the true collaborative nature of these studies will rest on the extent to which community residents' voices are heard *before* the genetics research ship sets sail.

Expressing a restrained optimism toward the current trend in community-based research on environmental health genetics, many activists wonder whether this new focus on *partnerships* might not respond to the concerns of the environmental justice movement. The success of these community-government partnerships depends, however, on the extent to which the NIEHS's research on 'gene-environment interactions' takes seriously the interactive role of both genetic *and* environmental factors. Indeed, many government geneticists contend that the EGP's stated focus on the *interaction* between genes and the environment mitigates against the narrow, genes-as-root-cause explanation that characterizes the genetic reductionist paradigm. Destabilizing the reductionist standpoint by emphasizing interaction would also challenge the underlying assumptions that are associated with it: 'inevitability' (of toxic pollution in the environment) and 'susceptibility' (of genetically-disadvantaged subpopulations). While this sounds like a plausible argument, during my interviews with EGP scientists references to the *environment* end of the equation rarely came up. The environment consistently falls out of the picture in EGP documents except as the naturalized substratum on which genetic variation operates.

To give one example, in 2003 researchers at Columbia University released the extraordinary statistic that 25% of Latino and African American children who live in Northern Manhattan (Harlem and Washington Heights) are diagnosed with asthma (Pérez-Peña 2003). The NIH has funded environmental genome researchers to work in partnership with community-based organizations in New York City to investigate this alarming statistic. In the name of promoting environmental justice and working in a scientific/community partnership, researchers have observed and monitored African American and Latino children as they walk to school and play in the school yard with 'diesel bus smoke blowing in their faces'. Then, in the laboratory, the researchers 'measure the concentration of 1-hydroxypyrene[29] in their urine and ultimately will study the effects that polymorphisms in their response genes might have on their ability to resist severe respiratory distress' (Pérez-Peña 2003). Despite the fact that Harlem houses 6 out of 8 of Manhattan's diesel bus depots and several waste

[29] 1-hydroxypyrene is a urinary metabolite that is used as biomarker of toxic exposure. It is produced when the body is exposed to aromatic hydro-carbons such as benzene and benzo (a) pyrene, both components of diesel exhaust.

transfer stations, and despite the fact that the EPA lists the hazardous components of diesel exhaust as probable asthmagenic and carcinogenic compounds, the thrust of the research persists in zeroing in on the genetic factors associated with high levels of respiratory disease in communities of color (Corburn 2005).

The women activists I interviewed argue that the single-minded focus on identifying a polymorphism that might make an African American child breathing diesel exhaust in Harlem more susceptible to asthma is not responding to the concerns of the environmental justice movement and does not represent a genuine 'science in the public interest', which is what the new community-based genetic research initiatives at the NIH claim to endorse. I think if the EGP is serious about using biotechnology to genuinely focus on gene-environment *interactions*, it could change the terms of the debate – it could show that susceptibility to environmental diseases is not an inherent quality, an inborn genetic disadvantage, but the result of the interaction between one's body and an unjustly contaminated environment. Therefore, the way to 'fix' the problem would not be primarily biotechnological, but would need to address the social, political, and environmental factors that produce disparities in environmental health.

These women activists instead ask different environmental health research questions: How do we remove pollutants from the environment and from people's bodies? How can we provide communities with adequate healthcare, nutrition, and clean air and water? How can we develop more sustainable industries and jobs that do not pollute the environment? For these women environmental justice activists whose children and communities are suffering the highest rates of environmental illnesses, the EGP is asking all the wrong questions.

11.8 Conclusion: Seeking Biotechnology in the Public Interest

The rhetoric of 'revolution', so widespread in popular science reporting on human genome research, demands that we take notice. But, what exactly are the promises of the genetics revolution to which we can look forward? The proclamations of an imminent 'golden age' (Olden 2002) of biomedicine that accompany recent trends yoking public health research to the new human genome bandwagon compel us to ponder just what is *golden* about this new age.

Environmental justice activists, most of whom are women of color, contest the flawed logic of genetic reductionism illustrated in the above examples the way it *forecloses* on and erodes commitment to solving environmental problems that are the result of social and economic injustices, *not* faulty genes.[30] Women activists like Debra Harry and Rose Augustine have become proactive technoscientific actors by

[30] Debra Harry and the Indigenous People's Council on Biocolonialism (IPCB) have published two booklets to educate Indigenous communities about the cultural, environmental, and health implications of the new wave of human genome research. See Harry et al. (2000, 2001).

increasing their scientific literacy and insisting on actively participating at the forefront, not the back of, decision making about scientific ‘revolutions’ that affect their lives. Refusing to passively accept the gender, class, and race reductionism that has historically restricted women and minorities from access to the upper echelons of science, they have challenged the genetic reductionism that tends to locate the source of environmental health problems in our *genes* rather than in our society’s environmentally-destructive activities that emit poisons into the earth and into the bodies of all living things.

Debra Harry and Rose Augustine both agree that, despite the government’s stated intentions to promote public acceptance of human genome research by instituting community-based dialogue groups, the greatest stride toward achieving *authentic* participation in the future directions of science and technology is enabling all communities the opportunity to develop diverse scientific literacies, a prerequisite for a genuinely democratic society committed to environmental justice at the turning of the ‘century of the gene’.[31]

References

Albers, J. W. (1997). Understanding gene-environment interactions. *Environmental Health Perspectives, 105*(6), 1–2.

Bullard, R. (Ed.) (1994). *Unequal protection: Environmental justice and communities of color.* San Francisco: Sierra Club Books.

Bullard, R. (Ed.) (2005). *The quest for environmental justice: Human rights and the politics of pollution.* San Francisco: Sierra Club Books.

Carrano, A. (1998). Genome resources help scientists explore life-environment interactions. *Human Genome News, 9*, 1–5.

Corburn, J. (2005). *Street science: Community knowledge and environmental health justice.* Cambridge, MA: MIT Press.

Di Chiro, G. (1996). Nature as community: The convergence of environment and social justice. In W. Cronon (Ed.), *Uncommon ground: Rethinking the human place in nature.* New York: W.W. Norton.

Di Chiro, G. (2004a). Local actions, global visions: Remaking environmental expertise. In R. Eglash, J. Croissant, G. Di Chiro, & R. Fouché (Eds.), *Appropriating technology: Vernacular science and social power.* Minneapolis, MN: University of Minnesota Press.

Di Chiro, G. (2004b). Producing “Roundup Ready®” communities?: Human genome research and environmental justice policy. In R. Stein (Ed.), *New perspectives on environmental justice: Gender, sexuality, and activism.* New Brunswick, NJ: Rutgers University Press.

Di Chiro, G. (2007). Indigenous peoples and biocolonialism: Defining the “science of environmental justice” in the century of the gene. In R. Sandler & P. Pezzullo (Eds.), *Environmental justice and environmentalism: The social justice challenge to the environmental movement.* Cambridge, MA: MIT Press.

Duster, T. (2002). *Human molecular genetics and the subject of race: Contrasting the rhetoric with the practices in law and medicine.* Paper delivered at the conference on Human Genetics, Environment, and Communities of Color, Columbia University, New York, February 4.

[31] A phrase coined by feminist science critic, Evelyn Fox Keller (2000).

Duster, T. (2003). The hidden eugenic potential of germ-line interventions. In A. R. Chapman & M. S. Frankel (Eds.). *Designing our descendants: The promises and perils of genetic modifications* (p. 159). Baltimore, MD: Johns Hopkins University Press.

Foucault, M. (1978). *The history of sexuality. Vol. 1: The will to knowledge*. New York: Vintage.

Harding, S. (1991). *Whose science? Whose knowledge: Thinking from women's lives*. Ithaca, NY: Cornell University Press.

Harry, D., Howard, S., & Shelton, B. (2000). *Indigenous peoples, genes and genetics*. Nixon, NV: Indigenous Peoples Council on Biocolonialism.

Harry, D., Howard, S., & Shelton, B. (2001). *Life, lineage and sustenance: Indigenous peoples and genetic engineering, threats to food, agriculture, and the environment*. Nixon, NV: Indigenous Peoples Council on Biocolonialism.

Heiman, M. (2004). Science by the people: Grassroots environmental monitoring and the debate over scientific expertise. In R. Eglash, J. Croissant, G. Di Chiro, & R. Fouché (Eds.), *Appropriating technology: Vernacular science and social power*. Minneapolis, MN: University of Minnesota Press.

Hofrichter, R. (Ed.) (2000). *Reclaiming the environmental debate: The politics of health in a toxic culture*. Cambridge, MA: MIT Press.

Keller, E. F. (2000). *The century of the gene*. Cambridge, MA: Harvard University Press.

Nickerson, D., Reider, M., Crawford, D., Carlson, C., & Livingston, R. (2005). An overview of the environmental genome project. *Environmental Health Perspectives*. Retrieved April 31, 2007, www.ehponline.org/docs/2005/7922/7922.html

Novas, C., & Rose, N. (2000). Genetic risk and the birth of the somatic individual. *Economy and Society, 29*, 485–513.

Olden, K. (2002). *The role of gene-environment interaction in health disparities* (p. 9). Paper presented at the conference on Human Genetics, Environment, and Communities of Color, Columbia University, New York, February 4.

Olden, K., & Guthrie, J. (2001). Genomics: Implications for toxicology. *Mutation Research, 473*, 9–15.

Olden, K., & Wilson, S. (2000). Environmental health and genomics: Visions and implications. *Nature Reviews in Genetics,* 1, 149–153.

Pérez-Peña, R. (2003). Study finds asthma in 25% of children in Central Harlem. *New York Times,* April 19, 2003, A-1.

Plows, A., & Boddington, P. (2006). Troubles with biocitizenship? *Genomics, Society and Policy, 2*(3), 115–135.

Prakash, S., Sze, J., & Shepard, P. (2003). Introduction: Human genetics, environment, and communities of color. *Human Genetics, Environment, and Communities of Color: Ethical and Social Implications: Conference Program and Resource Guide* (pp. 22–23). New York: West Harlem Environmental Action.

Roe, S. (1999a). Decades of risk: US knowingly allowed workers to be overexposed to toxic dust *Toledo Blade,* March 28–April 2, 1999, 7.

Roe, S. (1999b). Deadly alliance: How government and industry chose weapons over workers. *Toledo Blade,* March 28, 1999, 3–15.

Schmidt, C. (2000). Populations and polymorphisms: Building the new science of environmental genomics. *Genome News Network, Celera Corporation,* 1–3.

Schulz, A., Foster, S., Caldwell, C., Modell, S., & Singer, E. (2000). 1999/2000 focus group content report: Communities of color and genetics public policy project. Retrieved October 20, 2007, from http://www.sph.umich.edu/genpolicy/current/focus_report2.pdf

Sharp, R., & Barrett, C. (2000). The environmental genome project: Ethical, legal and social implications. *Environmental Health Perspectives, 108*, 279–281.

Shostak, S. (2003). Locating gene-environment interaction: At the intersections of genetics and public health. *Social Science & Medicine,* 56(11), 2327–2342.

Wang, Z. et al. (1999). Differential susceptibilities to chronic beryllium disease contributed by different Glu69 HLA-DPBI and –DPAI Alleles. *Journal of Immunology,* 163, 1647–1653.

Wilson, S., & Olden, K. (2004). The environmental genome project: Phase I and beyond. *Molecular Interventions, 4*(3), 147–156.

Chapter 12
Technological Challenges: Asbestos Past Experiences, Nanoparticles Future Developments

Qamar Rahman(✉)

Abstract: Science and technology are the main components of economic and social development, and globalization. They are, on the one hand, responsible for improvement in the quality of life, but on the other hand, if technological processes are applied prior to their toxicity evaluation and safety testing, they may also become unfriendly. It is therefore necessary to implement any new technology only after it has undergone complete risk assessment studies. The biotechnological revolution, also called the Life Science Revolution, and the new technologies available to it are claimed to have the potential to change everything about our society. Among the newest tools is the concept known as nanotechnology, the science of small particles. The advocates of nanotechnology predict that it will revolutionize the field of engineering, electronics, medicine, IT etc. However, present studies suggest that it also poses a number of threats to human health.

The use of nanotechnology techniques without prior human health evaluations faces society with the possibility that they could become the 'asbestos' of the twenty-first century. Asbestos was discovered in 1878 and became a common, highly desirable component in thousands of products and industrial applications all over the world. Only later was it realized that prolonged exposure to asbestos causes Asbestosis, Malignant Mesothelioma and Bronchogenic Carcinoma. These diseases were identified only after a long latency period. Unfortunately, many nanotechnology-related products are already on the market and in use without having undergone adequate safety evaluations. Serious adverse health effects can emerge when engineered NPs (nanoparticles), which in general were not properly characterized regarding their biohazard potential, are administered to humans intentionally (e.g., for medicinal purposes) or unintentionally (e.g., in the course of regular or accidental industrial processes).

Qamar Rahman
Research & Development, Integral University, Kurst road, Lucknow, India
qamar_15@sify.com

F. Molfino, F. Zucco (eds.) *Women in Biotechnology*,

Keywords: Asbestos, bronchogenic carcinoma, mesothelioma, nanotechnology, single and multi walled carbon nanotubes,, nanoparticles, new technologies

12.1 Introduction

In my approximately 40 years scientific journey, I have passed through various experiences. I have realized that at every stage a woman has to work harder than a man. In the male-dominated society, for a woman to achieve and make a name is very difficult. But I have never accepted defeat and have kept myself going in spite of many hurdles. I have always tried to work for the betterment of society and have never accepted or supported technologies that are harmful to the environment. In this presentation, I will discuss the problems caused by the use of asbestos, talk about my own findings with respect to technological change, and explore some of the possible future challenges of nanotechnology.

Particles and fibers have been present since the existence of human life. Their toxicity was first recognized at the time of Agricola and Paracelsus in the fourteenth century. During various revolutions, such as agricultural, industrial and biotechnological, these fibers and particles have become a matter of great concern. All the revolutions engendered by various technologies were thought to be for the benefit of human life. The basic idea was always to make life more beautiful, more comfortable and nobler. It was unfortunate that we never thought about the potential for adverse consequences of these technologies.

The biotechnological revolution, also called the Life Science Revolution, and the new technologies leading their development are claimed to have the potential to change everything about our society. Among these technologies, at this moment, is nanotechnology. This novel and rapidly growing field, using engineered and artificially synthesized nanoparticles (NPs), is currently expanding to another cardinal source of man-made toxic exposure. Manipulation of materials and processes on a nanometer scale is opening a world of creative possibilities, and as a growing applied science, nanotechnology has considerable global socioeconomic value. The nanotechnology-related economy is estimated to reach a value of US$1 trillion by 2015 (Roco 2005). The unique behavior and properties of materials on a nanoscale has revolutionized technology, producing an estimated 1,300 materials, either in use or being manufactured for uses in industry and other commercial uses. Enhanced strength, durability, flexibility, performance, and inimitable physical properties associated with the nanosize of these materials have been exploited in a multitude of areas in industries, several treatment modalities, including early detection of tumors, targeted site-specific drug delivery of anticancer drugs, and prognostic monitoring of therapy using visual aids (Gwinn and Vallyathan 2006). With these applications, come unprecedented avenues of exposure to humans directly to nanoparticles. Their growth and use in consumables and medical applications, without prior human health evaluations, challenge society with the possibility that they could become the ‘asbestos’ of the twenty-first century.

Asbestos was discovered in 1878, and commercially used in thousands of products and industrial applications all over the world. Because of its fire resistance, high tensile strength and durability, asbestos was used in the construction industry and its breakdown products created a major health hazard even in public schools. Several diseases including the malignant mesothelioma of the pleura and peritoneum, and cancer of the lung are caused by asbestos exposure. However, these diseases caused by asbestos were identified only after a long latency period.

In an International Meeting on fibers/particles/nanoparticles, Brooke Mossman in her presidential remarks stated, 'Naturally occurring minerals such as asbestos and silica have been studied epidemiologically and experimentally for many years, it is now established that inhaled ambient particles also are associated with adverse health effects in man and animals, and their compositional components and biologic effects are now under intense investigation. Because synthetic nanomaterials often occur in size ranges similar to ultrafine airborne particles that are considered to be a major factor contributing to adverse health effects of air pollution, information on the biological reactivity of these particulates is also necessary to allow their safety evaluation' (Mossman et al. 2007).

12.2 Asbestos! Past Experiences

Asbestos is an established solid phase carcinogen (Mossman and Gee 1989; Mossman et al. 1996). It is a generic term for a group of naturally occurring hydrated inorganic mineral silicates that possess a crystalline structure and are incombustible in the air and separable into filaments. From ancient times this mineral has been called the 'Magic Mineral' for its property of high tensile strength and heat resistance. In ancient Rome it was used as the wick of the lamp. It's tensile strength, undegradable structure and special chemical properties make it commercially an important mineral fiber. Asbestos is used as building material, with cement (roofing, shingles, cement pipes, cement tiles), friction resistant material (brake lining and clutch pads), suits for fire fighters and astronauts, insulation of electric wires in furnaces, and fireproof materials for the building industry and on ships as a high quality fire proof and soundproof material. There are two types of asbestos; serpentine and amphibole. Chrysotile with a curly stranded structure accounts for 95% of the world's asbestos consumption and is mainly used in the construction industry. Crocidolite, amosite tremolite and anthophylite, with straight rod like structures, are amphibole varieties (Kamp and Weitzman 1999).

Exposure to asbestos causes lung fibrosis known as asbestosis, effusion and pleural plaques and also two types of malignancies, bronchogenic carcinoma and malignant mesothelioma (Rahman 1995). Usually 15–40 years after the first exposure are the latency period for the development of these diseases (Khan et al. 1992; Kamp and Weitzman 1997). All the varieties of asbestos induce pulmonary diseases. Studies provide information about the implications of fiber structure, its chemical and physical properties in the induction of free radicals, both in cellular as well as in cell free systems, which, in turn, are responsible for its pathogenesis (Fubini 1997). Other predisposing factors such as biomass fuel, kerosene soot and vapours and cigarette smoke in the presence of asbestos exposure accelerate the disease process (Rahman et al. 2000a).

Asbestosis is a slowly progressive, diffused pulmonary fibrotic process caused by the inhalation of asbestos fibers. The development of asbestosis is directly associated with both multitude and duration of asbestos exposure (Mossman et al. 1990; Kamp and Weitzman 1997). The earliest symptom of asbestosis is usually insidiously progressive exertional dyspnoea (Kamp and Weitzman 1999). Patients with asbestosis may have a bibasilar fine end inspiratory crackles, clubbing and in advanced stages signs of cor-pulmonale (Fraser et al. 1990).

Development of human *malignant mesothelioma* is strongly associated with asbestos exposure (Craighead and Mossman 1982; Zeren et al. 2000). It arises preliminarily from the internal lining of the pleural and peritoneal body cavities. Its incidence had steadily increased world wide and accounted for approximately 20 deaths per million males in the western countries (Davies 1980). The exposure dose of asbestos may be extremely small, and the latency period between exposure and onset of disease ranges from 30 to 40 years. *Bronchogenic carcinoma* is a tumor, arising from the tracheobronchial epithelial or alveolar epithelial cells. The average latency period of disease is from 20 to 30 years. Bronchogenic carcinoma is common among asbestos exposed smokers (Lohani et al. 2002). A study conducted in an asbestos-cement factory in India showed increased pulmonary function abnormalities and respiratory impairments in the workers simultaneously exposed to cigarette smoke, kerosene soot along with asbestos, in comparison to asbestos alone (Rahman et al. 2000a). It was also suggested, by another study, that co-exposure of kerosene-soot and cigarette smoke enhances the genotoxicity of asbestos (Lohani et al. 2000).

The data on asbestos cement products was controversial. According to one group of researchers in the field, asbestos in cement pipes or sheeting is not a risk to the general public as the fibers are in a matrix and cannot emerge, even with weathering as small fibers. The other group believes that asbestos cement sheets, pipes etc. with wear and tear of the weather and aging breaks down and the asbestos fibers become free. In this regard we have conducted experiments using old asbestos cement sheets. Our results clearly showed that Asbestos Cement sheet powder is genotoxic and free asbestos fibers may come out from the sheets subjected to wear and tear. This is the first study to experimentally prove that in asbestos cement sheets, clean asbestos fibers can emerge (Dopp et al. 2005).

Our experimental studies demonstrated that asbestos contains redox active iron, which is pathogenic. It attacks heme proteins, depletes soluble antioxidants, and induces free radicals and oxidative stress, which are responsible for a cascade of reactions causing lung fibrosis and ultimately malignancies (Rahman et al. 1997). Asbestos-induced Oxy-radicals play an important role in its pathogenecity (Shukla et al. 2003). The studies conducted at the genetic and chromosomal levels revealed that asbestos induces structural damages in DNA and causes chromosomal breakage and aneuploidy (Rahman et al. 2000b). Studies using asbestos exposed human mesothelial cells as targets, for the first time provided strong evidence that Oxy-radicals are the key factors in asbestos-induced genotoxicity. The studies further indicate that the mechanism of free radical formation is different with different fibers (Poser et al. 2004).

The use of asbestos is now banned in most countries because of its toxic nature. Unfortunately India is still using asbestos and its use has increased 7% in the last decade.

India produces approximately 30,000 MT of asbestos per year and imports more than 85,000 MT of asbestos per year from Canada and Russia (Kulkarni 2001). Direct and indirect employment in asbestos related industries and mines is around 100,000. Male workers are engaged in under-ground mining and opencast mining regions, whereas female workers are involved in milling and processing units, where fiber concentration is very high.

In India, asbestos occurs in the states of Andhra Pradesh, Rajasthan, Bihar, Karnataka, Tamil Nadu and Manipur (Ramanathan and Subramanian 2001). There are 673 small-scale asbestos-based units out of which 45% exist only in Rajasthan. Rajasthan contributes nearly 95% of the total Indian production of asbestos, which is industrially processed within the State itself. The state of Rajasthan has the world's largest asbestos deposits of amphibole, mostly tremolite (Mansinghka and Ranawat 1996). Tremolite has been reported to be more fibrogenic to miners and millers than are other varieties (McDonald et al. 1999).

12.3 Epidemiological Studies Conducted in Asbestos-Based Units in India

As mentioned above, India is still using asbestos, therefore we have conducted an in-depth study in the organized as well as unorganized sectors in asbestos-based industries in India. The study aimed to monitor the asbestos fiber concentration, its identification, and size distribution along with the effectiveness and efficiency of control measures. Monitoring was carried out to assess asbestos fiber concentration in the occupational environment and in the vicinity of the industry. There are no early biomarkers for asbestos-induced diseases. Asbestos-associated diseases continue to be devastating worldwide, and new biomarkers for early detection of asbestos-induced cancers and fibrosis are sorely needed. The earliest symptom of asbestosis is usually insidiously progressive exertional dyspnea. Dyspnea typically progressed regardless of any further exposure to asbestos. As documented by others we have used markers such as asbestos bodies, pulmonary function test, serum markers and radiological examinations along with history to assess the exposure impact on the health of the workers occupationally exposed to asbestos as well as unexposed workers in the surroundings of the industry.

12.4 Monitoring and Identification of Asbestos in Different Asbestos-Based Industrial Units

Environmental monitoring of the air samples and raw material analysis were conducted in 18, unorganized sectors, involved in the mining, grinding, milling and manufacturing of asbestos-based cement products such as asbestos-cement pipes, pillars etc. (Ansari et al. 2007). These units were processing the indigenous asbestos available from nearby existing mines. The analysis of the samples after X-EDS and phase contrast microscopy showed that they were tremolite asbestos. The fiber concentration was 18–22 fiber/cc in the work place area, much higher than the prescribed Standard Limit 0.5 fiber/ml set by the Central Pollution Control Board (CPCB), New Delhi, 2005 at National level and the limit of 0.1fiber/cc set by the International Standard of Occupational Safety and Health Administration (OSHA).

In the unorganized sectors, workers were processing the asbestos with obsolete technologies, without using masks and gloves. Housekeeping was found to be very poor in these sectors. During the survey it was also noticed that children and pets were playing on the asbestos heaps and the workers were processing the asbestos while enjoying bidi smoking (Indian cigarettes) at the same time without using any safety and proper control measures, which further supports the contention that workers were not aware of the consequences of asbestos exposure.

In the organized sectors, the occupational asbestos concentration was lower by 1.71 fiber/cc than the unorganized sectors, which may be due to wet processing, but the housekeeping was found to be done poorly. Mostly the chrysotile variety of asbestos was in use for the making of asbestos-based products such as asbestos cement sheets, break-shoes, clutch plates etc. It was also noticed that workers did not change their working clothes and carried asbestos fibers to the indoor air environment in their houses. It was interesting that the owners of unorganized sectors claim that they are not using asbestos in their units but the collected samples as mentioned above were tremolite.

12.5 Comparative Clinical Findings of Asbestos Exposed and Unexposed Population

In the unorganized sectors 21% of the workers had asbestosis (Figs. 12.1 and 12.2). Among this 21%, 62% developed asbestosis in less than 5 years, 27% in 5–10 years, 15% in 11–20 years and 4% in 20–30 years.

Mostly these workers were on a contract basis and it was difficult to trace them. Their lung function test revealed high obstruction in the exposed workers. But significantly high restriction was observed in exposed as well as in unexposed workers (Fig. 12.3). Perhaps they were also exposed to asbestos either in the close vicinity or employed on a contract basis in the past.

Fig. 12.1 Women exposed to asbestos in the unorganized asbestos-based units in Rajasthan (India)

Fig. 12.2 Women exposed to asbestos in the unorganized asbestos-based units in Rajasthan (India)

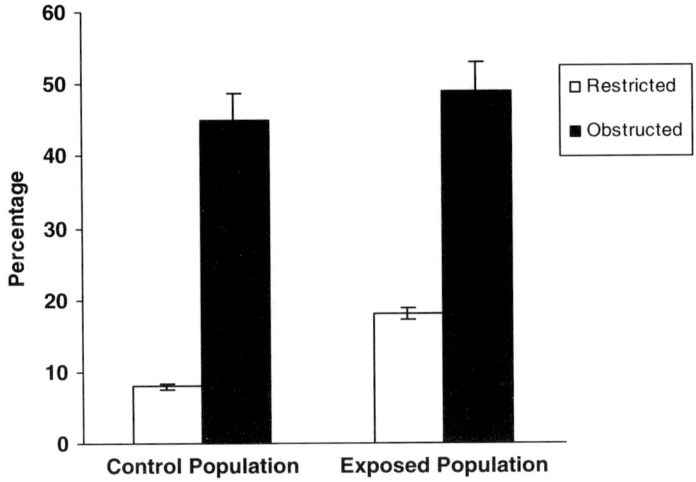

Fig. 12.3 High restriction due to asbestos in the PFT of exposed and unexposed workers

Fig. 12.4 Domestic cooking fuel exposure to asbestos workers

All these workers were domestically exposed to unprocessed cooking fuel smoke. The reason for developing asbestosis earlier than the scheduled time may be either high exposure of tremolite asbestos or double exposure: domestically, cooking fuel smoke and occupationally, asbestos (Fig. 12.4). Further predisposing factors can also not be ruled out.

In the organized sectors, 26% of the workers developed asbestosis in mostly 25–40 years. Most of the workers were using clean gas.

Several studies clearly show that all the varieties of asbestos are fibrogenic, co-carcinogenic and carcinogenic. Most countries have therefore banned the use of asbestos and are trying hard to remove from their buildings the asbestos that was used in the past. It is now very important for countries where asbestos is still in use to get rid of this mineral fiber, as they are not only exposing their workers but posing a threat for the coming generation also.

12.6 Nanoparticles and Health Effects

The novel and rapidly growing field of nanotechnology, using engineered and artificially synthesized NPs, currently expanding to create another cardinal source of man-made toxic exposure. With the rapid industrial revolution in nanotechnology, engineered nanoparticles (NPs) with diverse physical and chemical properties such as nanotubes, fullerenes, quantum dots or metal oxide-based particles are being manufactured worldwide. They are being commercially used in multiple fast-expanding industrial fields including automobiles, optics, microelectronics, cosmetics, medicine, diagnostic equipment, therapeutic products and drug delivery. Manipulation of materials and processes on a sub-100 nm scale is opening a whole new world of creative possibilities. Engineered NPs mainly are carbon-based, but silicon and titanium are also in use. Other synthesized NPs are made of metals or metal oxides, including zinc, iron, cerium, zirconium, gold, silver, copper, lead, cadmium, germanium, and selenium. The four key NP classes which are currently most frequently in use are fullerenes, carbon nanotubes, quantum dots, and metal oxide-based nanoparticles (Powell and Kanarek 2006a, b).

One of the most obvious differences between unintentionally formed and intentionally synthesized anthropogenic NPs (typically displaying size ranges 10–100 nm) is the poly-dispersed and chemically quite complex nature of the former, in contrast to the mono-dispersed and chemically precisely defined characteristics of the latter ones. Because of their extremely small size of <100 nm and a very high volume ratio, with high reactivity potential, the physicochemical properties of engineered NPs differ substantially from those of their native bulk materials.

Their unique properties like, small size, large surface area, specific surface properties, durability etc. could also lead to unexpected and sometimes harmful interactions, causing serious adverse health effects in humans and the ecosystem. With this suspicion, several studies have attempted with cellular and animal studies to explore whether some of the commercially used NPs can be toxic and potentially infiltrate lungs and migrate to other parts of the body to inflict injury. Adverse NPs-induced toxicity to different organelles (e.g., mitochondria) and organs (e.g., lymph nodes, spleen, heart, liver, pancreas, kidney, bone marrow, and even brain) is an important apprehension in the absence of biological evidence.

However, at the present time, public opinion still does not regard nanotechnology as a substantial threat to human health. Nonetheless, this momentary stance could change rapidly with further progression and spread of nanotechnology and parallel increasing evidence in NPs' adverse health effects. Therefore, even with limited evidence on the biological effects of NPs, now is the right time to inform the public about the numerous potential bio-hazards of engineered NPs. It is also the right time to enforce new basic safety guidelines in alliance with industries and establish toxicity test procedures that will ensure protection of workers and control release of NPs to the environment. In order to successfully tackle all the issues associated with nanotechnology, it is important to assemble a multidisciplinary team of experts in industry, engineering, industrial hygiene, toxicology, molecular biology and federal agencies to institute round-robin studies on toxicity and formulate guidelines on exposure.

12.7 Uptake and Somatic Distribution of Nanoparticles

Three different cardinal routes mediate exposure to NPs by humans: inhalation (respiratory tract), ingestion gastrointestinal tract-(GI), and dermal uptake (skin). The routes are:

1. Respiratory tract: when inhaled, NPs are efficiently deposited and after that distributed by diffusional mechanisms to all regions of the respiratory tract. Their small size then facilitates uptake into alveolar cells by phagocytosis/endocytosis and subsequent transcytosis across epithelial and endothelial cells. This finally results in the release of NPs into the blood and lymph circulation. At the end, the particles reach sensitive target sites such as lymph nodes, spleen, heart, liver, kidney, bone marrow, and even brain (Ferin et al. 1991; Ferin and Oberdörster 1992; Ferin et al. 1992; Warheit and Hartsky 1993; Nikula et al. 1997; Nemmar et al. 2002; Oberdörster et al. 2002; Semmler et al. 2004). The specific biokinetics of the phagocytosis/endocytosis and transcytosis processes of the NPs are also dependent on their respective surface chemistry (coating) and their possible subsequent in vivo surface modifications (Bazile et al. 1992; Kreuter 2004). Another important mechanism of endocytosis involves the NP uptake by sensory nerve endings embedded in airway epithelia, followed by the axonal translocation to ganglionic and even CNS structures. Therefore, the olfactory nerve pathway should also be considered as a critical portal of NP entry to the central nervous system of humans, especially under conditions of elevated environmental or occupational NP exposure (Gianutsos et al. 1997; Fechter et al. 2002). The access of NPs to the central nervous system and ganglia via a translocation along axons and dendrites of neurones was observed more than 60 years ago (Howe and Bodian 1940; Bodian and Howe 1941). Lockman et al. 2004 investigated the effect of neutral, anionic and cationic charged NPs on the blood brain barrier (BBB) integrity and NP brain permeability. Neutral NPs and low concentrations of anionic NPs were found to have no effect on BBB integrity, whereas high concentrations of anionic NPs and cationic NPs disrupted the BBB structure. Especially cationic NPs displayed an immediate

toxic effect regarding BBB integrity. Nanoparticles unambiguously have been shown to induce the production of reactive oxygen species and oxidative stress. Oxidative stress has been implicated in the pathogenesis of neurodegenerative diseases such as Parkinson's and Alzheimer's disease. More evidence for such effects is also presented by studies of biopsies from big city inhabitants, displaying Alzheimer's- like pathology as demonstrated by increased markers for inflammation and AB42-accumulation in the frontal cortex and the hippocampus in association with the presence of nanoparticles (Calderón-Garcidueñas et al. 2002; Borm et al. 2006).

2. Gastrointestinal tract: fewer studies have investigated the uptake of NPs by the GI tract. They have shown that the main fraction of the NPs rapidly pass through the GI tract and is fast eliminated via faeces. However, in fact a minor fraction is taken up by the gastrointestinal mucosa and finally is translocated to systemic organs (Jani et al. 1990, 1994).
3. Dermal exposure: dermal exposure to synthesized particulate materials occurs regularly during the use of sunscreen products, for example, TiO_2 and ZnO nanoparticles, that are often coated for minimizing their skin reactivity while maintaining their outstanding UV absorption properties. In healthy skin, the epidermis provides excellent protection against particle spread to the dermis. However, already heavy flexing of normal skin facilitates the penetration of micrometer-size fluorescent beads to the dermis, and damaged skin and allows micrometer-size particles to reach the dermis and the regional lymph nodes (Tinkle et al. 2003). In vivo imaging using intradermally injected quantum dots has been used to confirm particle trafficking to regional lymph nodes in animals. Such trafficking could deliver the particles to paracortical areas in the lymph nodes where macrophages and dendritic cells (DC), which are specialized in the uptake of particulate matter, are located. NP uptake then could lead to substantial effects on the immune system (e.g., interactions of certain proteins with particles may change their antigenicity and hence initiate autoimmune responses) (Kim et al. 2004).

Because of the worldwide prospering of nanotechnology, more and more studies are exploring their weird properties, but also the frequently associated adverse biological effects of the numerous newly designed nanotechnology-associated NPs. First of all, and analogous to the naturally occurring nanosized particles, the unusual physicochemical properties of synthesized NPs are mainly attributable to their extremely small size. For instance, 2 g of a 100 nm-diameter nanoparticle contains enough material to provide every human worldwide approximately with 300,000 particles each. Exactly these tiny dimensions of the designed NPs, together with the inevitably associated high surface-to-volume ratio, and predominantly in alliance with coupled high concentrations and purities, are responsible for their superior reaction capacities (also with biological systems). Furthermore, their extremely small size again creates the chance for increased uptake, fast body distribution and final interaction with target tissues. Secondarily, the individual physicochemical properties of different synthesized NPs are further determined by their respective, mostly unusual, specific physical shape (e.g., tubes, rings or spheres) and specific chemical composition (e.g., carbon, minerals or metals), both of which set up cardinal material characteristics like rigidity, durability, crystallite structure, aggregation status, solubility, and surface reactivity (e.g., electrical or catalytic qualities). In certain cases, cardinal surface structures are supplemented and further modified by additional surface groups or inorganic/organic coatings. Such additional surface groups can function as extra reactive sites in biological systems, for instance making NPs hydrophilic or hydrophobic, lipophilic or lipophobic, or catalytically more active. Although impressive from a physicochemical and technical perspective, the frequently unusual and novel properties of the artificial NPs also increasingly raise concerns about the probable adverse side effects of NPs on biological systems and human health (Brumfiel 2003; Giles 2003; Dreher 2004; Maynard et al. 2004; Matsudai and Hunt 2005; Maynard and Kuempel 2005; Oberdörster et al. 2005; Seaton and Donaldson 2005; Borm et al. 2006; Nel et al. 2006; Moore 2006; Peters et al. 2006). Indeed, a great number of studies already suggest that many synthesized NPs are not inherently benign and actually can affect biological activities at the cellular, sub-cellular, and molecular levels.

Eva Oberdörster has reported the toxicity of fullerence in largemouth bass. The exposure to pure fullerenes caused lipid peroxidation in their brains and glutathione depletion in their gills. Interestingly, in this study, the fullerenes also increased the water clarity in the fish tanks, an indication for their strong oxidizing and antibacterial properties (Oberdörster 2004). Yamawaki and Iwai (2006) investigated the direct effects of a specific fullerene derivative on endothelial toxicity. They have also suggested that fullerenes' cytotoxic effects may be due to lipid peroxidation of cell membranes and the resulting 'leakiness' of membranes. Fullerenes are highly lipophilic and tend to localize to cell membranes (Sayes et al. 2005).

Single-walled carbon nanotubes (SWNTs) induce oxidative stress, in keratinocytes and bronchial epithelial cells *in vitro* as evidence by the formation of free radicals, accumulation of peroxidative products, and depletion of cell antioxidants (Shvedova et al. 2004a, b). Similarly, multi-walled carbon nanotubes (MWNTs) showed pro-inflammatory effects due to ROS formation after their internalization in keratinocytes (Monteiro-Riviere et al. 2005). Because of their unique structure, carbon nanotubes (CNTs) simultaneously display features of NPs and conventional fibers, and also crucial potential to act as needle-like fibers (similar to asbestos). CNTs are essentially graphitic and so biologically extremely biopersistent. The length of single-walled nanotubes-SWNT is 1–3 nm and of multi-walled nanotubes-MWNT 2–50 nm. Several studies using intratracheal instillation of high doses of nanotubes in rodents demonstrated chronic lung inflammation, including foreign-body granuloma formation and interstitial fibrosis. The presence of metal impurities (e.g., Fe) also often accounts for further elevated toxicity (Warheit et al. 2004; Lam et al. 2004). Regardless of the synthesis mode, CNTs contain a number of toxic metals that clearly must be viewed as contaminants, because they are not required for the desired final CNT functions. These metals include for instance Co, Fe, Ni, and Mo, all of which have been documented to exhibit toxic effects.

Toxic effects on human health and wildlife could also be generated from the release of carbon nanotubes from car tires and vehicle plastic mouldings after these are disposed off by incineration. They could also easily pass into the soil and groundwater and finally enter our food-chain (Kreyling et al. 2006).

Several studies suggest Quantum dots (QD) cytotoxicity to be due to photolysis or oxidation. Especially under oxidative and photolytic conditions, QD core–shell coatings have been found to be very labile. This leads to degradations and thus exposure to potentially toxic 'capping' material or intact core metalloid complexes, resulting in dissolution of the core complex to QD core metal components (e.g., Cd, Se) (Derfus et al. 2004). Several studies have shown that QDs can be systemically distributed and can accumulate in organs and tissues including lymph nodes, kidneys, liver, lung, bone marrow and spleen (Oberdörster et al. 2005). Cadmium and selenium, two of the most widely used constituent metals in QD core metalloid complexes, are known to cause acute and chronic toxicities in vertebrates and are of considerable human health and environmental concern. For instance, Cd, a probable carcinogen, has a biologic half-life of 15–20 years in humans. It bio-accumulates, can cross the blood–brain barrier and placenta, and is systemically distributed to all body tissues, with liver and kidney being cardinal target organs of toxicity (Hamilton 2004; Henson and Chedrese 2004; Satarug and Moore 2004).

Exposure to ultrafine titanium dioxide particles has been associated with a variety of pulmonary effects in rats, including inflammation, pulmonary damage, fibrosis, and lung tumours (Bermudez et al. 2002; Bermudez et al. 2004). Furthermore, it was demonstrated that ultra-fine titanium dioxide particles can impair macrophage function and increase pulmonary retention, enter the epithelium faster, and translocate to the sub-epithelium space more readily than fine particles (Churg et al. 1998). We have shown that ultrafine titanium dioxide could also cause mitotic disturbances, DNA damage, and apoptosis in Syrian hamster embryo fibroblasts (Rahman et al. 2002).

The oxidative stress resulting from exposures to quartz, carbon black, or TiO_2 nanoparticles at higher levels, can result in pronounced cytotoxic effects like interstitial fibrosis and airway inflammation. It was also reported that radio-labeled ultrafine carbon black translocates through the respiratory epithelial layer to reach the lung interstitium or the blood and lymph circulations, involving alveolo-capillary translocation via transcytosis or endocytosis. In general, there is a direct relationship between the surface area, ROS-generating capability, and pro-inflammatory effects of nanoparticles (Nel et al. 2006).

A recent study demonstrated the potential of nanomaterials to cause a phytotoxic response in the ecosystem. In the case of alumina NPs, one of the US market leaders for nanosized materials, 99.6% pure nanoparticles with an average particle size of 13 nm were shown to cause root growth inhibition in five plant species (Yang and Watts 2005).

In sharp contrast to the many research efforts, which are accomplished with the aim to explore the desirable qualities of artificially constructed NPs, are so far very limited attempts for a simultaneous evaluation of their undesirable and harmful properties. Unfortunately, nowadays numerous commercial developments in nanotechnology are running ahead of a substantial global ethical understanding regarding the multiple potential adverse effects of nanoparticles on biological systems. For example very few data are available for impacts of engineered nanomaterials on environmentally relevant species. Almost no studies to date have been done on protists, fungi, plants, birds, reptiles, or amphibians. Only limited information is available for bacteria and the mammalian-related studies, which have been carried out so far, mainly used typical laboratory species. Considering that invertebrates constitute up to 97% of all known animal species, there is a considerable lack of information on ecological endpoints. So far filter-feeding organisms represent another unique non-investigated target group for nanoparticle toxicology. It should be mentioned that in the economically important aquatic ecosystems, zooplankton and filter feeding invertebrates make up the main basis of the food web.

Unfortunately, many nanotechnology-related products are already on the market without having undergone adequate safety evaluations.

Serious adverse health effects can also emerge when novel artificial NPs, which were not properly characterized regarding their biohazard potential, are administered to humans intentionally (e.g., for medicinal purposes) or unintentionally (e.g., in the course of regular or accidental industrial processes).

Consequently, future considerations for biohazard assessments of engineered NPs should include a careful selection of standardized tests as well as the use of relevant nanoparticle species doses/concentrations. Key elements of a future NP toxicity screening strategy should also include a detailed physicochemical characterization of the used NPs. Consequently, NP material should be precisely characterized with respect to size (size distribution, surface area, volume-to-surface ratio), chemical composition (purity, crystallinity; and magnetic, electronic, oxidative or catalytic properties, etc.), surface structure (surface reactivity, surface modifications, inorganic/organic coatings, etc.), solubility, shape and aggregation state. The physicochemically precisely characterized NPs should be distributed on demand to respective differently specialised laboratories. Furthermore, standardized reference materials (e.g., TiO_2, carbon black, quartz) are essential to compare NP material behaviour. Also standardized and highly accurate *in vitro* assays (cellular and non-cellular), and more comprehensive in vivo studies are obligatory. Cellular assays should reflect the portal-of-entry toxicity in lungs, skin, and mucus membranes as well as the toxic effects on target organ tissue derived from endothelium, blood cell elements, spleen, bone marrow, liver, nervous system, brain, heart, pancreas, kidneys or others. Non-cellular assays could include protein interactions and pro-oxidant activity. Examples include assays for oxidative stress (e.g. ROS formation, lipid peroxidation), C-reactive protein, immune and inflammatory responses, and cytotoxicity (e.g., release of liver enzymes and glial fibrillary acidic protein). The in vivo studies can make use of disease-specific animal models in order to measure portal of entry and target organ injury in living organisms.

As a whole, the principal objective of all biohazard assessments should be to find out which NP species are clearly dangerous for humans and the environment and which are undoubtedly not.

For that reason, a future interdisciplinary teamwork approach (e.g., an intense collaboration between the scientific fields of toxicology, biochemistry, physics, materials science, medicine, molecular biology, and bioinformatics, to name a few) is mandatory for a successful research effort, in order to create an appropriate and reliable risk assessment. This approach ultimately will lead to the final establishment of really suitable safety measures in nanotechnology and the firm creation of the new scientific field of nanotoxicology. In this regard, nanotoxicology can also be defined as a new branch of toxicology, which is mainly concerned with the proper safety evaluation of engineered artificial nanoparticles, including more complex nanostructures and interactive nanodevices, in order to support biologically safe technological developments in future nanotechnology.

I feel, being a woman scientist, that new technologies and human health issues should be addressed together and based rationally on scientific findings.

I would like to emphasise that, we must help to get rid off past environmental damages, correct present environmental problems, prevent future environmental impacts, and help sustain the planet for the next generation.

References

Ansari, F. A., Ahmad, I., Ashquin, M., Yunus, M., & Rahman, Q. (2007). Monitoring and identification of airborne asbestos in unorganized sectors, India. *Chemosphere, 68*(4), 716–723.

Bazile, D. V., Ropert, C., Huve, P., Verrecchia, T., Marlard, M., Frydman, A., Veillard, M., & Spenlehauer, G. (1992). Body distribution of fully biodegradable [14C]-poly(lactic acid) nanoparticles coated with albumin after parenteral administration to rats. *Biomaterials, 13*(15), 1093–1102.

Bermudez, E., Mangum, J. B., Asgharian, B., Wong, B. A., Reverdy, E. E., Janszen, D. B., Hext, P. M., Warheit, D. B., & Everitt, J. I. (2002). Long-term pulmonary responses of three laboratory rodent species to subchronic inhalation of pigmentary titanium dioxide particles. *Toxicological Sciences: An Official Journal of the Society of Toxicology, 70*(1), 86–97.

Bermudez, E., Mangum, J. S., Wong, B. A., Asgharian, B., Hext, P. M., Warheit, D. B., & Everitt, J. L. (2004). Pulmonary responses of mice, rats, and hamsters to subchronic inhalation of ultrafine titanium dioxide particles. *Toxicological sciences: An Official Journal of the Society of Toxicology, 77*(2), 347–357.

Bodian, D., & Howe, H. A. (1941). Experimental studies on intraneural spread of poliomyelitis virus. *Bulletin of the Johns Hopkins Hospital, 68,* 248–267.

Borm, P. J. A., Robbins, D., Haubold, S., Kuhlbusch, T., Fissan, H., Donaldson, K., Schins, R. P. F., Stone, V., Kreyling, W., Lademann, J., Krutmann, J., Warheit, D., & Oberdörster, E. (2006). The potential risks of nanomaterials: A review carried out for ECETOC. *Particle and Fibre Toxicology, 3,* 11.

Brumfiel, G. (2003). Nanotechnology: A little knowledge. *Nature*, 424(6946), 246–248.

Calderón-Garcidueñas, L., Azzarelli, B., Acuna, H., Garcia, R., Gambling, T. M., Osnaya, N., Monroy, S., Del Tizapantzi, M. R., Carson, J. L., Villarreal-Calderon, A., & Rewcastle, B. (2002). Air pollution and brain damage. *Toxicologic Pathology, 30*(3), 373–389.

Churg, A., Stevens, B., & Wright, J. (1998). Comparison of the uptake of fine and ultrafine TIO2 in a tracheal explant system. *The American Journal of Physiology, 274*(Pt 1), L81–L86.

Craighead, J. E., & Mossman, B. T. (1982). The pathogenesis of asbestos associated diseases. *The New England Journal of Medicine, 306*(24), 1446–1455.

Davies, R. (1980). The effect of dust on enzyme release from macrophages. In: R. C. Brown, I. P. Gormley, M. Chamberlain, & R. Davies (Eds.), *In vitro effects of mineral dust* (pp. 67). London: Academic.

Derfus, A. M., Chan, W. C. W., & Bhatia, S. N. (2004). Probing the cytotoxicity of semiconductor quantum dots. *Nano Letters, 4*(1), 11–18.

Dopp, E., Yadav, S., Furquan, A. A., Rödelsperger, K., Von Recklinghausen, U., Rauen, U., Florea, A.-M., & Rahman, Q. (2005). ROS-mediated genotoxicity of asbestos-cement in mammalian lung cells in vitro. *Particle and Fibre Toxicology, 2*, 9.

Dreher, K. L. (2004). Health and environmental impact of nanotechnology: Toxicological assessment of manufactured nanoparticles. *Toxicological Sciences: An Official Journal of the Society of Toxicology, 77*(1), 3–5.

Fechter, L. D., Johnson, D. L., & Lynch, R. A. (2002). The relationship of particle size to olfactory nerve uptake of a non-soluble form of manganese into brain. *Neurotoxicology, 23*(2), 177–183.

Ferin, J., & Oberdörster, G. (1992).Translocation of particles from pulmonary alveoli into the interstitium. *Journal of Aerosol Medicine: The Official Journal of the International Society for Aerosols in Medicine, 5*(3), 179–187.

Ferin, J., Oberdörster, G., & Penney, D. P. (1992). Pulmonary retention of ultrafine and fine particles in rats. *American Journal of Respiratory Cell and Molecular Biology, 6*(5), 535–542.

Ferin, J., Oberdörster, G., Soderholm, S. C., & Gelein, R. (1991). Pulmonary tissue access of ultrafine particles. *Journal of Aerosol Medicine: The Official Journal of the International Society for Aerosols in Medicine, 4*(1), 57–68.

Fraser, R. G., Pare, P. D., Fraser, R. S., & Genereux, G. P. (Eds.) (1990). Pleuro-pulmonary disease caused by inspiration of inorganic dust (pneumoconiosis). *Diagnosis of Disease of the Chest* (Vol. III, pp. 2346). Philadelphia: W.B. Saunders.

Fubini, B. (1997). Surface reactivity in the pathogenic response to particulates. *Environmental Health Perspectives, 105*(Suppl. 5), 1013–1020.

Gianutsos, G., Morrow, G. R., & Morris, J. B. (1997). Accumulation of manganese in rat brain following intranasal administration. *Fundamental and Applied Toxicology, 37*(2), 102–105.

Giles, J. (2003). Nanotechnology: What is there to fear from something so small? *Nature, 426*(6968), 750.

Gwinn, M. R., & Vallyathan, V. (2006). Nanoparticles: Health effects, pros and Cons. *Environmental Health Perspectives, 114*(12), 1818–1825.

Hamilton, S. J. (2004). Review of selenium toxicity in the aquatic food chain. *The Science of the Total Environment, 326*(1–3), 1–31.

Henson, M. C., & Chedrese, P. J. (2004). Endocrine disruption by cadmium, a common environmental toxicant with paradoxical effects on reproduction. *Experimental Biology and Medicine (Maywood, N.J.), 229*(5), 383–392.

Howe, H. A., & Bodian, D. (1940). Portals of entry of poliomyelitis virus in the chimpanzee. *Proceedings of the Society for Experimental Biology and Medicine, 43*, 718–721.

Jani, P., Halbert, G. W., Langridge, J., & Florence, A. T. (1990). Nanoparticle uptake by the rat gastrointestinal mucosa: Quantitation and particle size dependency. *The Journal of Pharmacy and Pharmacology, 42*(12), 821–826.

Jani, P. U., McCarthy, D. E., & Florence, A. T. (1994). Titanium dioxide (rutile) particle uptake from the rat GI tract and translocation to systemic organs after oral administration. *International Journal of Pharmaceutics, 105*(2), 157–168.

Kamp, D. W., & Weitzman, S. A. (1997). Asbestosis: clinical spectrum and pathogenic mechanisms. *Proceedings of the Society for Experimental Biology and Medicine, 214*(1), 12–26.

Kamp, D. W., & Weitzman, S. A. (1999). The molecular basis of asbestos induced lung injury. *Thorax, 54*(7), 638–652.

Khan, S. G., Ali, S., & Rahman, Q. (1992). Interaction of mineral fibers with lung cytochrome P-450 system: impairments in the activities of it's dependent and independent drug metabolising enzymes. *Chemosphere, 24*(8), 959–968.

Kim, S., Lim, Y. T., Soltesz, E. G., De Grand, A. M., Lee, J., Nakayama, A., Parker, J. A., Mihaljevic, T., Laurence, R. G., Dor, D. M., Cohn, L. H., Bawendi, M. G., & Frangioni, J. V. (2004). Near infrared fluorescent type II quantum dots for sentinel lymph node mapping. *Nature Biotechnology, 22*(1), 93–97.

Kreuter, J. (2004). Influence of the surface properties on nanoparticle-mediated transport of drugs to the brain. *Journal of Nanoscience and Nanotechnology, 4*(5), 484–488.

Kreyling, W. G., Semmler-Behnke, M., & Moller, W. (2006). Ultrafine particle-lung interactions: does size matter? *Journal of Aerosol Medicine: The Official Journal of the International Society for Aerosols in Medicine, 19*(1), 74–83.

Kulkarni, G. K. (2001). Asbestos: ban or not to ban. *Indian Journal of Occupational and Environmental Medicines, 5*(1).

Lam, C. W., James, J. T., McCluskey, R., & Hunter, R. L. (2004). Pulmonary toxicity of single-wall carbon nanotubes in mice 7 and 90 days after intratracheal instillation. *Toxicological Sciences: An Official Journal of the Society of Toxicology, 77*(1), 126–134.

Lockman, P. R., Koziara, J. M., Mumper, R. J., & Allen, D. D. (2004). Nanoparticle surface charges alter blood-brain barrier integrity and permeability. *Journal of Drug Targeting, 12*(9–10), 635–641.

Lohani, M., Dopp, E., Weiss, D. G., Schiffmann, D., & Rahman, Q. (2000). Kerosene soot induces genotoxicity and enhance the effect on co-exposure with chrysotile asbestos in Syrian Hamster Embryo Fibroblast. *Toxicology Letters, 114*(1–3), 111–116.

Lohani, M., Dopp, E., Becker, H. H., Seth, K., Schiffmann, D., & Rahman, Q. (2002). Smoking enhances asbestos-induced genotoxicity, relative involvement of chromosome 1: a study using multicolor FISH with tandem labeling. *Toxicology Letters, 136*(1), 55–63.

Mansinghka, B. K., & Ranawat, P. S. (1996). Mineral Economics and occupational health hazards of the Asbestos Resources of Rajasthan. *Journal of the Geological Society of India, 47*, 375–382.

Matsudai, M., & Hunt, G. (2005). Nanotechnology and public health. *Nippon kōshū eisei zasshi, 52*(11), 923–927.

Maynard, A. D., Baron, P. A., Foley, M., Shvedova, A., Kisin, E. R., & Castranova, V. (2004). Exposure to carbon nanotube material: Aerosol release during the handling of unrefined single-walled carbon nanotube material. *Journal of Toxicology and Environmental Health. Part A, 67*(1), 87–107.

Maynard, A. D., & Kuempel, E. (2005). Airborne nanostructured particles and occupational health. *Journal of Nanoparticle Research, 7*(6), 587–614.

McDonald, J. C., McDonald, A. D., & Hughes, J. M. (1999). Chrysotile, tremolite and fibrogenecity. *The Annals of Occupational Hygiene, 43*(7), 439–442.

Monteiro-Riviere, N. A., Nemanich, R. J., Inman, A. O., Wang, Y. Y., & Riviere, J. E. (2005). Multi-walled carbon nanotube interactions with human epidermal keratinocytes. *Toxicology Letters,155*(3), 377–384.

Moore, M. N. (2006). Do nanoparticles present ecotoxicological risks for the health of the aquatic environment? *Environment International, 32*(8), 967–976.

Mossman, B. T., Bignon, J., Corn, M., Seaton, A., & Gee, J. B. L. (1990). Asbestos: scientific developments and implications for public policy. *Science, 247*(4940), 294–301.

Mossman, B. T., Borm, P. J., Castranova, V., Costa, D. L., Donaldson, K., & Kleeberger, S. R. (2007). Mechanisms of action of inhaled fibers, particles and nanoparticles in lung and cardiovascular diseases. *Particle and Fibre Toxicology, 4*, 4.

Mossman, B. T., & Gee, J. B. L. (1989). Asbestos-related diseases. *The New England Journal of Medicine, 320*(26), 1721–1730.

Mossman, B. T., Kamp, D. W., & Weitzman, S. A. (1996). Mechanisms of carcinogenesis and clinical features of asbestos-associated cancers. *Cancer Investigation, 14*(5), 464–78.

Nel, A., Xia, T., Mädler, L., & Li, N. (2006). Toxic potential of materials at the nanolevel. *Science, 311*(5761), 622–627.

Nemmar, A., Hoet, P. H. M., Vanquickenborne, B., Dinsdale, D., Thomeer, M., Hoylaerts, M. F., Vanbilloen, H., Mortelmans, L., & Nemery B. (2002). Passage of inhaled particles into the blood circulation in humans. *Circulation, 105*(4), 411–414.

Nikula, K. J., Avila, K. J., Griffith, W. C., & Mauderly, J. L. (1997). Lung tissue responses and sites of particle retention differ between rats and cynomolgus monkeys exposed chronically to diesel exhaust and coal dust. *Microscopy Research and Technique, 37*(1), 37–53.

Oberdörster, E. (2004). Manufactured nanomaterials (fullerenes, C60) induce oxidative stress in the brain of juvenile largemouth bass. *Environmental Health Perspectives, 112*(10), 1058–1062.

Oberdörster, G., Oberdörster, E., & Oberdörster, J. (2005). Nanotoxicology: An emerging discipline evolving from studies of ultrafine particles. *Environmental Health Perspectives, 113*(7), 823–839.

Oberdörster, G., Sharp, Z., Atudorei, V., Elder, A., Gelein, R., Lunts, A., Kreyling W., & Cox, C. (2002); Extrapulmonary translocation of ultrafine carbon particles following whole-body inhalation exposure of rats. *Journal of Toxicology and Environmental Health. Part A, 65*(20), 1531–1543.

Peters, A., Veronesi, B., Calderón-Garcidueñas, L., Gehr, P., Chen, L. C., Geiser, M., Reed, W., Rothen-Ruthishauser, B., Schürch, S., & Schulz, H. (2006). Translocation and potential neurological effects of fine and ultrafine particles: A critical update. *Particle and Fibre Toxicology, 3*, 13.

Poser, I., Rahman, Q., Lohani, M., Yadav, S., Becker, H. H., Weiss, D. G., Schiffmann, D., & Dopp, E. (2004). Modulation of genotoxic effects in asbestos-exposed primary human mesothelial cells by radical scavengers, metal chelators and a glutathione precursor. *Mutation Research, 559*(1–2), 19–27.

Powell, M. C., & Kanarek, M. S. (2006a). Nanomaterial health effects. Part 1, Background and current knowledge. *WMJ: Official Publication of the State Medical Society of Wisconsin, 105*(2), 16–20.

Powell M. C., & Kanarek, M. S. (2006b). Nanomaterial health effects. Part 2, Uncertainties and recommendations for the future. *WMJ: Official Publication of the State Medical Society of Wisconsin, 105*(3), 18–23.

Rahman, Q. (1995). Asbestos: an occupational and environmental carcinogen. In B. B. Dhar & D. N. Thakur (Eds.), *Mining and environment* (pp. 549–564). New Delhi: Oxford & IBH Publishing.

Rahman, Q., Dopp, E., & Schiffmann, D. (2000a). Genotoxic effects of asbestos fibers. In G. A. Peters & B. J. Peters (Eds.), *Sourcebook on asbestos diseases* (Vol. 21, pp. 223–242). New Hampshire: Butterworth Legal Publishers.

Rahman, Q., Dopp, E., Lohani, M., & Schiffmann, D. (2000b). Occupational and environmental factors enhancing the genotoxicity of asbestos. *Inhalation Toxicology, 12*, 157–163.

Rahman, Q., Lohani, M., Dopp, E., Pemsel, H., Jonas, L., Weiss, D. G., & Schiffmann, D. (2002). Evidence that ultrafine titanium dioxide induces micronuclei and apoptosis in Syrian hamster embryo fibroblasts. *Environmental Health Perspectives, 110*(8), 797–800.

Rahman, Q., Mahmood, N., Khan, S. G., Arif, J. M., & Athar, M. (1997). Mechanisms of asbestos mediated DNA damage: Role of heme and heme proteins. *Environmental Health Perspectives, 105*, 1109–1112.

Ramanathan, A. L., & Subramanian, V. (2001). Present status of asbestos mining and related health problems in India: a survey. *Industrial Health, 39*(4), 309–315.

Roco, M. C. (2005). Environmentally responsible development of nanotechnology. *Environmental Science & Technology, 39*(5), 106A–112A.

Satarug, S., & Moore, M. R. (2004). Adverse health effects of chronic exposure to low-level cadmium in foodstuffs and cigarette smoke. *Environmental Health Perspectives, 112*(10), 1099–1103.

Sayes, C. M., Gobin, A. M., Ausman, K. D., Mendez, J., West, J. L., & Colvin, V. L. (2005). Nano-C60 cytotoxicity is due to lipid peroxidation. *Biomaterials, 26*(36), 7587–7595.

Seaton, A., & Donaldson, K. (2005). Nanoscience, nanotoxicology, and the need to think small. *Lancet, 365*(9463), 923–924.

Semmler, M., Seitz, J., Erbe, F., Mayer, P., Heyder, J., Oberdörster, G., & Kreyling, W. G. (2004). Long-term clearance kinetics of inhaled ultrafine insoluble iridium particles from the rat lung, including transient translocation into secondary organs. *Inhalation Toxicology, 16*(6–7), 453–459.

Shukla, A., Gulumian, M., Hei, T. K., Kamp, D., Rahman, Q., & Mossman, B. T. (2003). Multiple roles of oxidants in the pathogenesis of asbestos-induced diseases. *Free Radical Biology & Medicine, 34*(9), 1117–1129.

Shvedova, A., Kisin, E., Keshava, N., Murray, A. R., Gorelik, O., Arepalli, S., Gandelsman, V. Z., & Castranova, V. (2004a). Cytotoxic and genotoxic effects of single wall carbon nanotube exposure on human keratinocytes and bronchial epithelial cells [Abstract]. *227th American Chemical Society National Meeting, 27 March-1 April 2004, Anaheim, CA*, American Chemical Society, IEC 20, Washington, DC.

Shvedova, A., Kisin, E., Murray, A., Schwegler-Berry, D., Gandelsman, V., Baron, P., Maynard, A., Gunter, M., & Castranova, V. (2004b). Exposure of human bronchial cells to carbon nanotubes caused oxidative stress and cytotoxicity. *Proceedings of the Meeting of the SFRR Europe 2004, Ioannina, Greece,*: Taylor & Francis, Philadelphia, pp. 91–103.

Tinkle, S. S., Antonini, J. M., Rich, B. A., Roberts, J. R., Salmen, R., & DePree, K. (2003). Skin as a route of exposure and sensitization in chronic beryllium disease. *Environmental Health Perspectives, 111*(9), 1202–1208.

Warheit, D. B., & Hartsky, M. A. (1993). Role of alveolar macrophage chemotaxis and phagocytosis in pulmonary clearance responses to inhaled particles: Comparisons among rodent species. *Microscopy Research and Technique, 26*(5), 412–422.

Warheit, D. B., Laurence, B. R., Reed, K. L., Roach, D. H., Reynolds, G. A. M., & Webb, T. R. (2004). Comparative pulmonary toxicity assessment of single-wall carbon nanotubes in rats. *Toxicological Sciences: An Official Journal of the Society of Toxicology, 77*(1), 117–125.

Yamawaki, H., & Iwai, N. (2006). Cytotoxicity of water-soluble fullerene in vascular endothelial cells. *American Journal of Physiology. Cell Physiology, 290*(6), C1495–502.

Yang, L., & Watts, D. J. (2005). Particle surface characteristics may play an important role in phytotoxicity of alumina nanoparticles. *Toxicology Letters, 158*(2), 122–132.

Zeren, E. H., Gumurdulu, D., Roggli, V. L., Zorludemir, S., Erkisi, M., & Tuncer, I. (2000). Environmental Malignant Mesothelioma in Southern Anatolia: A study of fifty cases. *Environmental Health Perspectives, 108*(11), 1047–1050.

Chapter 13
Affective Implications of GM Food on Social and Individual Integrity: An Ethical Approach

Susanne Uusitalo(✉)

Abstract: Genetically modified organisms are a current topic that gives rise to various reactions across different forums ranging from academic discussions to public forums. Recent advances in biotechnology are not only a matter of concern to scientists but to lay people as well. Genetically modified food is one of the issues of biotechnology that consumers may encounter in their daily lives. The reactions that genetically modified organisms or more specifically genetically modified food bring about may stem from various reasons. What is important is that these reactions should be taken into account in decision making. In this paper, I argue for the importance of respecting consumers' integrity by providing them with a sufficient amount of information for making meaningful choices concerning their food purchases. The context of this paper is consumer autonomy in the European Union and the focus is more specifically on labelling. As long as consumers regard the difference between genetically modified food and traditional food significant, labelling provides at least one means for respecting the consumer's integrity.

Keywords: Genetically modified food, consumer, ethics, autonomy of choice, information

13.1 Introduction

The concept of genetically modified organisms is a topic that touches people in many ways all over the world. We are all being faced to an increasing extent with these kinds of products of biotechnologies. Needless to say, biotechnologies provide us with more than genetically modified organisms, but in this paper I concentrate on genetically modified organisms or more specifically on food that is produced from

Susanne Uusitalo
Department of Philosophy, University of Turku, Finland
Assistentinkatu 7, Department of Philosophy, 20014 University of Turku, Finland
susuus@utu.fi

F. Molfino, F. Zucco (eds.) *Women in Biotechnology,*

or contains genetically modified organisms or genetically modified ingredients. Furthermore, the context of this paper is that of the European Union in which genetically modified food products are accepted in the market even if at the moment they may be difficult to find in practice (see for instance GMO Compass 2007).

One of the most concrete ways of facing biotechnologies in everyday life concerns genetically modified food products, as purchasing food is a very common everyday chore. Genetically modified food has given rise to various discussions ranging from food safety to environmental and economic issues. My focus, however, is on consumers. Scientists' understanding of biotechnologies is sometimes contrasted with lay people's reactions to the latest applications. Eurobarometer in 2005 on social values, science and technology has indicated that European respondents are reluctant to consume genetically modified food. Leaving the question of the roots (and legitimacy) of this reluctance aside, we need to consider in what ethically acceptable ways people are to deal with products that have been realised with biotechnologies. In this paper I look specifically at the issue of genetically modified food, as food generally is an issue that can bring about many strong responses ranging from disgust and fear to rationalisations. Moreover, I suggest that labelling genetically modified food is one of the ways in which people's personal integrity can be respected in relation to genetically modified food.

By integrity I mean an individual's possibility to follow her authentic beliefs and desires when choosing the products she wants to buy.[1] It is not my intention to provide a thorough and deep conceptualisation of the notion of integrity, but to use it only in this restricted context. Basically this aspect of integrity means that the consumer is capable of making an autonomous choice.[2] In particular it means that first, the individual has competence, i.e., psychological and physical capabilities needed for that choice, second, that she has authentic beliefs and desires and third that she has power to execute her desires at least in terms of the choice[3] (Räikkä 1999). There is no requirement to have critical reflection on these beliefs and desires (Hyun 2001), rather there has to be a possibility for that kind of reflection. In this paper a paradigm for the consumer's autonomous choice is purchasing food products from her local supermarket. An important point is to notice that autonomy of one's choice is a matter of degree. The consumer may have more or less autonomy in making that choice. Labelling is one of the aspects that may increase the consumer's possibility for making an autonomous choice, i.e., choosing the product according to her authentic beliefs and desires.

---

[1] It has to be acknowledged that consumers in the European Union cannot avoid the constant flow of commercials and adverts. This can make choosing according to one's authentic beliefs and desires difficult. This interesting question is worth further analysis even if it is beyond the scope of this paper.

[2] It is good to keep in mind that autonomy is only one value among others in discussions concerning genetically modified food.

[3] This does not imply that she can carry out an action that follows the choice.

Social integrity refers to a society in which differences between different groups and individuals are identified and respected. In a pluralist society the public is taken into account, i.e., on a personal level people's integrity – or we could talk of consumer autonomy in this case – is not undermined in the context of genetically modified food products. Different consumers prefer to choose different products and enabling them to make those choices according to their beliefs and desires is respecting the consumers' integrity. In this paper the notions of integrity and autonomy overlap and can thus be regarded as synonyms in this respect.

In what follows, I will briefly discuss the significance of food to an individual's integrity and then discuss the moral relevance of affective reactions to genetically modified food. Finally I argue that labelling of genetically modified food products is a means for respecting consumers' integrity.

13.2 Genetically Modified Food and Moral Relevance

Food is relevant to everyone. In fact, it is self-evident to say that people's survival is dependent on it. However, food is not merely a means for survival, but it is also an important part of the ways in which people express themselves, or as Elsbet Probyn (Probyn 2000) puts it, it provides the ways in which people eat into cultures, into identities and into themselves. Food carries biological, social, and cultural-symbolic significance (Pascalev 2003). It should, thus, be no surprise that the amount of intimacy that is involved when one eats something, i.e., ingests something, may give rise to strong responses.

I argue that emotional outbursts (or rationalisations for that matter) should not be directly dismissed as irrelevant or trivialised at first hand, rather they should be weighed and analysed whether they are expressions of something morally relevant. Moral relevance is understood as related to moral values and moral norms. A thing may have value-adding properties and value-subtracting properties. The thing can have the properties simultaneously and thus the properties should be weighed and analysed in relation to the other properties the thing has. Having a certain property does not make the thing directly morally acceptable or unacceptable. A similar case can be presented for actions. An action may have wrong-making and right-making properties and these should be weighed and analysed in relation to the other properties the action has. Similarly to the value case, having a particular property that is, for instance, a right-making property does not automatically result in the action being morally right, but the property should be considered in relation to other morally relevant properties the action has. Moral relevance means that a morally relevant thing can and should affect people's judgements of what is a morally right or wrong action. (Siipi 2005) An example of moral relevance would be a case of *Golden Rice*, which is genetically engineered rice that contains A-vitamin. At least in the areas in which individuals suffer from A-vitamin deficiency, this rice has a morally valuable or desirable property; it provides the individuals with something that promotes their health (see Dawe et al. 2002).

British philosopher Mary Midgley (Midgley 2000) argues that feeling is an essential part of people's moral life. Disgust can be a sign of something being seriously wrong. However, not all feelings of disgust and outrage are expressions of moral relevance. Genetically modified food is not automatically morally relevant, but depending on its constitution and the situation it can be. For some people, for instance, it does not make a difference whether the brand of oatmeal they intend to buy is genetically modified or not. In these cases those consumers' integrity is not violated even if there were not any labels indicating genetically modified food products. Affective reactions that genetically modified food can bring about may not be morally relevant either. Feelings of repulsion may turn out to be merely an aesthetic or some other response. It may well be that some individual responses based on some so-called intrinsic objections, for instance on unnaturalness or naturalness claims, might not have any justified grounds *per se*.[4] Claims of unnaturalness can be understood as unfamiliarity. Disgust can come about because the object of the feeling is so unfamiliar to the subject. For instance, many people may experience revulsion when they are introduced to the idea of eating rotten shark meat, an Icelandic delicacy. This type of unnaturalness is not morally relevant. On the other hand, these gut feelings may be morally relevant too. Revulsion can be an indicator of beliefs and fears for instance and thus it should be taken into account. Furthermore, genetically modified vegetables that have been modified with porcine genes, for instance, may make the products undesirable for Muslims or for vegans who reject the use of animals in the product.

13.3 Labelling as a Practice of Respecting People's Integrity in the Context of Purchasing Genetically Modified Food Products

Relevant responses, i.e., responses based on people's desires and beliefs concerning genetically modified food, have to be considered and taken into account not only on a social decision-making level, but also on an individual level. Social decision-making processes concerning genetically modified food products have to be constructed in such a way that consumers' integrity is taken into account. Basically this means at least providing the consumers with enough information about the products they have available to them to make a meaningful decision of what to purchase. Providing information that is understandable and that does not mislead the consumer is crucial for that consumer's integrity. With the help of this kind of information, consumers are able to make choices and hopefully act according to their beliefs and desires (within individual and social parameters). It is important to identify the criteria and

[4] However, the quantity of such responses may make them something that is to be taken into account on a social level.

justification for these choices for decision makers and the relevancy of individuals to the issue. In this case, an important criterion is access to a sufficient amount of information to allow people's choices to be justified by relying on the notion of integrity, i.e., people's consumer autonomy should be respected. In practice this could mean that in addition to the labels on the genetically modified food products, there could be leaflets and more information on genetically modified food next to the products or at some other obvious location in the stores. In some cases a consumer's integrity may be undermined due to local conditions. For instance, at the moment a Finnish consumer who wishes to buy only genetically modified food cannot act accordingly as there are hardly any genetically modified food products in the market. In other words, even if the Finnish consumer is able to make a meaningful decision, i.e., she has freedom of choice concerning her food purchases, she cannot act according to her choice as the state of affairs. The fact that there is hardly any, if any at all, genetically modified food available in the market at the moment, casts limitations on her ability to act according to her choice. However, this kind of limitation can be regarded as unavoidable, for consumers do not act in a vacuum but within the market with other agents such as sellers and farmers.

It must also be taken into account that not all responses are protests that focus on the possible risks of the products and/or on lack of knowledge (i.e., on the so-called extrinsic consequentialist issues). Reluctance to consume genetically modified food does not have to stem from ignorant emotional reactions (which can perhaps be soothed by increasing the reactor's knowledge of the issue in question). Because people can assume that, as long as these risks are taken into account and appropriately seen to and information is provided to the ignorant, reluctance will cease, it is important to point out that there may be other reasons for reluctance.[5] Thus it is crucial to think of the ways in which individuals' integrity is not undermined by recent developments of biotechnologies. This can be achieved in this context by labelling the genetically modified foods, to name but one means. The European Union requires that food products be appropriately labelled if they contain genetically modified organisms or are produced from or contain ingredients produced from genetically modified organisms (see Article 12 in European community 2003).

At the moment, however, the technology available cannot guarantee measurement of the percentage of a product containing genetically modified ingredients that is less than 0.9%. This, of course, means that products that contain less than 0.9% of genetically modified ingredients cannot be identified and thus labelled. This fact cannot be considered as an unacceptable violation of consumers' integrity, since at the moment there is simply no means to modify this distortion. Thus the more

[5]This is to say that no matter what the reasons behind the reactions are, the reactions are those of consumers and their integrity needs to be respected. By highlighting the point that not all reactions are due to extrinsic issues, I want to emphasise the need to have respect for people's integrity; my argument is that assuming the reactions stem from merely one reason, the situation can seem to be solved by taking action regarding that one point, and thus the consumers whose reason for reluctance to consume is not due to that reason may have their integrity undermined.

technology developments advance in this respect (measuring the amount of genetically modified ingredients), the more labelling can be seen as respecting consumers' integrity (given that the consumers take the labels into account in making their purchasing choices).

All in all, as long as consumers regard the difference between genetically modified food and traditional food significant, labelling provides at least one means for respecting the consumer's integrity.[6] This may, however, change if for instance non-genetically modified food products eventually disappear from the market.[7]

References

Dawe, D., Robertson, R., & Unnevehr, L. (2002). Golden rice: What role could it play in alleviation of vitamin A deficiency? *Food Policy, 27*, 541–560.

European Community (2003). Regulation (EC) No. 1829/2003 of the European Parliament and of the Council of 22 September 2003 on genetically modified food and feed. Retrieved June 11, 2007, from http://europa.eu.int/eur-lex/pri/en/oj/dat/2003/l_268/l_26820031018en00010023.pdf

GMO Compass (2007, January 23). GMO labelling: Labelled goods hard to find. Retrieved July 10, 2007, from: http://www.gmo-compass.org/eng/regulation/labelling/92.gmo_labelling_labelled_goods.html

Hyun, I. (2001). Authentic values and individual autonomy. *The Journal of Value Inquiry, 35*, 195–208.

Midgley, M. (2000). Biotechnology and monstrosity: Why we should pay attention to the Yuk Factor. *The Hastings Center Report, 30*, 7–15.

Pascalev, A. (2003). You are what you eat: Genetically modified food, integrity, and society. *Journal of Agricultural and Environmental Ethics, 16*, 583–594.

Probyn, E. (2000). *Carnal appetites: Food sex identities* (p. 2). London/New York: Routledge.

Räikkä, J. (1999). On the morality of avoiding information. In V. Launis, J. Pietarinen, & J. Räikkä (Eds.), *Genes and morality: New essays* (pp. 63–75). Amsterdam: Rodopi.

Siipi, H. (2005). *Naturalness, unnaturalness and artifactuality in bioethical argumentation* (pp. 15–16). Dissertation, University of Turku.

---

[6] This, however, entails that the consumers know what the label 'genetically modified' means and they know about the labelling practice in the European Union.

[7] I am deeply grateful to Alexandra Plows, and to Helena Siipi and Juha Räikkä for their valuable comments at the WONBIT conference and at the Department of Philosophy, University of Turku, Finland, respectively.

Chapter 14
Women's Perceptions of Biotechnologies: The Case of Genetically Modified Foods in Switzerland

Fabienne Crettaz von Roten(✉) and Elvita Alvarez

Abstract: The applications of biotechnology, in particular genetically modified foods, have been the object of considerable hopes and debate. Analyzing the public's perceptions of biotechnology, studies have found 'publics' rather than a single 'general public'. Among other social classifications, women constitute a definable public to analyze, in particular in a feminist perspective. GM foods lie at the juncture of two essential questions of gender perspective: the relationship of attitudes toward science and the responsibility and management of domestic life. This paper proposes to analyze the differences in perceptions of GM foods between women and men by emphasizing social sex roles over traditional explanations of differences of attitudes toward science on the basis of data contained in the Swiss Eurobarometer Biotechnology 2002.

Women are more concerned about the quality of food and the risks inherent to GM foods and are more critical toward their benefits. Arguments in favour of GM foods seem to be less appealing to women than to men. These differences are not explained by a lack of knowledge about genetics, but they may partially be explained by trust and values variations. However, the latter explanatory factors are related to the different socialization patterns. Thus the myth of the nurturing woman still remains deep-rooted in our social spirit and in the posture of the privileged managers of the domestic universe, which makes GM foods of principal concern to women.

Keywords: Women, biotechnologies, GM food, Switzerland

Fabienne Crettaz von Roten
Observatoire Science, Politique et Société, Université de Lausanne, SSP, Bâtiment Vidy, CH-1015 Lausanne, Switzerland
Fabienne.Crettazvonroten@unil.ch

Elvita Alvarez
Faculté des Sciences Economiques et Sociales, Uni Mail, Bd du Pont-d'Arve 40, CH-1211 Genève, Switzerland
Elvita.Alvarez@politic.unige.ch

F. Molfino, F. Zucco (eds.) *Women in Biotechnology,*

14.1 Introduction

The applications of biotechnologies have been the object of considerable hopes and debate in most European countries in the past decades. Experience has shown that the introduction of a new technology requires careful attention to the interactions between technology and society (Bauer 1995). This implies a need to analyze and follow the *public's* perceptions of biotechnologies, as studies have identified a plurality of 'publics' rather than a single 'general public' (Einsiedel 2000). Since 1991, the European Commission has financed six surveys on the public's perception of biotechnology.[1] The abundance of these surveys provides a rich empirical basis for understanding the public's attitudes towards biotechnology, with the following main results.

The heterogeneity of the perception of biotechnological applications: medical and industrial applications have been received positively, but the agricultural applications, in particular genetically modified foods, have engendered general criticisms and anxieties (Pardo et al. 2002).

The perception among the population is very heterogeneous: supporters of biotechnologies are more likely to be male, younger, and better educated (Crettaz von Roten and Alvarez 2005). However, summaries support the natural tendency to stereotype others by ignoring the details of the decisions, emotions, and opinions that influences their choices (Fischhoff and Fischhoff 2001).

Analysis highlights the various roles which the public may endorse within the framework of his relation to biotechnology: citizens, patients, consumers, etc. (Michael 1998). The public may be:

- Citizens, who have to vote on the regulation of biotechnology. In Switzerland, where the political system of direct democracy allows the public to launch a campaign to collect signatures for a so-called *initiative* on any subject, biotechnology was the object of six referenda since 1987.
- Patients, who can benefit from GM medicine or who can become co-constructor of knowledge in taking part in researches (patient groups).
- Consumers, who decide to buy or not to buy GM foods, and whose choice can influence research and innovation strategies.

Finally, surveys show great differences in public perception of biotechnology among countries (Gaskell et al. 2002). In Switzerland, biotechnology is a key economic issue, because the country hosts a number of large multi-national companies in the pharmaceutical, agricultural and food industries. What is more, biotechnology has been high on the Swiss political agenda with six referenda on this issue, creating intensive public debates about biotechnology with a strong social stratification: the 1998 Gene Protection initiative opposed a male-dominated scientific establishment

---

[1] Surveys were realised in 1991, 1993, 1996, 2000, 2002, and 2005. For details about these surveys, see http://ec.europa.eu/public_opinion/archives/eb_special_en.htm

to a coalition led by educated, successful women (Schatz 1998). The Swiss biotechnology regulations are restrictive: even if GMOs are not prohibited, an authorization is necessary for the commercialisation of products consisting of or containing genetically modified organisms and, in the shops, labelling is obligatory. Finally, in 2005, the Swiss have voted in favour of a five-year moratorium on GMOs.

To study women and biotechnology in Switzerland, we have chosen, among the various applications of biotechnology, genetically modified food because this application has engendered general criticisms and anxiety and because it highlights most of the social roles that the public can, or can refuse to, endorse.

Moreover, the question of GM foods opens an innovative entry, which allows adopting a gender perspective (Alvarez and Crettaz von Roten 2006). The GM foods lie at the crossing of two essential questions of gender perspective. Firstly, the relationship toward science and technology: following Amancio, 'modern science was born as an exclusively masculine activity' (Amancio 2005), therefore 'lack of women in science reflects the gendered nature of the culture of science and scientific institutions' (Amancio 2005). Secondly, the responsibility and the management of the domestic sphere: women remain the primary childcare providers and household managers,[2] food purchase and cooking are a large part of their activities (Delphy 1998).

This paper proposes to analyze the differences in the perceptions of GM foods between women and men by emphasizing gender over traditional explanations of differences of attitudes toward science and technology (*deficit model*, trust, etc.).

14.2 Data

This paper is based on the Eurobarometer (EB) 58.0 realised in 15 countries, plus Switzerland. The EB surveys, financed by the European Commission, have been conducted between two and five times per year since 1971. They measure the state of public opinion towards the European Union and other issues, including science and biotechnology. The national samples come from a multistage random design of the adult population, with approximately 1,000 face-to-face interviews conducted.

The Swiss EB 58.0 was financed by the Swiss National Science Foundation at the end of the year 2002. The survey was conducted face-to-face on 1,025 people, aged 15 and over, from the three linguistic regions of Switzerland: French, German and Italian speaking regions.[3] At national level, the error rate is 3%.

[2] In Switzerland, women spend almost twice as much time as men on housework (on average, 30 hours per week against 17 hours for the men); women in couples with young children spend on average 54 hours per week on domestic and parental work (21 hours for men), this exceeds the labour market weekly hours of work (Département Fédéral de l'Intérieur 2004).

[3] See www.sidos.ch for details about the survey design, methodology and response rate.

14.3 Results

Before analysing perceptions of biotechnologies in Switzerland, it is interesting to set the framework, in particular the level of engagement toward science and technology. We have measured engagement with three indicators: interest, information and understanding. Compared to Europeans,[4] the Swiss have a higher level of engagement (see Table 14.1): 41% of the Swiss are most of the time interested in science (only 33% in Europe), 27% are informed about science (25% in Europe) and 38% understand science stories in the news (only 30% in Europe).

These indicators show a significant[5] difference between women and men: women say that they feel less interested, less informed and understand science stories less often, as noticed before by numerous studies. *She Figures 2006* highlights gender imbalance in the science and technology fields: Switzerland has a lower proportion of female PHDs and a lower proportion of female researchers than most European countries (European Commission 2006).

Gender studies have illustrated the fact that scientific and technological competences are less attributed to women (Dhavernas 1992; Wajcman 1995), and also that women feel less competent in scientific fields (Kimball 1989; Fox and Firebaugh 1992). Women end up excluding themselves from the scientific field, for example by paying less attention to information about science and by trying less to understand it.

14.3.1 Perceptions of GM Foods

Food safety has always been a concern (hormone-treated meat, mad cow disease, foot-and-mouth disease) (Ferrières 2005). It's a small wonder that an overwhelming majority of Swiss people reject the safety of GM foods (see Table 14.2): 24% believed that it is safe to eat GM foods while 76% believed it is not safe to eat GM foods. There is a significant difference between women and men: if 33% of men find GM foods safe, only 15% of women find it safe.

However, if a majority of respondents are sure about their opinions about GM (59%) and find it important to have an opinion on GM foods (86%), only 43% find

Table 14.1 Engagement toward science and technology by sex (% « Most of the time »)

	Total	Women	Men
Interest in science and technology	40.7	24.4	57.0
Informed about science and technology	27.0	14.1	39.8
Understand science stories in the news	38.3	26.9	49.3

[4] See http://ec.europa.eu/public_opinion/archives/ebs/ebs_224_report_en.pdf for European results.

[5] p-value less than 0.05.

it easy to form an accurate judgment on GM foods. Among these last three items, there is no sex difference. In other words, if women find GM foods less safe than men, it is not because they are more unsure, nor because they find it less important to form a judgment, nor because they have more difficulty in forming a judgment. Therefore, what explains this difference in perception?

14.3.2 Deficit Model

A well-known explanation of negative attitudes toward science is the deficit model, which postulates that a deficit in scientific knowledge explains the public's doubts about the value of scientific progress (Allum et al. 2002). Therefore, the EB surveys have included items to measure knowledge of biology and genetics. Table 14.3 shows that the percentage of correct answers varies between 88% and 26%, with a higher percentage of correct answers about bacteria and pre-natal screening and a lower percentage of correct answers about ordinary tomatoes and the transfer of animal genes into plants. On the whole, the relation between knowledge and attitude toward biotechnology is not significant. This result is consistent with former studies.[6]

Table 14.2 Perceptions of GM foods by sex (% « Agree »)

	Total	Women	Men
I think it is safe for me to eat GM foods	24.1	15.5	33.0
I am sure about my opinions about GM foods	58.9	n.s.[a]	n.s.
It is important for me to have an accurate judgment on GM foods	86.3	n.s.	n.s.
It is easy for me to form an accurate judgment on GM foods	43.3	n.s	n.s.

Category « Disagree »: difference to 100%
[a]n.s. means that the sex difference isn't significant at p-value = 0.05

Table 14.3 Knowledge of biology and genetics by sex (%« Correct »)

Selection of affirmations	Total	Women	Men
There are bacteria which live from waste water (T)	88.4	85.0	91.6
It is possible to find out in the first months of pregnancy whether a child will have Down's syndrome, Trisomy, Mongolism (T)	80.1	86.2	74.1
Yeast for brewing beer consists of living organism (T)	74.3	n.s.	n.s.
By eating a genetically modified fruit, a person's genes could also become modified (F)	59.2	n.s.	n.s.
Ordinary tomatoes do not contain genes, while genetically modified tomatoes do (F)	47.5	43.2	51.8
It is not possible to transfer animal genes into plants (F)	26.4	n.s.	n.s.
Mean score (nb. of correct answers among 12 items)	7.5	n.s.	n.s.

False and don't know answers: difference to 100%

[6]Bucchi and Nerisini (2002) showed that exposure to information does not always lead to a greater approval of biotechnologies; Gaskell (1997) found that more knowledge about biotechnologies makes people more sceptical or polarised.

If we analyse the pertinence of the deficit model to explain the difference of perception of GM foods, we observe that half of the items did not show differences between women and men (yeast, genetically modified fruit, gene transfer). Among the significant items, the percentage of correct answers is higher among men for bacteria and tomatoes (92% and 52%) but the percentage of correct answers is higher among women for prenatal screening (86%). Regarding the mean score calculated from the 12 items of knowledge,[7] there is no significant difference between women and men. Therefore the deficit model cannot explain this difference in perception.

14.3.3 Evaluative Schemes of GM Foods

To evaluate a biotechnological application, respondents were asked whether they thought the application was useful for society, risky for society, morally acceptable and whether it should be encouraged.[8] The evaluative schemes are quite different among the biotechnological applications (Gaskell et al. 2000): the risk attached to gene therapy – a useful application – is tolerable, whereas for GM foods – an application perceived as less useful – it is unacceptable.[9] Figure 14.1 reports the mean scores of men and women for GM foods on each criterion.

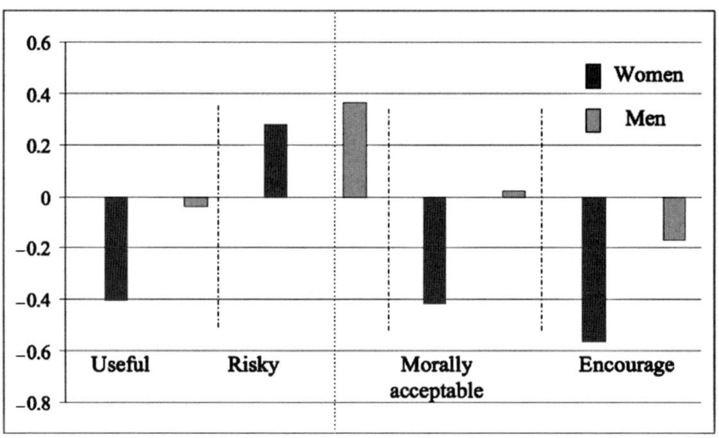

Fig. 14.1 Evaluative schemes of GM foods by sex

[7] The survey contains 12 items of knowledge, but, to make this simpler, we only include items related to GM foods in Table 14.3, see Crettaz von Roten and Alvarez (2005) for details.

[8] Following the literature (http://ec.europa.eu/public_opinion/archives/ebs/ebs_177_en.pdf), the data have been recoded from 1 = « Definitively agree » to 4 = « Definitely disagree » (don't know responses excluded) into +1.5 to −1.5 in order to show the midpoint of zero; then mean scores have been calculated.

[9] As the judged usefulness of technologies declines so there is an increase in perceived risk, along with a decline in perceptions of moral acceptability and overall levels of support (Gaskell et al. 2000).

The evaluative schemes for women and men are quite different[10]: women find GM food less useful than men, equally risky, less morally acceptable, and less to be encouraged. According to the height of the absolute value of the mean scores, we see that women think principally about GM foods in terms of usefulness and moral acceptability, whereas men think in terms of risk.

14.3.4 *Perceptions of Risks of GM Foods*

Historically, risk science has mostly dealt with high-risk industries (nuclear power industry, chemical industry, aviation industry, etc.), but the risk issue has also become important for the biotechnology industry. Analyzing the perceptions of risks of GM foods, we find – as no surprise – that 67% of respondents believed that GM foods are a risk for society, with no difference between women and men, as stated before (see Table 14.4).

Precisely, the risks concern future generations (79%), the environment (72%) and health (68%). There is a sex difference on two items: women stress more the risk related to future generations and to health. In fact, these differences are related to socialization: taking care of the children and of one's health is still mostly socially attributed to women.

Results show that the worries about foods are not peculiar to 'food related with science': 44% of respondents are concerned most of the time about the quality of food, indicating that the public is worried about the side effects of the mass-production of food (Bray 2003). Moreover, there is also a sex difference: 52% of women are concerned about the quality of food, but only 37% of men.

14.3.5 *Perception of Benefits of GM Foods*

The other criterion of evaluation, usefulness, is studied more deeply in the survey with questions about the benefits related to the introduction of GM foods. A majority

Table 14.4 Risks related to the introduction of GM foods by sex (% « Agree »)

	Total	Women	Men
GM foods application is a risk for society	67.1	n.s.	n.s.
GM foods pose threats to future generations	78.7	85.8	71.7
Growing genetically modified crops will be harmful to the environment	72.0	n.s	n.s.
Eating GM foods will be harmful to my health and my family's health	68.3	79.4	57.4

Category « Disagree »: difference to 100%

[10] On the whole, GM foods are not useful (–0.22), risky (0.32), not morally acceptable (–0.19), and not to be encouraged (–0.37).

of respondents believe that GM foods could benefit industry (61%) and the fight against third world hunger (53%), but not the economy (38%) and the consumer (27%) (Table 14.5).

Figures indicate three significant sex discrepancies. First, women consider it more likely that GM foods will only be good for industry. Then, women believe it less likely that GM food will be useful in the fight against third world hunger and for the consumer. In fact, these differences are deeply linked with the role of nurturer, which is still socially attributed to women (Delphy 1998).

The proponents of GM foods have formulated various arguments to sell their products. Among all the reasons suggested as a motivation to buy GM foods, there are more Swiss people saying that they would not buy them than those saying they would (see Fig. 14.2). Moreover, 46% reject all the reasons.

Let's analyse the results reason by reason. Somewhat surprisingly, price is the least incentive for buying GM foods (only 11%). The most persuasive reason is the environmental benefit (41%) and the health benefit of lower pesticide residues (38%). Among all the reasons offered, there are fewer purchasing intentions among women than among men (about 10% fewer women, expect an improvement in the price and in the taste; for the latter, the differences are smaller).

Nevertheless, it is important to remember that what people say and what people do are sometimes rather different. This is related to *cognitive dissonance* (Akerlof

Table 14.5 Benefits related to the introduction of GM foods by sex (% « Agree »)

	Total	Women	Men
GM foods and crops will only be good for industry and not for the consumer	60.6	67.9	53.4
GM foods will be useful in the fight against third world hunger	52.8	46.7	58.9
In the long run, a successful Swiss GM foods industry will be good for the economy	38.0	n.s	n.s.
GM foods will be useful for me and other consumers	27.1	18.9	35.2

Category « Disagree »: difference to 100%

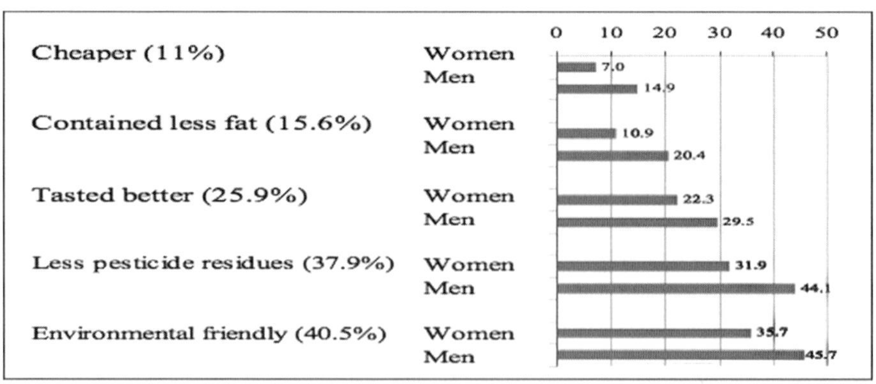

Fig. 14.2 Reasons to buy GM foods by sex (% « Agree »)

and Dickens 1982): even if people are aware, for example, of the horrible conditions in the mass production of poultry, they often buy cheap chicken produced in such conditions.

14.3.6 Trust in Key Actors in Biotechnology

Some authors (Hornig Priest et al. 2003; Siegrist 2000) postulate that trust in actors or institutions are more important than scientific knowledge in predicting levels of supports for biotechnology. Their hypothesis is that without confidence, people are likely to have exaggerated perceptions of risks.

Swiss people have especially confidence[11] in doctors, patients' organisations, university scientists, consumers' organisations, farmers and environmental groups (more than 80% of respondents, Table 14.6).

Trust explains few of the sex differences in perception of GM foods because only two actors show a significant difference: women trust environmental groups more than do men and men trust industry more than do women. However, this analysis highlights key actors: according to Hornig Priest et al. (2003) 'for food biotechnology it is clear that it is the difference between how much a nation's population trusts industry and how much they trust environmental groups that is the key predictor of encouragement, rather than either factor alone'.

14.3.7 Values

We ended our analysis with values, as many researches have stressed the gender differences in fundamental values orientations: compassion, materialism and meaning

Table 14.6 Trust in key actors by sex (% « Doing a good job »)

	Total	Women	Men
Medical doctors keeping an eye on the health implications of biotechnology	90.1	n.s.	n.s.
Organisations of patients or their relatives looking after patient's interests	89.6	n.s.	n.s.
University scientists doing research in biotechnology	88.4	n.s.	n.s.
Consumer organisations checking products of biotechnology	85.2	n.s.	n.s.
Farmers deciding which types of crops to grow	82.3	n.s.	n.s.
Environmental groups campaigning against biotechnology	82.2	86.3	78.3
Newspapers and magazines reporting on biotechnology	76.6	n.s.	n.s.
Scientists in industry doing research in biotechnology	74.4	n.s.	n.s.
Our government making regulations on biotechnology	72.4	n.s.	n.s.
Shops making sure our food is safe	71.9	n.s.	n.s.
Industry developing new products with biotechnology	61.7	54.0	68.2

[11] The indicator used – « Doing a good job » – constitutes a proxy measure of trust; it views the actor as both competent and behaving in a socially responsible way.

(Beutel and Marini 1995); equality, inner harmony, true friendship (Di Dio et al. 1996). Gender-based differences in values are likely to emerge because the differentiated socialization imposes different gender roles.

Women are more likely than men to express concern for the environment (see Table 14.7): 73% of women agreed that nature is fragile (12% less among men), 89% of women believed that technology has upset the balance of nature (12% less among men), 76% of women refused to exploit nature (5% less among men). The scale built on these three items[12] shows significant sex difference.

On the other side, women are less likely than men to express free-market liberal economics value: 51% of women believed that economic growth brings better quality of life (15% more among men), 43% believed that private enterprise can solve the Swiss' problems (10% more among men). The scale built on these three items[13] shows significant sex difference.

Finally, we tested the difference of approval of GM foods[14] among women and men, controlling for these two values. The results of the analysis of covariance[15] (see Table 14.8) show two significant effects: the environmental value and sex. The environmental value helps to explain the difference of approval of GM foods among women and men: the parameter estimate is negative, which means that an increase of the environmental value implies a decrease of approval of GM foods. The free-market liberal economics value is not a right choice of confounding variable for approval of GM foods. Finally there is a sex effect, with a parameter estimate of 0.28 for men and 0 for women (corresponding to a higher level of approval of GM foods among men) but the percentage of total variability attributable to this factor is half the percentage of the environmental value.[16]

Table 14.7 Environmental and free-market liberal economic values by sex

% « Agree »	Women	Men
Nature is fragile and easily damaged by human actions	73.0	61.0
Modern technology has upset the balance of nature	88.6	77.1
Exploiting nature is not unavoidable if humankind is to progress	75.7	70.6
Environmental value	2.41	2.10
Economic growth brings better quality of life	50.7	66.0
What is good for business is good for the citizens	7.0	14.0
Private enterprise is the best way to solve the Swiss' problems	43.3	53.4
Free-market liberal economics value	1.02	1.36

[12] The items were dummy coded 1 if « Agree » on the first item and coded 1 if « Disagree » on the second and third items, then the recoded items were summed.

[13] The items were dummy coded 1 if « Agree » on the three items, and then the recoded items were summed.

[14] 'It is safe to eat GM foods' from 1 « Agree » to 3 « Disagree ».

[15] The assumptions of the analysis of covariance are fulfilled as the two covariates are significantly related to approval of GM foods. The environmental value is negatively linked with the approval of GM foods ($r = -0.28$). The free-market liberal economics value is positively linked with the approval of GM foods ($r = 0.13$).

[16] Partial eta-squared is 0.032 for sex and 0.065 for the environmental value.

Table 14.8 Sex difference in approval of GM foods and biotechnology by values (F statistic, p value and parameter estimate in parenthesis)

	Approval	
	GM foods	Biotechnology
Intercept	$208.70^{s.}$	$425.74^{s.}$
	(1.86)	(5.33)
Environmental value	$24.77^{s.}$	$27.01^{s.}$
	(−0.25)	(−0.51)
Free-market liberal economics value	$0.63^{n.s.}$	$16.15^{s.}$
	(0.04)	(0.40)
Sex	$11.87^{s.}$	$10.62^{s.}$
	(H = 0.28 F = 0)	(H = 0.54 F = 0)

$^{s.}$ significant at level 0.05, n.s. – non-significant

To extend our analysis, we performed the same analysis for biotechnology, by using a general indicator of approval of biotechnologies.[17] For the general attitude, the three effects are significant[18]: the environmental value, the free-market liberal economics value and sex. The environmental value is significant with a negative parameter estimate as in the former analysis. The free-market liberal economics value helps to explain the difference of approval of biotechnology among women and men: the parameter estimate is positive, which means that an increase of the free-market liberal economics value implies an increase of approval of biotechnology. Finally, the parameter estimates of sex are 0.54 for men and 0 for women (corresponding to a higher level of approval of biotechnology among men) but the percentage of total variability attributable to this factor is much smaller than the percentages of the environmental and free-market liberal economics value.[19]

The environmental value is important to understand the perception of biotechnology and GM foods. Biotechnological developments have supplemented and sometimes supplanted 'natural ways', for example in food and reproduction, with accusations of 'playing God' (Pretty 2001; Peters et al. 2007). Some scientific developments have had severe environmental impacts and the population has now concerns of the unforeseen consequences of interfering with nature, and in particular are reluctant to accept an increasing (bio)technological domination of nature. Conversely, the free-market liberal economics value is important to understand the perception of biotechnology, which is not surprising because biotechnology is a key economic issue in Switzerland with many biotechnology companies.

[17] 'What is your personal attitude towards biotechnology' from 1 « Strongly in favour » to 10 « Strongly against».

[18] The assumptions of the analysis of covariance are fulfilled. The environmental value is negatively linked with the approval of biotechnology (r = −0.24). The free-market liberal economics value is positively linked with the approval of biotechnology (r = 0.23).

[19] Partial eta-squared is 0.016 for sex, 0.039 for the environmental value and 0.024 for the free-market liberal economics value.

14.4 Conclusion

The goal of this paper was to analyze the differences in the perception of GM foods between women and men. Our analyses show that the traditional explanations of negative attitudes toward science – the deficit model, engagement, trust, etc. – are not pertinent to understand why women have a more negative perception of GM foods.

To summarise, women find GM foods less useful than men, equally risky, less morally acceptable, and less to be encouraged. We find significant sex differences in the items concerning safety of GM foods, risks for the family and the children's health of consuming GM foods. On the other hand, we found that women feel more concerned by the negative effects of biotechnology and GM foods on the environment.

This last result is often explained by the particular sensitivity and relation between women and the environment that would be inherent to the values of women, related to the reproductive function. This function would make them more sensitive to life and to future generations. Other explanations are related to their function of housewife, still largely widespread in the patriarchal societies, which would give them a more integrated vision of the relations between human beings, their activities of consumption and their natural environment.

In fact, women are socially responsible for domestic management: food, reproduction, children and health. The work in the family sphere, such as meals and the care includes the responsibility for others' well-being and the mental, emotional, and physical work in the course of fulfilling that responsibility (Hochschild 2003). The authors suggest adopting a gender perspective, which allows focusing on social sex roles and on sexual division of labour. Family involvement is therefore often a better indicator than biological sex, as such, for understanding the differentiated valorisation of GM foods. This focus is necessary to understand women's specific perception of biotechnology and GM foods.

References

Akerlof, G., & Dickens, W. (1982). The economic consequences of cognitive dissonance. *American Economic Review*, *72*(3), 307–319.

Allum, N., Boy, D., & Bauer, M. (2002). European regions and the knowledge deficit model. In M. Bauer & G. Gaskell (Eds.), *Biotechnology: The making of a global controversy* (pp. 224–243). Cambridge: Cambridge University Press.

Alvarez, E., & Crettaz von Roten, F. (2006). Science et rapports sociaux de sexe: les femmes face aux aliments génétiquement modifiés. In J.-P. Leresche, M. Benninghoff, F. Crettaz von Roten, & M. Merz (Eds.), *La fabrique des sciences. Des institutions aux pratiques* (pp. 329–347). Lausanne: PPUR.

Amancio, L. (2005). Reflections on science as a gendered endeavour: Changes and continuities. *Social Science Information*, *44*(1), 65–83.

Bauer, M. (1995). Resistance to new technology and its effects on nuclear power, information technology and biotechnology. In M. Bauer (Ed.), *Resistance to new technology* (pp. 1–41). Cambridge: Cambridge University Press.

Beutel, A., & Marini, M. (1995). Gender and values. *American Sociological Review, 60*, 436–448.

Bray, F. (2003). Genetically modified foods: Shared risk and global action. In B. Harthorn & L. Oaks (Eds.), *Risk, culture, and inequality* (pp. 185–2007). Westport, CT: Praeger.

Bucchi, M., & Neresini, F. (2002). Biotech remains unloved by the more informed. *Nature, 416*, 261.

Crettaz von Roten, F., & Alvarez, E. (2005). Perception des biotechnologies en Suisse: Perspectives longitudinales et de genre. *Les Cahiers de l'Observatoire, 11*, 1–59.

Delphy, C. (1998). *L'ennemi principal, Economie politique du patriarcat*. Paris: Syllepse.

Département Fédéral de l'Intérieur (2004). *Rapport sur les familles 2004*. Bern: DFI.

Dhavernas, M. J. (1992). Je ne suis pas celle que vous pensez. In F. Collin (Ed.), *Le sexe des sciences* (pp. 129–142). Paris: Autrement.

Di Dio, L., Saragovi, C., Koestner, R., & Aube, J. (1996). Linking personal values to gender. *Sex Roles, 34*, 621–636.

Einsiedel, E. (2000). Understanding "publics" in the public understanding of science. In M. Dierkes & C. Grote (Eds.), *Between understanding and trust: The public, science and technology* (pp. 205–215). Reading: Harwood Academic Publishers.

European Commission (2006). She Figures 2006: Women and science statistics and indicators. Retrieved January 22, 2007, from http://ec.europa.eu/research/science-society/pdf/she_figures_2006_en.pdf

Ferrières, M. (2005). Risque alimentaire et conférence du consensus: l'expérience de 1669. In C. Lahellec (Ed.), *Risques et crises alimentaires* (pp. 3–20). Paris: Lavoisier.

Fischhoff, B., & Fischhoff, I. (2001). Publics' opinions about biotechnologies. *AgBioForum, 4*(3–4), 155–162.

Fox, M. F., & Firebaugh, G. (1992). Confidence in science: The gender gap. *Social Science Quarterly, 73*, 101–113.

Gaskell, G. (1997). Europe ambivalent on biotechnology. *Nature, 387*, 845–847.

Gaskell, G. et al. (2000). Biotechnology and the European public. *Nature Biotechnology, 18*, 935–938.

Gaskell, G., Thompson, P., & Allum, N. (2002). Worlds apart? Public opinion in Europe and the USA. In M. Bauer & G. Gaskell (Eds.), *Biotechnology: The making of a global controversy* (pp. 351–375). Cambridge: Cambridge University Press.

Hochschild, A. R. (2003). *The commercialization of intimate life, notes from home and work*. Berkeley, CA: University of California Press.

Hornig Priest, S., Bonfadelli, H., & Rusanen, M. (2003). The "trust gap" hypothesis: Predicting support for biotechnology across national cultures as a function of trust in actors. *Risk Analysis, 23*(4), 751–766.

Kimball, M. (1989). A new perspective on women's math achievement. *Psychological Bulletin, 105*, 198–214.

Michael, M. (1998). Between citizen and consumer: multiplying the meanings of the "public understanding of science". *Public Understanding of Science, 7*, 313–327.

Pardo, R., Midden, C., & Miller, J. D. (2002). Attitudes toward biotechnology in the European Union. *Journal of Biotechnology, 98*, 9–24.

Peters, H. P., Lang, J., Sawicka, M., & Hallman, W. (2007). Culture and technological inovation: Impact of institutional trust and appreciation of nature on attitudes towards food biotechnology in the USA and Germany. *International Journal of Public Opinion Research, 19*(2), 191–220.

Pretty, J. (2001). The rapid emergence of genetic modification in world agriculture: Contested risks and benefits. *Environmental Conservation, 28*(3), 248–262.

Schatz, G. (1998). The Swiss vote on gene technology. *Science, 281*, 1810–1811.

Siegrist, M. (2000). The influence of trust and perceptions of risks and benefits on the acceptance of gene technology. *Risk Analysis, 20*(2), 195–203.

Wajcman, J. (1995). Feminist theories of technology. In S. Jasanoff, G. E. Markle, J. C. Petersen, & T. Pinch (Eds.), *Handbook of science and technology studies* (pp. 189–203). Thousand Oaks, CA: Sage.

Chapter 15
A Transdisciplinary Approach to Face the Plant Gene Transfer Technique: From Laboratory to Society

Lucia Martinelli(✉) and Floriana Marin

Abstract: We present a transdisciplinary project aiming to face the controversial issue of transgenics. An analysis of the attitude of the scientific community, stakeholders and citizens, towards the techniques for gene transfer and for genetically modified food and feed is the core of our study. The most innovative aspect of our research is a survey that specifically focuses on the scientific community where scientists' attitudes concerning the techniques for gene transfer and their perception of related risks is assessed. In addition, the meaning of the development of 'sustainable' practices for gene transfer is discussed. We also report the results of a research conducted in the Trentino region (Italy) aiming to assess the risk perception of several actors and consumers' willingness to buy genetically modified food. Finally, we point out the role of women scientists as crucial actors in the decision-making process concerning research, development and application.

Keywords: Gene transfer, genetically modified organisms (GMOs), marker genes, public perception, communication

15.1 Unlocking the Doors of the Ivory Tower

The demand of our society to participate in the decisions concerning scientific research is pressing the scientists to take into account bioethical and economic issues, social perceptions, and political priorities in their research lines (Beck 2001; Slovic 1999).

Lucia Martinelli
Fondazione Edmund Mach – IASMA Research Centre, Genetics and Molecular Biology Department, Via E. Mach, 1, 38010 San Michele all'Adige (TN), Italy
Lucia.Martinelli@iasma.it

Floriana Marin
Fondazione Edmund Mach – IASMA Research Centre, Genetics and Molecular Biology Department, Via E. Mach, 1, 38010 San Michele all'Adige (TN), Italy
Floriana.Marin@iasma.it

F. Molfino, F. Zucco (eds.) *Women in Biotechnology,*

Nowadays the direction of research has become mostly dependent on society rather than on academia (and often in competition with it) as a process described as 'the post-academic era of science' (Ziman 2000; Greco 2005). This means that the idea of the researcher as an inaccessible holder of privileged knowledge, who lives isolated in an 'ivory tower' and communicates exclusively with 'the peers', does not fit the role of the contemporary scientific community (Castelfranchi and Pitrelli 2007). Accordingly, caring about public communication and understanding of science is progressively becoming a must for researchers (Gregory and Miller 1998a, b; Scholderer and Frewer 2003).

This concern, however, still needs to be fully acknowledged by the scientific community, as is well represented by the controversial debate involving agrobiotechnology, and in particular the use of Genetically Modified Organisms (GMOs) in everyday lives (Gaskell et al. 2004; Savadori et al. 2004).

In this framework, after years of active laboratorial research in plant biotechnology, we felt the need to analyze gene transfer techniques as a part of a whole complex context, where laboratory research, social sciences, bioethics, and communication come together in a multidisciplinary approach. We are working in a stimulating research area that – in our experience – is still not fully familiar to the scientific community.

In this paper we present (i) our research on the development of a more agreeable technique for transferring genes into plants, and (ii) a study of the risk perception related to genetically modified organisms.

In our research, which started in 2000 and is still in progress, besides the laboratorial aspects, the cooperation with researchers of other disciplines is essential. Concepts of social science enable us to investigate the scientists' consciousness of the importance of bridging the gap between science and society, while bioethics focuses on the responsibility of the subjects involved in agrobiotechnology.

Finally, analysis of the risk perceptions of various actors involved in the field of GMOs applications leads us to combine our scientific background with an improved feeling for social concerns.

15.1.1 Choosing a (Politically) Correct Technique for Gene Transfer into Plants

In considering gene transfer techniques we decided, as part of the scientific community, to analyze the possible effects of our research on society at large, and to take into account the increasing social demand for knowledge and participation. As a result, we embarked on a research line which aims to employ 'sustainable' strategies for plant gene transfer. We define 'sustainable' those techniques based on the use of constructs that are exploited to allow the transfer of low environmental impact marker genes to plant tissues, and even only of the target gene without the marker gene, through co-transformation, transposable elements, or site-specific recombination (Martinelli et al. 2006).

15.1.1.1 Marker Gene Removal and Alternative Marker Genes in Grape Gene Transfer

According to our established experience on grapevine gene transfer (Martinelli and Mandolino 1994, 2001; Martinelli et al. 2002) we decided to apply some 'sustainable' techniques to this plant.

In grapevine – as for many other plants – the gene transfer technique is mostly based on *in vitro* co-culture of plant tissues with agrobacteria carrying the gene to be transferred. The efficiency of the genetic transformation depends on the possibility to rely on an efficient tissue culture and to discriminate the cells that have inserted the foreign gene from those that don't. These latter have to be removed from the cultures (Martinelli and Mandolino 2001).

Usually, the capability conferred to a cell by an exogenous marker gene to survive in a medium to which has been added an otherwise toxic antibiotic is the strategy most often applied for a proper selection among the cells. For this reason marker genes are typically associated to the gene of interest during its transfer to the plant.

Antibiotic-resistance, however, is a crucial issue that has addressed public concerns against genetically modified plants, so much so that at least some of those genes are now going to be banned (Dir. 2001/18/CE). Moreover, it being the exclusive function of marker genes to help during the gene transfer process, their presence in transgenic plants becomes undesirable.

These two drawbacks require either alternative strategies that allow removal of the marker genes after gene transfer or the use of new marker genes.

We evaluate two approaches for the gene transfer in grapevine the constructs for the marker gene removal and the use of marker genes alternative to the ones conferring the antibiotic resistance trait (Dale and Ow 1991; Reed et al. 2001; Zuo et al. 2001).

To eliminate the marker gene after insertion of the target gene, the self-excision can be obtained with a site-specific recombination, a strategy that today is suitable also for plants with vegetative propagation, such as grapevine. In our activity, embryogenic calli or somatic embryos of *V. vinifera* (cultivars, Chardonnay, Brachetto), 110 Richter and *V. rupestris* were co-cultured with *Agrobacterium tumefaciens* carrying the chemically-inducible site-specific cre/loxP pX6 vector with the gene for the Green Fluorescent Protein (GFP), and the gene for the neomicin phosphotransferase (NPTII). This latter confers resistance to the antibiotic kanamycin. In this construct, the expression of the cre recombinase is regulated by the 17-B-estradiol. The construct was kindly provided by The Rockefeller University of New York (USA), Professor Nam-Hai Chua (Hare and Chua 2002; Martinelli et al. 2006). We have already regenerated some transgenic plant lines, and the assays for fully evaluating the efficiency of the gene removal are currently under way.

As for the use of marker genes alternative to the antibiotic resistance, among the possible strategies, we tested the genes that enable cells to metabolize specific carbon sources, such as mannose as an alternative to sucrose (Posytech, Sygenta licence). This strategy, which is based on the capability conferred to the cells by the phosphomannose isomerase gene to grow on culture media containing mannose instead of sucrose, proved to be successful with various plants (Reed et al. 2001); however it gave doubtful results in grapevine (Kieffer et al. 2004). As a first step, we subcultured wild-type calli and embryogenic calli of *V. vinifera* cultivar Brachetto on different formulations of the same basal medium, respectively added with mannose, sucrose or free of these carbohydrate sources. Cultures grown on the substrate containing mannose or sucrose showed a similar growth rate, even after long times of subcultures. No growth was observed in calli placed on the medium without carbohydrate source, which progressively turned brown and died (Martinelli et al. 2006).

According to our results, we can here anticipate some considerations on the possibility of applying these two 'sustainable' strategies for plant gene transfer to grapevine.

The use of the phosphomannose isomerase gene proved to be unsuitable since remarkably long times are required for discriminating transgenic from non-transgenic tissues, and the cost of the mannose is extremely high compared to the cheap sucrose that usually is added to the culture media.

In addition, concerning the choice of gene transfer techniques, we believe that a relevant attribute of 'sustainability' is related to the option to reduce the genetic material transferred to a plant to the strictly required traits. For this reason, the cre/loxP system seems to better suit this feature, since insertion of the marker gene allows us to take advantage of an effective selection method, while its subsequent removal eliminates the non-desired genetic material that was inserted during the gene transfer process.

15.1.1.2 Evolution of the Technique and Attention to Societal Concern

In his/her research activity, a researcher becomes accustomed to modifying a technique to achieve better results and empirical research is based on this flexible attitude.

As for gene transfer into plants, the current technology is susceptible of improvement owing to the progress of molecular biology and of tissue culture. In our study, we aim to understand how this technical evolution may work together with the non-technical aspects proposed by the society. Our assumption is that the attempt to exploit more 'sustainable' practices for gene transfer into plants would represent a conciliation effort between technology progress and attention to the public concern.

In this framework, we investigate the needs and the capabilities of the scientific community to manage their relationship with the society at large, considering the specific field of gene transfer into plants.

Perceived risks of this technology and the use of 'sustainable' strategies were investigated during specific surveys conducted in selected European laboratories. Qualitative research techniques, such as ethnographic observation and interviews of the persons in charge of the laboratories and of the research guidelines, were applied in a methodological approach. We assessed the scientists' perception of exogenous gene transfer in general, and in particular we focused on the reasons to eventually apply 'sustainable' technology.

In our sample, the scientists seemed to be conscious of the fact that public opinion seriously affects institutional priorities and policies with an immediate impact on funding schemes (Greco 2005). Accordingly, mostly due to practical reasons, our researchers acknowledged the importance of applying less impacting technologies as an effective response to public diffidence. However, the reasons leading scientists to apply 'sustainable' technologies in gene transfer in most cases related to motivations that are functional to their research goals, rather than to a true consideration of public needs and concerns.

This latter finding while disappointing us – being ourselves members of the scientific community – proves the urgency of acting in a reliable manner with a view to filling the gap between science and society. This is why, accordingly, bridging science with society is one of the priorities of the research policy of the European Community (http://cordis.europa.eu/science-society/).

15.1.2 Studying the Risk Perception Related to Genetically Modified Organisms (GMOs)

Besides the aspects related to the technology of gene transfer into plants discussed above, we conducted in the Trentino region an extensive study concerning the use of genetically modified food and feed. Hazards perceived and willingness to buy such products were the objectives of our survey.

15.1.2.1 Analysis of the Hazard Perception Concerning GMOs

We addressed our study to stakeholders (delegates of public institutions, of private and public analysis laboratories, of farmers, breeders and industry associations and trade-unions, and representatives of consumers and environmental associations) and lay citizens. During focus groups, we investigated their opinions on the capability of European regulations on food and feed labeling (EC1829/2003 and 1830/2003) to assure consumers and to prevent risks (Martinelli et al. 2005a, b).

A general aversion towards the use of genetically modified products was observed. However, while stakeholders admitted that their reserve was mostly based on market motivations, citizens expressed a concern based on a complex combination of emotional, political, cultural and ethical aspects. Stakeholders faced the debate with a problem-solving approach and were able to respect the agenda of the debate. Citizens, on the other hand, tended to take part in the discussions with a more passionate attitude, and in some cases the facilitator's role was quite critical in managing the debate. Moreover, laypersons tended continuously to divert the discussion from the principal theme of the debate (EU regulation on food labeling) and to direct it to wider issues concerning all of humanity (i.e., world famine, biodiversity defense, multinational resources exploitation, long term consequences of GMOs spread, etc). Furthermore, they showed a strong feeling of outrage against the GM technology application (Sandman 1999).

The results of the focus groups emphasized the relevance of the ethical component in the debate on genetically modified food and feed. Besides, the need of the society to be involved in the decision-making process was tangible. A crucial factor in the opposition to agrobiotechnology seemed to be a distrusting feeling of the decision-making process that leads to the permission to produce, grow and sell GM products, in particular when industries are involved. Finally, both the stakeholders and the citizens expressed a considerable feeling of incommunicability and distance by the scientific community.

15.1.2.2 Analysis of the Willingness to Buy GM Food

We investigated the consumers' willingness to buy food containing genetically modified ingredients. A Discrete Choice (DC) approach (McFadden 1974a, b, 1981, 2000; McFadden and Ruud 1994; McFadden and Train 2000) was applied to model consumer preferences for alternative GM yoghurts. Such yoghurts were characterized by hypothetical GM ingredients that promised three kinds of benefits: a higher agricultural yield, a lower environmental impact, or the prevention of serious health diseases (Marin and Martinelli 2006).

These GM yoghurts were inserted into a choice bundle, together with a non-GM one, and each product was assigned a price. People were asked to choose their preferred alternative. The choices were analyzed with the Multinomial Logit and Random Parameter Logit models (Train 2003), and were the chief part of a questionnaire that was administered to a random sample of 532 individuals, with a response rate of 60.5%.

Besides the individual attitude toward buying GM food – and in particular yoghurt – the questionnaire also checked the respondents' ethical values, priorities, habits, general knowledge on GMOs, and the attitudes toward different uses of gene technology. Two specific questions were addressed to investigate the risk-taking attitude and the willingness to support institutional initiative to characterize local agro-food production (namely, a Non-GM quality label).

In our study, the majority of respondents were averse to paying extra money for GM yoghurts even in the cases where these products promised benefits related to better health, to a prize discount or a lower environmental impact. Moreover, risk perception influenced the responses, and nullified the potential benefits offered by gene technology to the consumers, while the Non-GM label did not provide a proper tool for ensuring consumers' confidence in the products. Perceived risks as emotional factors, accordingly, have been already described as a crucial component of public perception of GMOs (Slovic 1999; Wolt and Peterson 2000; Sandman 1999).

Finally, respondents expressed the need for better information on the technological applications that involve their lives.

15.2 Conclusions

The work here presented describes our trans-disciplinary approach to face the controversial issue of transgenics. Our effort to understand the attitude of scientific community, stakeholders and citizens towards techniques for gene transfer and genetically modified food and feed are the core part of the research. Our analysis that specifically focuses on the scientists and considers their attitudes toward the techniques for gene transfer and their perception of related risks is the most innovative aspect of our research.

Considering the existence of 'sustainable' practices for gene transfer, we believe that the effort to combine improvement of a technique with the respect of the public concerns can make it possible for the scientists to communicate with the society and not only with 'the peers'.

Finally, in our project, within the partners, a group of women scientists with diverse formations and age contributed to the research. Women's teamwork represents a characteristic element of our activity, enabling us to widen our research horizons for a better result from all the work packages of the project.

We hope that the research here presented will also be seen as describing the effort of women scientists to consider a specific technical matter with a trans-disciplinary approach. Our final goal is to achieve a better consciousness of the reasons determining the gap between laboratory and society, and to provide ourselves with proper tools for managing this problem. As women scientists, in fact, we feel ourselves to be relevant actors in the society with the important role of contributing to the decision-making processes concerning science and technology transfer.

Acknowledgements The research presented here was supported by the Autonomous Province of Trento, in the framework of the projects *OSSERVA3* and *EcoGenEtic.Com* (scientific coordinator Lucia Martinelli). Besides all the partners in these projects who contributed with their useful discussions, the authors wish to expressly thank Lorenza Dalla Costa, Ilaria Vaccari and Valentino Poletti for their valuable work in the part of the research carried out in the laboratory, and Giuseppe Pellegrini for the valuable surveys of stakeholders and the scientific community.

References

Beck, U. (2001). *La società globale del rischio*. Trieste: Asterios.

Castelfranchi, Y., & Pitrelli, N. (2007). *Come si comunica la scienza?* Roma-Bari: Laterza.

Dale, E. C., & Ow, D. W. (1991). Gene Transfer with Subsequent Removal of the Selection Gene from the Host Genome. *PNAS USA, 88*, 10558–10562.

Gaskell, G. et al. (2004). GM foods and the misperception of risk perception. *Risk Analysis, 24*(1), 185–194.

Greco, P. (2005). Il Modello Venezia. La comunicazione nell'era post accademica della scienza. Retrieved September, 2006, from http://ics.sissa.it/conferences/csIntroduzione.pdf

Gregory, J., & Miller, S. (1998a). The public understanding of science. In G. Wilson et al. (Eds.), *Handbook of Science communication* (pp. 3–17). London: IOP.

Gregory, J., & Miller, S. (1998b). *Science in public: Communication, culture, and credibility*. New York: Plenum Press.

Hare, P. D., & Chua, N.-H. (2002). Excision of selectable marker genes from transgenic plants. *Nature Biotechnology, 20*, 575–580.

Kieffer, F. et al. (2004). Mannose and xylose cannot be used as selectable agents for Vitis vinifera L. transformation. *Vitis, 43*, 35–39.

Marin, F., & Martinelli, L. (2006). *The discrete choice analysis as a valuable tool for estimating consumer attitudes towards genetically modified food in Europe*. Paper presented at the 9th International Conference on Public Communication on Science and Technology (PCST-9), Seoul, May 17–20, 2006.

Martinelli, L., & Mandolino, G. (1994). Genetic transformation and regeneration of transgenic plants in grapevine (*Vitis rupestris* S.). *Theoretical and Applied Genetics, 88*, 621–628.

Martinelli, L., & Mandolino, G. (2001). Genetic transformation in Vitis. In: Y. P. S. Bajaj (Ed.), *Biotechnology in agriculture and forestry, transgenic crops* II (Vol. 47, pp. 325–338). Berlin: Springer.

Martinelli, L. et al. (2002). Stable insertion and expression of the movement protein gene of Grapevine Virus A (GVA) in grape lines (*Vitis rupestris* S.). *Vitis*, 41, 189–193.

Martinelli, L. et al. (2005a). *Genetically modified food and feed: Traceability and labeling in the public debate*. Paper presented at the 9th International Consortium on Agricultural Biotechnology Research (ICABR), Agricultural biotechnology, 10 years later, Ravello, July 6–10, 2005.

Martinelli, L. et al. (2005b). European regulation on traceability and labeling of genetically modified food and feed: Analysis on the public debate. *In Vitro Cellular & Developmental Biology – Animal, 2005*, 1133.

Martinelli, L. et al. (2006). *"Eco-friendly" genes: From scientific research to risk management, ethical issues and communication*. Paper presented at the 9th International Conference on Public Communication on Science and Technology (PCST-9), Seoul, May 17–20, 2006.

McFadden, D. (1974a). Conditional logit analysis of qualitative choice behavior. In P. Zarembka (Ed.), *Frontiers in econometrics*. New York: Academic.

McFadden, D. (1974b). The measurement of urban travel demand. *Journal of Public Economics, 3*, 303–328.

McFadden, D. (1981). Econometric models of probabilistic choice. In C. F. Manski & D. McFadden (Eds.), *Structural analysis of discrete data with econometric applications* (pp. 198–271). Cambridge, MA: MIT Press.

McFadden, D. (2000). Economic choices. Nobel Prize Lecture, December 8, 2000. Retrieved September 15, 2007, from http://nobelprize.org/nobel_prizes/economics/laureates/2000/mcfadden-lecture.pdf

McFadden, D., & Ruud, P. A. (1994). Estimation by simulation. *Review of Economics and Statistics, 76*, 4, 591–608.

McFadden, D., & Train, K. (2000). Mixed MNL models for discrete response. *Journal of Applied Econometrics, 15*, 447–470.

Reed, J. et al. (2001). Phosphomannose isomerase: An efficient selectable marker for plant transformation. *In Vitro Cellular & Developmental Biology - Plant, 37*, 127–132.

Sandman, P. M. (1999). Risk = Hazard + Outrage: Coping with controversy about utility risks. *Engineering News-Records, October 4*, A19–A23.

Savadori, L. et al. (2004). Expert and public perception of risk from biotechnology. *Risk Analysis, 24*(5), 1289–1299.

Scholderer, J., & Frewer, L. J. (2003). The biotechnology communication paradox: Experimental evidence and the need for a new strategy. *Journal of Consumer Policy, 26*, 125–157.

Slovic, P. (1999). Trust, emotion, sex, politics and science: Surveying the risk-assessment battlefield. *Risk Analysis, 19*(4), 689–701.

Train, K. (Ed.) (2003). *Discrete choice methods with simulation*. Cambridge: Cambridge University Press.

Wolt, J. D., & Peterson, R. K. D. (2000). Agricultural biotechnology and societal decision making: The role of risk analysis. *AgBioForum, 3*(1), 39–46.

Ziman, J. (2000). *Real science: What it is and what it means*. Cambridge: Cambridge University Press.

Zuo, J. et al. (2001). Chemical-regulated, site-specific DNA excision in transgenic plants. *Nature Biotechnology, 19*, 157–161.

Part IV
Facing Impact: Society and Biotechnology

Chapter 16
Gender and Justice in the Gene Age: The Challenges Presented by Reproductive and Genetic Biotechnologies

Marsha J. Tyson Darling(✉)

Abstract: This chapter examines some of the challenges created by the rapid emergence, and in some prosperous nations, unregulated development of new reproductive and genetic biotechnologies and biomedical protocols that use and impact women's reproductive capacity. It also reports on the ova trade in the United States, and attempts at enforceable regulatory oversight of reproductive biotechnologies and egg donation in the US. Lastly, this article considers the social justice challenges posed by some emerging biomedical technologies in the context of distributive justice issues like women's bodily integrity and 'red biotechnologies', intergenerational justice, existing health equity resource sharing, and the long-standing challenge to the women's movement to support reproductive justice for marginalized women.

Keywords: Biomedical technologies, ova trade, reproductive justice, women's bodily integrity, women's movement.

In terms of the emergence of reproductive and genetic biotechnologies, this essay focuses on several of the most important challenges to securing distributive justice, particularly, the exercise of the right to bodily integrity for all, not just some women and girls. Women's bodily integrity is inextricably linked with women's health and well-being. Hence, securing responsible and enforceable regulatory governance of reproductive and genetic technologies has appropriately emerged as a pressing civil society mandate, both in the North where an ova trade is in full stride, as well as in the South where a history of women being used for risky clinical trials, places poor women of color at risk for exploitation in the global race to secure women's ova.

Marsha J. Tyson Darling
Adelphi University, 1 South Avenue, 300 Alumnae Annex, Garden City, NY 11530, USA
darling@adelphi.edu

F. Molfino, F. Zucco (eds.) *Women in Biotechnology,*

Among the biotechnologies that have emerged as central to any meaningful discussion of women's and children's health and well-being are Assisted Reproductive Technologies (especially, In Vitro Fertilisation (IVF)), Pre-Implantation Genetic Diagnosis (PGD) (medical and non-medical use), Sex Selection and Disability De-Selection Technologies (prenatal screening use of Ultrasound and/or Amniocentesis, Testing Mother's DNA to determine sex of fetus, and Sperm Sorting), Somatic Cell Nuclear Transfer (SCNT), and recently proposed Altered Nuclear Transfer (ANT).

Some of the technologies that use women's bodies in risky and often invasive ways are not new, and date back to the 1970s and 1980s. Dating to several decades now, feminists have questioned, challenged, and in some cases opposed emergent assisted reproductive technologies that use women's bodies in often unregulated and experimental protocols. Then as now, the philosophical backdrop for feminist concerns about protecting and advancing women's health and well-being has been a focus on promoting a women centered analysis.

Feminists concerned with social justice and human rights protections for all women have advanced an analysis that conceptualizes women as individuals, and as a social group living social identities in differing organizational forms and structures. Significantly, a women centered analysis has emphasized that women have a right to bodily integrity, extending beyond the 'male science' view of women's organs as discrete parts useful for their biological function and for their value as the property of men in control of 'families'. This has become particularly important, as a male dominated marketplace, the male defined institution of the family and much of male controlled science, now intersect to present images of especially women's wombs and the reproductive organs and tissues that 'service' wombs in terms of their 'market potential', industrial 'capacity' and contribution to male controlled 'family property'.

In much of the global arena, fundamentalisms are driving patriarchal norms directed at elevating the moral status of unborn fetuses over the autonomy rights of pregnant women. Fundamentalisms impacting culture and the gender related traditions of the family, religious and political developments impacting women's bodies, and an 'enclosure' corporate initiative that seeks patent, production and market developments based on women's ovum, tissues, cells and genes now converge as a formidable challenge for public discourse and civil society engagement on behalf of women's health and bodily integrity (Darling 1999; Feinman 1992).

An industrial processing view of women's bodies and reproductive capacity has opened the door for what has become in recent decades a two-fold scientific quest. First, there is an effort to technologically 'manage' reproduction and relocate it from the 'randomness' and 'inefficiency' of nature's genetic selections at the point of fertilisation in a women's womb, to the predictability of intentional genetic combinations in a Petri dish. Many civil society stakeholders, including many scientists, are concerned that this development is the early stage of pursuing consumer driven genetic enhancements and varying manifestations of cloning. In this regard, in recent decades, some scientists have begun to create and patent new and novel configurations of life, creating genetically modified and transgenic organisms, insects, animals, seeds and plants. Obviously, humans are next.[1]

Secondly, a number of scientists and corporations are patenting human genetic sequences and life processes, thereby seeking to privatize the genetic 'knowledge commons' of all living organisms. In this vein, women's bodies, tissues, and genetic materials are at the center of industry plans for a 'red biotechnologies' genomics based industry (Fowler 1995; Kollek 1995; Mies and Shiva 1993). Hence, feminist expressions of concern for how often women's bodies have been viewed and treated as 'walking wombs' and 'experiment stations' have remained highly relevant now for several decades (Arditti et al. 1984; Callahan and Knight 1992; Corea 1991; Hynes 1991; Klein and Rowland 1988; Lorber 1992; Meleo-Erwin 2001; Raymond 1991, 1993; Rich 1977; Wertz and Fletcher 1992).

Spanning the decades in which feminists have focused on questioning and challenging the increasingly male dominated market and science profession's fascination with the commodification of the female body's womb, other feminists have closely examined the relationship between both conception and anti-conception technologies, and their role in international development, and global state pro-natal and anti-natal policies. Viewing women's bodies through a social justice prism, feminist activists and 'engaged' scholars and legal experts have noted the promotion of women's fertility among middle class whites in the global North, the suppression of women's fertility in the global South, and the commodification (pregnancy surrogacy) and criminalization of the bodies of poor women of color in the global North (Cohen and Taub 1989; Forte and Judd 1998).

Deeply concerned with protecting and advancing women's health, including women's reproductive health, feminists have called for regulatory mechanisms, legal agreements and signatory commitments to state level enforcement of women's human and reproductive rights. Even as progressive efforts on behalf of promoting women's health and reproductive rights remained steadfast through four United Nations Decade Conferences, and several UN inter-governmental conferences (including the International Conference on Population and Development and ICPD held in Cairo on 1994), an unregulated *in vitro* fertilisation industry (in the US), and an emerging biotechnology industry focused on providing 'gene based enhancements', has directed keen attention at expanding scientific exploration of women's tissues, eggs, and wombs (Cicero 2005; Knowles and Parens 2003). In recent decades, an emerging biotechnology industry promised enhanced fertility for some women, while many policy makers insisted on population control for many other women. Neo-Malthusian fertility limiting experimental protocols (like long acting contraceptives Depo and Norplant), anti-fertility vaccine trials (clinical trials in India), and 'crude' sterilization procedures (Quinacrine sterilization in Peru), were

[1] See public education materials on human reproductive and genetic technologies: http://www.genetics-and-society.org, http://www.cwpe.org, http://www.bwhbc.org, http://www.gene-watch.org, http://www.wgnrr.org, info@hgalert.org, http://www.thecornerhouse.org.uk, http://www.the humanfuture.com

integrated into development planning and international aid packages in global South countries. Then as now, access to reproductive autonomy and fertility enhancement for mostly white middle-class women in the North, stands in stark contrast to perceptions that poor women of color in the North and South are responsible for 'excess' fertility leading to the population explosion (CornerHouse 2000; Hartmann 1995; Roberts 1997; Rodriquez-Trias 1980; Tokar 2001).

Women of color and their progressive allies have worked assiduously to promote reproductive justice and not just reproductive autonomy for economically privileged women. This challenge has had the effect of dividing the women's movement, especially in the global North, into two differing positions namely, women who advocate for unrestricted reproductive autonomy, regardless of whether the exercise of such a right exacerbates existing social inequalities. Alternately, there are women who have insisted that all women, not just some women's reproductive rights are crucial to advancing women's health and human rights (Burrell 1995; Darling 2001; Quadagno 1996; Roberts 1999; Silliman et al 2004; Sparr 1994).

In recent decades, the issue of reproductive autonomy or 'choice' and reproductive justice have remained squarely at the center as one of several major tensions in the global women's movement. Largely unregulated reproductive and genetics based biotechnologies are for sale to whoever can buy them. In the United States this takes place within a market-based paradigm, often buttressed by pro-natalist state policy and economic incentives. Fearing a right wing attack on abortion rights, the major women's movement organizations in the United States, led by primarily White women, have dug in their heels and insisted that reproductive autonomy is the start and end of the discourse regarding women's reproductive choices and actions. Taking a step back and looking at the big picture, it is clear that a more far-reaching discourse and activism regarding the impact of risky reproductive and genetic technologies is being held captive to the politics of the abortion issue in the United States.

To be clear, North American women in support of reproductive 'choice' are right to be on the defensive about protecting abortion rights. But the silence of the US based Women's Movement's leadership about the impact of women's engagement with technologies that weaken women's right to bodily integrity, and which also will likely draw North American women into values and behaviors that support Eugenics laden consumer choices, ignores the real possibility that in the name of reproductive freedom women will engage in 'unwise' and 'unjust' choices and decisions. This is particularly important as silence about 'unwise' and 'unjust' choices, decisions and actions will further exacerbate social inequalities, and undermine the vigilance that is necessary to protect women's 'bodily integrity' from a genetic enclosure movement. Hence, the challenge for the women's movement in the US is to support a women's right to elective abortion while at the same time working to establish protective regulatory governance of risky and invasive genetic technologies that have clearly targeted women's wombs, tissues, ovum and DNA for industrial processing. In contrast, elsewhere, particularly in Western Europe, specifically in Germany, feminists involved in *ReproKult* have been able to negotiate both the challenge to reaffirm reproductive autonomy for women while simultaneously working to establish responsible regulatory governance of risky and invasive emerging biotechnologies (Cook 1991; Correa 1994; Degener 1990; Nelson 2003; ReproKult 2002; Ross et al. 2002; Sherwin 2000).

Transnational civil society engagement and feminist engagement in bioethics and justice discourses and activism are particularly important at this time because technology and scientific endeavor are never value free; they always serve ideological, political and economic interests. Often, technology and science serve class, cultural (racial/ethnic) and gendered interests. In recent decades, corporate interests in global North countries have sought to promote seed, plant, insect, animal and human genetic engineering. Women's body tissues, cells and especially ova and the processes these organs engage in are essential 'parts' and 'processes' (Knoppers 1999) for industry plans to sell genetically enhanced babies. Advocates for pursuing this particular direction in assisted reproductive technologies are scientists, philosophers, economists and social science scholars, many of whom occupy important professional positions.

Importantly and not so surprisingly, a number of advocates for human genetic enhancement through germline engineering either explicitly or implicitly envision the application of a Eugenics standard as integral to the utilization of reproductive genetic technologies. So either one has an 'enhanced' designer baby, or the *In-vitro* industry can sell you one; either way, one could chase perfection as an ideological goal for human reproduction (Darling 2006). Dating to the 1990s, a number of influential voices have gone on record as being committed to a consumer driven Eugenics agenda and a post-human future:

> What is called for here is not genocide, the killing off of the population of incompetent cultures. But we do need to think realistically in terms of the "phasing out" of such peoples…Evolutionary progress means the extinction of the less competent (Lynn 1994).

And:

> If the cost of reprogenetic technology follows the downward path taken by other advanced technologies like computers and electronics, it could become affordable to the majority members of the middle class in Western societies. And the already wide gap between wealthy and poor nations could widen further and further with each generation until all common heritage is gone. A severed humanity could very well be the ultimate legacy of unfettered global capitalism (Silver 2000).

And:

> The right to a custom-made child is merely the natural extension of our current discourse of reproductive rights. I see no virtue in the role of chance in conception, and great virtue is expanding choice…If women are allowed the 'reproductive right' or 'choice' to choose the father of their child, with his attendant characteristics, then they should be allowed the right to choose the characteristics from a catalog (Hughes 1996).

And:

> Many people love their retrievers and their sunny dispositions around children and adults. Could people be chosen in the same way? Would it be so terrible to allow parents to at least aim for a certain type, in the same way that great breeders…try to match a breed of dog to the needs of a family? (Pence 1998).

And:

> With a little marketing by IVF clinics, traditional reproduction may begin to seem antiquated, if not downright irresponsible. One-day people may view sex as essentially recreational, and conception as something best done in the laboratory (Stock 2002).

Interestingly, a much more guarded perspective has come from one of a very few women commentators on human genetic engineering:

And:

> Humans have long since possessed the tools for crafting a better world. Where love, compassion, altruism and justice have failed, genetic manipulation will not succeed (Maranto 2003).

Attempting to engineer 'perfection' into children has always been an essential Eugenics goal. Conventional genetic biotechnologies may offer the potential to chase that goal, not as a 20th century Eugenics influenced state mandated initiative, but rather as a consumer 'choice'. Then as now, Eugenicists have seduced progressives by appealing to the widely held goal of treating contagion, curing diseases, and preventing illnesses while improving public health. The reality connected with the other side of seeking an ideal of human perfection as a manifestation of Eugenics biological determinism, the darker side, is that Eugenics assumptions, goals, and behaviors have targeted human populations for the application of 'standards' of racial, gender, ethnic, class, ability, and religious 'genetic fitness and purity', and hence, superiority and inferiority (Darling 2003; Gould 1981; Haller 1963; Kevles 1985; Ordover 2003).

Importantly, in terms of our contemporary concerns, the most enduring influence that 20th century Eugenics exerted on many societies was the impact it had on reproductive justice. That is precisely the fundamental issue under-girding contemporary calls for regulating techno-fixes that offer the promise of using genetic enhancements to allow genetically 'superior' people to assert their dominion over less genetically 'worthy' people. Also, many find the Eugenics goal of using reprogenetic technologies to alter our human destiny in ways that pursue a post-human future dangerous and untenable. It is precisely this threat to human genetic biodiversity and human social development, including a tolerance for human imperfection that is the driving force behind much of civil society work on behalf of regulating reprogenetic technologies (Galpern and Darnovsky 2005; Gender and Justice in the Gene Age 2004; The Return of Eugenics 2007; Mooney 2002).

Public interest stakeholder concerns abound as reproductive cloning and embryonic stem cell cloning emerge at the center of highly contested debates about:

(a) The health and equity challenges embedded in pursuing egg donation for research.
(b) The ethics of using embryonic stem cells for research and development of regenerative medicines and biomedical procedures.
(c) The ethics of permitting scientists to privatize genetic sequences and to patent what a few years ago would have been unimaginable, namely, patenting a scientific fact, or fact of nature.[2] Many are concerned to create ethical parameters for

[2] There is growing concern about technological protocols that have not undergone Institutional Review Board 'IRB' oversight, FDA regulation, oversight by managed care providers, randomized clinical trials, rigorous data collection, and scholarly and scientific peer review input.

attempts to engage in species-altering manipulation of the body's germ cells (Andrews 2006; Beyond Cloning 2001; Norsigian 2005).

Indigenous peoples and their allies are deeply concerned, and have in recent decades organized to oppose the expropriation of their DNA by governments or private biotech ventures. Long accustomed to painful engagements with western scientists, social scientists, and government leaders, communities of indigenous peoples have been particularly on alert since the NIH patented the DNA of Indigenous Peoples without informed consent in the early 1990s. Most recently, the Indigenous Peoples Council on Biocolonialism (IPBC) went on record as stridently opposing the Genographic Project, sponsored by the National Geographic Society (NGS), the IBM Corporation, and the Waitt Family Foundation. If throwing open the door to the Human Genome has let the genie out of the bottle, the persistence of Eugenics values and beliefs could portend serious social development problems, especially for Indigenous Peoples who are along with stateless people like gypsies, are the most marginalized and subordinated peoples on the planet.

Speaking for the interests of global indigenous communities, spokesperson Debra Harry refers to the Genographic Project as the 'Vampire Project', and notes that, 'The historical practice of objectifying Indigenous Peoples in unethical research should be over'. Indeed, in May 2006, the United Nations Permanent Forum on Indigenous Issues recommended that the Genographic Project's goal of collecting 100,000 DNA samples from Indigenous Peoples 'be immediately suspended and report to the Indigenous peoples on the free, prior and informed consent of all the communities where activities are conducted or planned'.[3]

Clearly, Indigenous Peoples and their allies are concerned about the parameters in which scientific and biomedical research takes place, its impacts on vulnerable social groups, and the extent of accountability to existing standards of informed consent, fairness, and an equitable sharing of the benefits of technological developments. After all, if Indigenous Peoples furnish their DNA, and it is privatized by western for-profit corporations, how is the communal structure of Indigenous society served when the profits of innovations made possible by Indigenous DNA contributions by-pass Indigenous Peoples? What is in it for them, are they once again just 'useful' to western profitmaking, as was the case during colonialism; is this an attempt at biocolonialism? Furthermore, what becomes of 'capacity building,' or equitable partnerships in North/South development?

The disability rights, gay/lesbian rights, Indigenous and stateless peoples' rights movements have focused attention on their relationship to emerging biotechnologies, over the course of the past several decades. It has become clear that some of the emerging biotechnologies that impact different and diverse social groups unavoidably impact women's bodies, and hence, women's health and well-being. Against the backdrop of a focus on the status of an embryo, which for over two decades has been the fundamentalist

[3] See 'United Nations Recommends Halt to Genographic Project', and 'Indigenous Peoples Oppose National Geographic & IBM Genetic Research Project that Seeks Indigenous Peoples DNA', both retrieved November 20, 2007 from http://www.ipcb.org/issues/human_genetics/html/geno_pr.html

challenge to a woman's right to elective abortion in the global North as well as in much of the global South, a domestic and global trade in women's ova has emerged. The global ova trade has important and serious consequences for women's health and women's rights.

On the one hand, the highly lucrative trade in women's ova is linked with the infertility industry's interest in offering and promoting a cornucopia of expensive technologies and gene based protocols that interact with women's bodies, and that offer medical and non-medical reproductive interventions. On the other hand, scientists, biotech and pharmaceutical corporate leaders in the medical industrial complex are also interested in building a gene based personal wellness and health care industry based on embryonic stem cell therapies, products, and protocols. They desire women's embryos and ova as sources of stem cells, not for fertility, but for research applications towards what has been coined 'regenerative medicine' (Darling 2006; Dickenson 2001, 2002b; Radin 1996).

Either way, women's bodily tissues and cells associated with reproduction and regeneration, contain the genetic blueprint of all human knowledge; these cells and tissues represent for some the 'raw materials' of a new scientific frontier. Since the late 1990s when human embryonic stem cell laboratory cultivation became a reality, scientists, biotech and pharmaceutical companies in the United States, Germany, the UK, Canada, Australia, Spain and other European nations, India, Korea, China, and Japan have intensified a quest to acquire human embryos from *IVF* centers, abortion clinics, and increasingly by paying young women for their ova (Ahuja et al. 1999).

Feminists and their progressive allies in the global North have been in the forefront of efforts to protect and safeguard women's health in the emerging egg donation industry. Because the ova trade is global, thereby moving ova across state and country boundaries, it is particularly likely that issues like informed consent, psychological counseling and procedural safeguards for young women's health and well-being are impossible to oversee effectively. Balancing women's autonomy and legal rights in constitutional democracies, and in some cases just women's physical access to participation in the ova trade, the UK, Canada, Germany and others have enacted legislation that establishes a responsible regulatory mechanism to oversee the market trade in young women's ovum. Importantly, some of the legislation attempts to transfer the burden of proof regarding safety on to the fertility and biotech industries (Tauer 2004; Washenfelder 2002).[4]

---

[4] See, the German Embryo Protection Act of 1990, in UK. the Human Fertilisation and Embryology Act of 1990, Victoria Infertility Act of 1997; and in Canada the Assisted Human Reproduction Act (AHRA) of 2004. Also, see The Convention for the Protection of Human Rights and Dignity of the Human Being with Regard to the Application of Biology and Medicine: Convention on Human Rights and Biomedicine, the UNESCO Universal Declaration on the Human Genome and Human Rights, the International Declaration on Human Genetic Data, and the Universal Declaration on Bioethics and Human Rights, and the International Ethical Guidelines for Biomedical Research Involving Human Subjects of the Council for International Organizations of Medical Sciences. In the United States, see the reports by the National Bioethics Advisory Commission, *Ethical Issues in Human Stem Cell Research: Conclusions and Recommendations*, National Institute of Health, *Report of the Human Embryo Research Panel*, 'Reproduction and Responsibility: The Regulation of New Biotechnologies' is available at http://www.bioethics.gov. Also, the *Opinion of the European Group on Ethics in Science and New Technologies*, the Australian NHMRC *Guidelines*, *Proceed with Care: Final Report of the Royal Commission on New Reproductive Technologies*.

The global ova trade presents a daunting governance challenge. For instance, some poor Eastern European (the Ukraine and Cyprus) women are engaged in the ova trade, sometimes undergoing potentially dangerous greater doses of fertility hormones, egg stimulation and extraction. Young women in the unregulated egg extraction industry in the US enter into contracts with brokers and sometimes undergo multiple extraction cycles. Julia Derek, author of *Confessions of a Serial Egg Donor*, reports that she paid her college expenses by undergoing twelve egg donations cycles. In a climate of profit making from egg extraction and donation, ethical concerns abound about informed consent, psychological counseling, the declining importance of the 'precautionary principle' of do not harm, and education for women about the risks involved in participating in new, risky, and invasive technologies. It is particularly worrisome when the scientists and doctors who stand to benefit financially oversee consent issues and safety protocols (Hempel 2006; Papadimos and Papadimos 2004).

Clearly, women's well-being is best served by greater not less transparency, and by the enactment of enforceable governance that seeks responsible accountability in what in many places has been reduced to a market 'arrangement'. Professor of Medical Ethics and Humanities, Donna Dickenson of the UK, sums up the dynamics of egg extraction and market based egg donation arrangements, referred to as 'gift relationships'. Dickenson notes: 'If donors believe they are demonstrating altruism, but biotechnology firms and researchers use the discourse of commodity and profit, we have not "incomplete commodification" but complete commodification with a plausibly human face' (Dickenson 2002a; Grubb 1998).

Viewed in this manner, the paid and unpaid trade in women's ova situates women's bodies at the center of corporate interests in making ova derived products. A key question concerns whether women will maintain proprietary rights to their own reproductive processes, or will those genetic sequences, cell lines, and women's reproductive labor become the property of pharmaceutical and biotechnology companies and genetic researchers? What becomes of women's bodily integrity in an atmosphere of expropriation of women's cells? At the same time that the medical industrial complex and a growing global stem cell research agenda presses states, politicians, and the courts to allow for the development of a global embryonic stem cell trade, women are sought after as consumers of exploratory gene based protocols. To this end, the landmark US based court decision *Moore vs. Regents of the University of California (1990)* challenged the notion that we possess property in the organs, tissues and cells of our bodies (Andrews 1999).[5]

At present, the unregulated infertility industry in the United States is the engine that is driving egg donation in the US. Many women donors and egg recipients are unaware that they engage in a risky context in participating in ARTs. Most women assume that industries which provide medical services are regulated by state or federal officials responsible for protecting the public's health. Hence, many women enter the infertility industry as recipients or donors in the 'blind' so to speak. In place of

[5] See Moore vs. Regents of the University of California (1990) 51 Cal. 3d. 120.

transparency and accountability to publicly appointed officials, the ART industry self-reports and is not answerable to any institution charged with safeguarding women as members of a public constituency.

No doubt, many ART practitioners are skilled, honest and concerned not only with producing an 'outcome' (multiple eggs) in the context of egg donors, and a baby in terms of the expectations of being paid to produce a product. However, as has been noted by many, the absence of regulatory oversight over a multi-billion dollar a year industry should cause us concern about not only what is reported, but also what is not self-reported. At issue is the concern for how we as the public evaluate the conduct of scientists against an ethical barometer that values and monitors medical protocols that impact women's well-being.

In terms of new reproductive technologies (NRTs), ever since women's eggs became the production site for technological intervention into infertility, women's health advocates and feminist voices have raised concerns for the health impacts of emerging reproductive technologies, especially the core technology of Assisted Reproduction Technology (ART), and In-Vitro Fertilisation Embryo Transfer (IVF-ET). With little exception (Gena Corea's interviews w/women in the early 1980s), the public has relied on the fertility industry's self-reporting notes that since the birth of the first US born IVF baby in July, 1978, ART intervention cycles have exceeded 100,000, with over 30,000 successful pregnancies and over 40,000 babies born in the US. The fertility industry's emphasis is on how ARTs 'win' by producing live births. The 'win…win' spin conceals the unknown health risks of induced ovarian stimulation for women's reproductive health, as well as ART as related 'failures' (Our Bodies, Ourselves). Hence, too often we still lack 'reliable', peer reviewed qualitative and quantitative data on the experiences, and short and long term impacts of NRTs on women's and infant's health. We also need better reporting on the successes, mistakes and failures of IVF-ET (Sample 2005).[6]

Clearly, we are approaching a threshold point wherein what began as an unregulated attempt to provide a non-coital mechanism for combining ovum and sperm has evolved into an industry that increasingly seeks to separate conception, and reproductive and gestational phases of baby making, into discreet procedures and technology controlled protocols. IVF can now accompany contract maternal surrogacy not only for heterosexual couples, but also for lesbian, gay, bisexual, transgender, and queer persons (LGBTQ), and differently-abled singles or couples. Essentially, adults are being encouraged, even pressured to surrender reproduction over to a biotechnology industry, the 'section' of the medical industrial complex that will handle the reproduction

[6]There is little recognition of unethical conduct on the part of ART practitioners; fertility clinics sometimes publish inflated success rates in advertising materials; practitioners have used stored eggs for prospective parents for impregnating others; a physician used his own sperm for ART without the patient's consent; and egg donors are often not provided with accurate information about the risks involved with ARTs. See 'Issues of Ethical Conduct in Assisted Reproductive Technology', a Joint Report of the Council on Ethical and Judicial Affairs and the Council on Scientific Affairs, American Association of Medicine Task Force on Ethical Issues in Assisted Reproductive Technology.

of better, more 'perfect' babies. Conversely, low-income women only have access to population planning programs; even if some low income women suffer from pelvic inflammatory disorders that reduce or prevent fertility, they cannot access ARTs.

The developing mandate to produce 'perfect babies' comes along with an increasing emphasis on suggesting that fetuses are patients who should receive the best patient care the state can afford in the context of public health. De-selecting for disability is also high on the list of the uses of NRTs, leading many to question the consequences of encouraging intolerance for 'imperfection' and 'disability'. One wonders whether seeking 'perfect' children is becoming a women's responsibility. If so, what does the intersection of socio-economic access and NRTs mean for what human genetic perfection will look like?

Feminists are also particularly concerned about the non-medical uses of reproductive and genetic biotechnologies. Procedures like pre-implantation genetic diagnosis (PGD) were initially developed with a medical use, namely, as an embryonic pre-implantation procedure designed to identity the presence of genetic markers for debilitating and chronic disorders, or genetically transmitted diseases. Feminists and concerned civil society stakeholders are correct in being deeply concerned that currently the pre-conception, pre-pregnancy and prenatal screening technologies (each is used to undertake sex-selection and/or to de/select for disability) that are in use have serious implications for maternal health, and fetal well-being. Very significantly, the use of PGD as a sex selection technology has serious consequences for females, as its use to further male/son preference nearly exclusively eliminates female embryos. This in turn has serious consequences for the stability of male/female sex ratios, in light of the non-medical use of sex selection technology to socially mark and discard female embryos (Alexander 2000; Darnovsky 2004; Mahowald 2000).

In the global South, particularly in Asia, feminists have noted that the non-medical use of PGD continues to facilitate the deepening of public perceptions of female gender social inequality. Recent census numbers clearly indicate serious sex ratio disparities especially in some middle class Asian Indian suburbs. It is difficult to ascertain accurate numbers of sex ratio imbalances for regions in China, and sex selection abortion and the uses of PGD for non-medical purposes is prohibited by the letter of the law in both India and China. South Asian Indian feminists have been persistent in their efforts to document the extent of the use of sex selection technology. They have also been vigilant in seeking to prod the Indian government to enforce legal bans.

Research on the effectiveness of state enforcement measures and mechanisms in the abolition of sex selection technologies in India, China, etc. is needed, as is reporting on the effectiveness of state and/or civil society efforts to 'put teeth' into the laws. Also, unlike western democracies that are unfettered by continuing neo-colonial relationships with International Financial Institutions (IFIs), as we engage advocacy for improving women's health in global South countries, we have to deal with the 'conditionality priorities' of IFIs – the World Bank and International Monetary Fund.

While the use of technology to identify and abort female fetuses is most acute in China and India, full-page newspaper and magazine advertisements in the United

States promote sex-selection technology under the banner of 'family balance'. Pre-Implantation Genetic Diagnosis is aggressively marketed to Asian Indian and Chinese communities in the US by ethnic based newspapers and magazines. Feminists and social justice advocates have organized to oppose the aggressive marketing of PGD to Asian communities in the US and western Europe (Sachs 2001).

Feminists should be concerned whether poor women in the global South are coerced into providing ovum. The exponential growth of globalized clinical trials for research and development in low-income countries has important implications for the interest scientists have in securing large amounts of women's ova for research. There are already organizations that oversee commercialized clinical trial investigations in low-income countries where multiple and intersecting social inequalities are a reality. We should be concerned about allowing companies that gain materially from genomics based medical research to oversee informed consent and public education. Also, some bio-ethicists are concerned about diminishing peer review processes for experimental gene and nanobased protocols and drugs used in the global South. Hence, the fertility industry's interest in using global South poor women for egg donation for research purposes should concern social justice advocates, especially since third party intermediary companies are already in place, facilitating access to poor women in the South.

In conclusion, feminists should be exploring how the mystique of genetic manipulation leads to genetic reductionism and the geneticization of health disorders. Many disorders, whether social, physiological, or environmental, etc. are presented to the public as gene based, thereby justifying the searching for genetic 'fixes'. Feminists should be in the forefront of exposing the not so tacit pressure that is mounting to integrate ARTs into public health spending. To situate ARTs into the public health sector is to burden an already overtaxed system with a responsibility that does not serve the many. In addition, in the United States where an enormous number of persons lack basic healthcare, it is unthinkable that there are calls to fund ARTs from the public health sector. Some seek to influence public policy and state financing for research and development, and to direct state payment for technologies and biomedical protocols that were just yesterday experimental. Lest this sound unduly alarmist, consider when and how did technologies like ultrasound, amniocentesis, fetal monitors, and maternal serum testing become normalized?

Feminist scholars should continue to be engaged in research that explores the collapsing standards of care and the 'precautionary principle', that were the norm just a few decades ago. Given the pressing medical global healthcare needs we confront, we must engage in shaping the policies that influence expenditures for research on behalf of public health. Feminists should make our voices heard about society's obligations to intergenerational justice: exactly what kind of genetic legacy are we leaving to the generations of children who will follow us; what are our rights, obligations and responsibilities to them? Do we owe children of the future planetary biodiversity? Will we by our actions ensure that they inherit more and not fewer rights than we inherited? Finally, what are the longstanding challenges to the global women's movement to support reproductive justice for marginalized women?

References

Ahuja, K., Simons, E., & Edwards, R. (1999). Money, morals and medical risks: conflicting notions underlying the recruitment of egg donors. *Human Reproduction, 14*(2), 279–284.

Alexander, W. (2000). Fatal daughter syndrome. In R. Tong et al. *Globalizing feminist bioethics* (pp. 179–197). Boulder, CO: Westview Press.

Andrews, L. (1999). *The clone age*. New York: Henry Holt.

Andrews, L. (2006). Patently wrong: The first privately owned law of nature. *Gene Watch, 19*(3), 16–17.

Arditti, R., Klein, R., & Minden, S. (Eds.) (1984). *Test-tube women: What future for motherhood?* London: Pandora Press.

Beyond cloning: protecting humanity from species-altering procedures (2001). Boston, MA: Boston University School of Public Health.

Burrell, D. E. (1995). The Norplant solution: Norplant and the control of African American motherhood. *U.C.L.A. Women's Law Journal, 5*, 401.

Callahan, J., & Knight, J. (1992). Women, fetuses, medicine and the law. In H. Holmes & L. Purdy (Eds.), *Feminist perspectives in medical ethics* (pp. 224–239). Bloomington, IN: Indiana University Press.

Cicero, K. (2005). The 10 best fertility centers. *Child magazine*, 65–75. Retrieved October 20, 2007, from http://infertility-ivf.com/ChildMagazine.pdf

Cohen, S., & Taub, S. (Eds.) (1989). *Reproductive laws for the 1990s*. Clifton, NJ: Humana Press.

Cook, J. (1991). *New reproductive technologies: International legal issues and instruments*. Ottawa, ON: Royal Commission on New Reproductive Technologies.

Corea, G. (1991). How the new reproductive technologies will affect all women. In H. P. Hynes (Ed.), *Reconstructing babylon: Essays on women and technology* (pp. 41–60). Bloomington, IN: Indiana University Press.

CornerHouse (2000, July). *The malthus factor: Poverty, politics and population in capitalist development: briefing Paper #20*. Dorset: The CornerHouse.

Correa, S. (1994). *Population and reproductive rights: Feminist perspectives from the south*. London: ZED Press.

Darling, M. J. T. (1999). The state: friend or foe?: Distributive justice issues and African-American women. In J. Silliman & Y. King (Eds.), *Dangerous intersections: Feminist perspectives on population, environment, and development* (pp. 214–241). Boston, MA: South End Press.

Darling, M. J. T. (2001). Gendered globalization, state interests: Women of color and marginalized women. *N. Paradoxa International Feminist Art Journal, Economies/Exchanges, 8*, 65–73.

Darling, M. J. T. (2003). Eugenics unbound: Race, gender and genetics. In V. Agnew (Ed.), *Women's health, women's rights: Perspectives on global health issues*. Toronto, ON: York University.

Darling, M. J. T. (2006). Gender, new technologies and development. *Development, 49*, 23–28.

Darnovsky, M. (2004). Hi-tech sex selection: A new chapter in the debate. *Genewatch, 17*(1).

Degener, T. (1990). Female self-determination between feminist claims and "voluntary" eugenetics, between "rights" and ethics. *Issues in Reproductive and Genetic Engineering, 87*, 3.

Dickenson, D. (2001). Property and women's alienation from their own reproductive labour. *Bioethics, 15*, 205–217.

Dickenson, D. (2002a). Commentary – Commodification of human tissue: Implications for feminist and development ethics. *Developing World Bioethics, 2*, 55–63.

Dickenson, D. (2002b). Who owns embryonic and fetal tissue. In D. Dickenson (Ed.), *Ethical issues in maternal-fetal medicine* (pp. 233-246). Cambridge: Cambridge University Press.

Feinman, C. (Ed.) (1992). *The criminalization of a woman's body*. New York: Haworth Press.

Forte, D. J., & Judd, K. (1998). The south within the north. In R. P. Petchesky & K. Judd (Eds.), *Negotiating reproductive rights: Women's perspectives across countries and cultures* (pp. 256–294). London: ZED Press.

Fowler, C. (1995.) Biotechnology, patents and the third world. In V. Shiva & I. Moser (Eds.), *Biopolitics: A feminist and ecological reader on biotechnology* (pp. 214–225). London: ZED Books.

Galpern, E., & Darnovsky, M. (2005). Eggs vs. ethics in stem cell debate. *Nation*. Retrieved November 29, 3005 from http://www.geneticsandsociety.org/article.php?id=199.

Gender and justice in the gene age: a report from a feminist meeting on new reproductive and genetic technologies (2005). Retrieved November 20, 2007 from http://www.gjga.org

Gould, S.J. (1981). *The mismeasure of man*. New York: W.W. Norton.

Grubb, A. (1998). I, me, mine: Body parts and property. *Medical Law International, 3*, 299–313.

Haller, M. (1963). *Eugenics: Hereditarian attitudes in american thought*. New Brunswick, NJ: Rutgers University Press.

Hartmann, B. (1995). *Reproductive rights and wrongs: The global politics of population control* (Rev. ed.). Boston, MA: South End Press.

Hempel, C. (2006, June 25). Golden Eggs. *Boston Globe Sunday Magazine*.

Hughes, J. (1996). Embracing change with all four arms. *Eubios Journal of Asian and International Bioethic, 6*, 94–101.

Hynes, P. H. (1991). Biotechnology in agriculture and reproduction: The parallels in public policy. In P. H. Hynes (Ed.), *Reconstructing babylon: Essays on women and technology* (pp. 103–130). Bloomington, IN: Indiana University Press.

Kevles, D. J. (1985). *In the name of eugenics: Genetics and the uses of human heredity*. Cambridge, MA: Harvard University Press.

Klein, R., & Rowland, R. (1988). Women as test sites for fertility drugs. *Reproductive and Genetic Engineering: International Feminist Analysis, 1*, 251–73.

Knoppers, B. M. (1999). Status, sale and patenting of human genetic material: An international survey. *Nature Genetics, 22*, 23–26.

Knowles, L. P., & Parens, E. (2003). Reprogenetics and public policy: Reflections and recommendations. *The Hastings Report, 33*.

Kollek, R. (1995). The limits of experimental knowledge: A feminist perspective on the ecological risks of genetic engineering. In V. Shiva & I. Moser (Eds.), *Biopolitics: A feminist and ecological reader on biotechnology* (pp. 95–111). London: ZED Books.

Lorber, J. (1992). Choice, gift, or patriarchal bargain?: Women's consent in *in vitro* fertilization in male infertility. In H. Holmes & L. Purdy (Eds.), *Feminist perspectives in medical ethics* (pp. 169–180). Bloomington, IN: Indiana University Press.

Lynn, R. (1994, January). *Interview in Newsday*.

Mahowald, M. (2000). Cultural differences and sex selection. In Tong, R. et al. (Eds), *Globalizing feminist bioethics* (pp. 165–178). Boulder, CO: Westview Press.

Maranto, G. (2003) *Quest for perfection: The drive to breed better human beings* (p. 278). New York: Scribner Books.

Meleo-Erwin, Z. C. (2001). Reproductive technology: Welcome to the brave new world. In B. Tokar (Ed.), *Redesigning life?: The worldwide challenge to genetic engineering* (pp. 195–204). London/New York: ZED Books.

Mies, M., & Shiva, V. (1993). *Ecofeminism*. London: ZED Books.

Mooney, P. (2002). Making well people better. *World Watch*, 13–16.

Nelson, J. (2003). *Women of color and the reproductive rights movement*. New York: New York University Press.

Norsigian, J. (2005). Stem cell research and embryo cloning: Involving laypersons in the public debate. Symposium issue on stem cell research and human cloning: Where do we draw the line? *New England Law Review, 39*, 527–534.

Ordover, N. (2003). *American eugenics: Race, queer anatomy, and the science of nationalism*. Minneapolis, MN: University of Minnesota Press.

Papadimos, T., & Papadimos, A. (2004). The student and the ovum: The lack of autonomy and informed consent in trading genes for tuition. *Reproductive Biology and Endocrinology, 2*, 56. Retrieved November 20, 2007 from http://www.rbej.com/content/2/1/56

Pence, G. (1998). *Who's afraid of human cloning*? (p. 168). Lanham, MD: Rowman & Littlefield.
Quadagno, J. (1996). *The color of welfare: How racism undermined the war on poverty*. New York: Oxford University Press.
Radin, M. J. (1996). *Contested commodities: The trouble with trade in sex, children, body parts and other things*. Cambridge, MA: Harvard University Press.
Raymond, J. G. (1991). Of eggs, embryos, and altruism. In P. H. Hynes (Ed.), *Reconstructing babylon: Essays on women and technology* (pp. 61–69). Bloomington, IN: Indiana University Press.
Raymond, J. G. (1993). *Women as wombs: Reproductive technologies and the battle over women's freedom*. Melbourne: Spinifex.
ReproKult (2002). *Reproductive medicine and genetic engineering: women between self-determination and societal standardization*. Berlin: ReproKult, Women's Forum for Reproductive Medicine Conference.
The Return of Eugenics (2007). *Genewatch, 20*(1).
Rich, A. (1977). The theft of childbirth. In C. Dreyfus (Ed.), *Seizing our bodies: The politics of women's health* (pp. 146–163). New York: Vintage Press.
Roberts, D. (1997). *Killing the black body: Race, reproduction, and the meaning of liberty*. New York: Pantheon Books.
Roberts, D. (1999). Welfare's ban on poor motherhood. In G. Mink (Ed.), *Whose welfare*? Ithaca, NY: Cornell University Press.
Rodriquez-Trias, H. (1980). Sterilization abuse. In R. Arditti, P. Brennan, & S. Clark (Eds.), *Science and liberation* (pp. 113–127). Boston, MA: South End Press.
Ross, L. J. et al. (2002). Just choices: Women of color, reproductive health and human rights. In J. Silliman & A. Bhattacharjee (Eds.), *Policing the national body: Race, gender and criminalization* (pp. 147–174). Boston, MA: South End Press.
Sachs, S. (2001, August 15th). Clinics' pitch to indian émigrés: It's a boy. *The New York Times*.
Sample, I. (2005, November). Genetic flaws in IVF embryos. *The Guardian*.
Sherwin, S. (2000). Normalizing reproductive technologies and the implications for autonomy. In R. Tong, G. Anderson, & A. Santos (Eds.), *Globalizing feminist bioethics: cross cultural perspectives* (pp. 96–113). Boulder, CO: Westview Press.
Silliman, J. Fried, M. G., Ross, L., & Gutierrez, E. R. (2004). *Undivided rights: Women of color organize for reproductive justice*. Boston, MA: South End Press.
Silver, L. (2000). *Reprogenetics: How do a scientist's own ethical deliberations enter into the process? Humans and genetic engineering in the new millennium: How are we going to get "gen-ethics" just in time?*. Copenhagen: Danish Council of Ethics.
Sparr, P. (Ed.) (1994). *Mortgaging women's lives: Feminist critiques of structural adjustment*. London: ZED Books.
Stock, G. (2002). *Redesigning humans: our inevitable genetic future* (p. 55). New York: Houghton Mifflin.
Tauer, C. (2004). International policy failures: Cloning and stem-cell research. *The Lancet, 364*, 9429.
Tokar, B. (Ed.) (2001). *Redesigning life: The worldwide challenge to genetic engineering*. London: ZED Books.
Washenfelder, C. (2002). Regulating a revolution: the extent of reproductive rights in Canada. *Health Law Review*, *12*(2), 44–52.
Wertz, D.C., & Fletcher, J. (1992). Sex selection through prenatal diagnosis: A feminist critique. In H. B. Holmes & L. Purdy (Eds.), *Feminist perspectives in medical ethics* (pp. 240–256). Bloomington, IN: Indiana University Press.

Chapter 17
Health Risks and Benefits from Biotechnology

Amalia Bosìa(✉)

Abstract: The revolution in DNA technology has permitted the development of otherwise unavailable and novel biological medicines; has provided new and efficient methods for the large-scale production of existing substances; is the basis for novel, highly sensitive and specific diagnostic tests; has developed new techniques in genetic engineering; is the basis for the development of new and safer vaccines, which are more effective and can be produced in larger quantities than by any other means; is a major key to new fundamental understanding of normal and disease processes.

As far as risks are concerned, a few examples will be mentioned, related to biotechnology-derived products (biologicals), human embryonic stem cells, gene therapy and application of functional genomics. As a final step, the so-called precautionary principle will be introduced, since it raises a cluster of questions about how prudently to engage in risk-taking.

Keywords: Health risks and benefits, biotechnology, biological medicines, cell therapy, genetic enhancement

17.1 DNA Technology

The revolution in *DNA technology* has opened up a new and exciting vista for global health care:

(A) It has permitted the development of otherwise unavailable and novel biological medicines and it has provided new and efficient methods for the large-scale production of existing substances.

Amalia Bosìa
Dipartimento di Genetica, Biologia e Biochimica, Università degli Studi di Torino, Via Santena, 5 bis, 10126 Torino, Italy
amalia.bosia@unito.it

Biotechnology-derived products (biologics) hold a great deal of promise among the therapeutic interventions for a wide range of disorders, including cancer, inflammatory diseases and atherosclerosis. Millions of people worldwide have benefited from the many biologic products and vaccines available today, and more than 370 new products are currently in various stages of clinical testing. Biologics differ from traditional pharmaceutical drugs. They are generally large and complex proteins, which might or might not be glycosylated, and can be of non-human origin, contain a partial human-sequence or can comprise a complete human-sequence. All of these types of products have the potential to generate immune responses or immunogenicity through the production of anti-drug antibodies (ADA) or cell-based immune responses.

ADA also have the potential to neutralize the endogenous counterpart of the biologic and worsen the disease that the biologic was intended to treat. Neutralizing ADA that are cross-reactive with the endogenous protein can cause significant safety concerns because they inhibit the inherent physiological pathway and can lead to a deficiency syndrome.

In order to assess immunogenicity in non-clinical studies, animal models can be useful for understanding the physiological sequelae that might occur upon the binding of ADA, particularly neutralizing ADA that cross-react with endogenous protein. Such data can reveal the level of redundancy of a natural molecule, such as induction of thrombocytopenia in the presence of anti-thrombopoietin antibodies.

In order to assess immunogenicity in clinical studies, it is a common policy to include immunogenicity as part of the review of clinical safety assessments for biologic license applications. Immune responses to therapeutic proteins have been defined almost exclusively by the detection of circulating antibodies. To understand the true induction of ADA and their potential clinical effects, samples need to be taken before and during the clinical study.

(B) It is the basis for novel, highly sensitive and specific diagnostic tests, that are:

- Biomarker discovery with omics technologies: in the post-genome era, efforts are focused on biomarker discovery and the early diagnosis (e.g.) of cancer through the application of various omics technologies – transcriptomics, proteomics, metabonomics, peptidomics.... on tissue samples and body fluids. Currently, the biological samples analyzed include blood, urine, sputum, saliva, nipple-aspirate fluid, breath, tear fluid, cerebrospinal fluid and tissue samples. The diversity of components in these samples (e.g., amino acids, peptides, proteins or metabolites) can further increase the analytical complexity.
- Transcriptomics: the transcriptome is the complete set of RNA products transcribed in a given organism, and transcriptomics is the study of the transcriptome. Microarray technology is a powerful tool for transcriptomics analysis that has been used to identify biomarkers associated with some tumor types – the patterns of gene expression can be used to classify

types of tumors and predict the outcome. Numerous reports have demonstrated the potential power of expression profiling for the molecular diagnosis of human cancers. In particular, using large scale meta-analysis of cancer microarrays some common cancer biomarkers have been identified; these biomarkers could be broadly used to increase the sensitivity and accuracy of cancer diagnosis.

- Proteomics: proteomics is the large-scale identification and functional characterization of all the expressed proteins in a given cell or tissue, including all protein isoforms and modifications. The frequently used tools for proteomic investigations include two-dimensional gel electrophoresis and mass spectrometry (MS). MS-based proteomics technology has been considered as a promising approach for the early diagnosis of cancers. Up to now, this technology has been applied to many types of cancers for biomarker discovery and diagnosis, including ovarian, prostate, breast, bladder, renal, lung, pancreas and astroglial tumors.
- Metabonomics: In this context, two terms are frequently used: metabolomics and metabonomics. Metabolomics refers to the measurement of all metabolite concentrations in cells and tissues. Metabonomics is the quantitative measurement of the metabolic responses of multicellular systems to pathophysiological stimuli or genetic modification. Nuclear magnetic resonance (NMR) spectroscopy and MS, both in combination with modern separation approaches, are two primary analytical methods for conducting metabonomic measurements.
- Bioinformatics: no matter which omics technology is used in biomarker development, bioinformatics tools are required to extract the diagnostic or prognostic information from the complex data. Based on pattern recognition technologies, discriminatory patterns (a panel of gene, protein or peptide patterns) can be identified for the diagnosis.

17.1.1 Outstanding Questions in Omics-Based Biomarker Discovery:

(i) MS-based omics technology is biased towards high-abundance proteins in biological samples, and has low sensitivity to low-abundance molecules.
(ii) Experimental bias exists throughout the entire process, from sample collection to data generation; this calls for the standardization of experimental protocols and automation of all sample preparation steps for each omic platform.
(iii) Data overfitting is a substantial problem in omics research, where the number of parameters in a model is too large relative to the number of samples available.
(iv) Small numbers of samples are used in omics studies; consequently, different data analysis methods for the same omics data generate different predictive results.

17.1.2 Other Messages

- Use of all these pieces of information at the personal level, possibly integrated into the individual patient's electronic health record, presents real concerns, as well as offering opportunities for real health benefits.
- Benefits anticipated from genetic research include: improved diagnosis of disease and earlier detection of a genetic predisposition to disease; pharmacogenetics, where drugs are designed to interact with a person's particular genetic make-up; and gene therapy.
- Pharmacogenetics is expected to lead in the long term to more powerful and customized medicines and a significant decrease in the costs of health care. But the technology, and the ability to identify patients with a need for more expensive drugs, is again a risk area for certain ethnic groups, and individuals, in relation to the costs of health care and health insurance.
- The most commonly expressed fear is that genetic information could be used in ways that could harm people, for example, to deny them access to health insurance, employment, education and even loans.

Implications for practice:

- As genomic information becomes more readily available in health care at the individual and the public level, and genetic testing more routine, health professionals and health consumer groups will need to start addressing these issues more openly.
- Promotion of public discussion of the ethical and social issues related to genetic research at this level in all countries is urgently needed.
- There is an important role that health information professionals should play in promoting wider public knowledge of genetic research and its implications, in fostering public debate on key issues, and in seeking clearer policies in the health sector concerning the use of genetic information about individuals.

17.1.3 The Risk of Disease Mongering

Moreover, owing to the market expansion of new diagnostic tools and medications, the problem of disease mongering is attracting increasing attention, as it turns healthy people into patients, wastes precious resources, and causes iatrogenic harm.

Convincing healthy people that they are sick and in need of medicines creates an enormous market for drugs and medicines. Medicalisation is the process of turning ordinary life events and its customary ups and downs into medical conditions. Does a five-year-old who is unable to concentrate and finds it very difficult to sit still suffer from 'attention deficit hyperactivity disorder' (ADHD)? The concept of what is and what is not a disease at times can be extremely slippery. Many of life's normal processes like birth, ageing, sexuality, unhappiness and death can be medicalised. The term, 'medicalisation', became prominent in the 1970s with a central theme that definitions of diseases were being broadened to increase demand for medical services and new drugs and other products. Disease mongering can turn ordinary troubles into medical problems, see mild symptoms as serious, treat personal problems as medical ones, see risks as diseases, and frame prevalence estimates to increase potential markets. Disease mongering is the selling of sickness that widens the boundaries of illness and increases the market for medicines.

17.2 Nanotechnology

A revolution in health care and medical technology looms large on the horizon on the basis of the discipline of nanotechnology.

Applications of nanotechnology may include the use of nanosystems or nanoparticles for the detection of early disease and the delivery of therapeutic agents. The vision is that nanoparticles may be able to seek out a target within the body (e.g., a cancer cell) and perform treatment. The treatment delivered by the nanoparticles may be that of releasing a drug in a localised area, thus minimising the potential systemic side effects of generalised drug therapy as in, for instance, chemotherapy. There are numerous engineered constructs, assemblies, architectures and particulate systems used for diagnostics and targeted drug delivery, whose unifying feature is their nanometre-scale size range (from a few to 250 nm). Therapeutic and diagnostic agents can be capsulated, covalently attached, or adsorbed on to nanocarriers. These approaches can overcome drug solubility issues, particularly in view of the fact that large numbers of the new drug candidates emerging from high-throughput drug screening initiatives are water insoluble. Second, by virtue of their small size and by functionalising their surface with synthetic polymers and appropriate ligands, nanoparticulate carriers can be targeted to specific cells and locations within the body after intravenous and subcutaneous routes of injection. Such approaches may enhance therapeutic effectiveness and decrease side effects. Some of these carriers can be engineered in such a way that they can be activated by changes in the environmental pH, chemical stimuli, or by the application of an external heat source. Such modifications offer control over particle integrity, drug delivery rates, and the location of drug release, for example, within specific organelles. Some are being designed with the focus on multifunctionality; these carriers target cell receptors and deliver simultaneously drugs and biological sensors.

But if nanoparticles are to be used for targeted drug delivery, we need to be aware of the toxicity of nanoparticles resulting from their nanoscale size. Materials in this size range may approach the length scale where their properties differ substantially from those of bulk materials of the same composition, allowing them to perform exceptional feats of reactivity, for instance. Possible undesirable results of these capabilities are harmful interactions with biological systems and the environment with the potential to generate toxicity. So we need to perform a risk-benefit analysis. Although some of the nanosystems used in drug delivery may be pre-manufactured, many may need to be created by self-assembly. The scientific challenge will then be to control these self-assembling processes. Current research into the self-assembling of nanostructures deals with the self-assembly of, for instance, carbon nanotubes and rodcoil polymers.

Nanoparticles used in drug delivery may get ‘out of control’ in the absence of feedback mechanisms to control their function. To take this into account, it may be possible to develop nanoparticles that are biodegradable or composed of naturally occurring substances which can be eliminated from the body through the natural mechanisms of metabolism and excretion. Alternatively, nanoparticles could have ‘homing’ devices which would allow them to be collected and removed after performing the desired function.

17.3 Cell Therapy, Cell Transplantation and Regenerative Medicine

Recent developments in the identification, *in vitro* culture and differentiation of stem cells point to the unprecedented potential of these cells, or their derivatives, to cure degenerative disorders.

Several somatic cell therapy products have been in clinical use for many years and have proven to be safe. Human embryonic stem cells (hESC) are attractive new source-material for cell therapy but have their intrinsic risks when used clinically. Translating their advantages into clinical benefits faces many challenges, including efficient differentiation into the desired cell type, maintaining genetic stability during long-term culture and ensuring the absence of tumorigenic cells from the final product.

There is considerable debate about the respective benefits and disadvantages of adult or embryonic stem (ES) cells, with both technical and ethical arguments being used to promote each set of claims. So far, all proposed cell therapy applications need to consider several safety concerns: (i) the immunogenicity of non-autologous cell transplants; (ii) the general possibility of infectious agents spreading from graft to recipient; (iii) the genomic instability of stem cells in long-term cell culture, which has been investigated extensively in the case of hESC but less so with adult stem cells; (iv) the largely unexplored migratory potential of transplanted cells *in vivo*; and (v) the inherent or culture-induced tumorigenic potential of transplanted cells, particularly hESC. It is useful to start by asking whether any long-term detrimental effects have been observed after the clinical transplantation of successful cell therapy products.

17.3.1 The Safety of Transplanted Somatic Cell Therapy Products

The use of *in vitro*-expanded human autologous or allogeneic cell populations for cell therapy use in humans is progressing rapidly. When cells are cultured for less than

24 h before their homologous clinical use (e.g., bone marrow stem cells), the procedure is not regulated as a 'drug' but as a 'surgical procedure'. The first cell therapy product approved by the Food and Drug Administration (FDA) under an accelerated approval procedure was Carticel1 in 1997: autologous chondrocytes are expanded *in vitro* before being reintroduced into a lesioned or degenerated cartilage area, with the goal of treating severe cartilage defects. This approach proved highly successful, even for large full-thickness symptomatic chondral injuries. More than 10,000 patients have been treated with Carticel1 during the past ten years without reports of serious adverse effects, and patients are reimbursed for the treatment by their health care providers in the USA. The Food and Drug Administration also approved several cell therapy products for skin replacements used for the treatment of acute (e.g., burns) or chronic (e.g., diabetes ulcer) inflammatory skin disorders.

Porcine and human hepatocyte preparations are currently in clinical trials for handling of acute liver failure. The most promising application is the extracorporeal liver-assist device (ELAD), incorporating immortalized human hepatocyte (C3A) cells to replace liver function for at least 30 days. This approach has been proven safe and highly useful to bridge the time from acute liver failure diagnosis to liver transplantation.

Skeletal myoblasts have been expanded *in vitro* for as long as two months before allogeneic transplantation into muscular dystrophy patients. No efficacy has been demonstrated, mainly owing to cell survival issues, but neither have tumors resulted. Similarly, patients with ischemic heart failure have been injected with as many as 2×108 autologous myoblasts, expanded *in vitro* over a 17-day period. In this stand-alone pilot study, enrolling only five subjects, a significant improvement in several heart function parameters, and no adverse effects, were reported after a six months observation period. However, more careful follow-up studies revealed that arrhythmia and tachycardia can arise after myoblast transplantations, and there is considerable interest in the outcome of the MAGIC (myoblast autologous grafting in ischemic cardiomyopathy) phase I/II clinical trial at centers in Europe, involving the treatment of 97 randomized patients with cells (myoblasts) or placebo. The above examples involve the use of terminally differentiated cells; however, there are many ongoing clinical trials involving the direct application of stem cells. Most prominent (and provocative) among these are trials involving the injection of unpurified or purified populations of autologous bone marrow stem cells into the hearts of patients suffering acute infarctions. In recent clinical trials performing intra-coronary infusion of bone marrow derived cells, one study did not observe any improvement, whereas another study reported a moderate improvement in left ventricular function. In no case is there any evidence that cardiomyocytes were formed from the transplanted cells, and the interpretation of data, in many cases, has been difficult because trials have often lacked appropriate controls. The list of recent developments in cell therapy is certainly an incomplete collection but it illustrates the considerable activity in the area of cell transplantation and regenerative medicine. Notably, and in contrast to early gene therapy trials, these applications of somatic cells have mostly been proven safe and effective.

17.3.2 Generation and Directed Differentiation of Human Embryonic Stem Cells

The first embryonic stem cells (hESC) were generated from mice in 1981, and hESC were established much later from surplus early pre-implantation *in vitro* fertilisation blastocysts. These cells showed a potentially unlimited proliferation capacity combined with a pluripotent differentiation capacity, while maintaining karyotypic integrity in culture These unique qualities of hESC make them an attractive source material for studying the events of early human embryogenesis *in vitro* and for the development of cell therapy applications. Many laboratories have focused on the differentiation of mouse embryonic stem cells (mESC), and the literature in this area is extensive In general, uncontrolled differentiation can be induced spontaneously by the formation of embryoid bodies. Alternatively, a more controlled and directed differentiation of mESC or hESC towards cells of particular lineages can be obtained by culturing the cells in defined culture conditions containing specific growth factors, matrix proteins or other instructive molecules. Furthermore, transgenic methods have been developed to select for the survival of the desired cell type. However, it is important to note that genetic manipulation might introduce yet another level of regulatory complexity to the clinical application of any cell therapy product. Using these methods, cell populations can be generated that are close to 100% homogenous in the case of mESC, or up to 20% for cardiomyocytes or definitive endoderm in hESC. Generally, mESC differentiation strategies tend to be five to ten years ahead of their hESC counterparts: hematopoietic precursor cells were first differentiated from mESC in 1993, but from hESC in 2003, with similar delays for lymphoid and cardiomyocyte lineages. It is worth noting that, in many cases, differentiation strategies developed for mESC can differ substantially from methods applied to hESC. Therapeutic cloning may represent a way of producing cells that can differentiate into all cell types and replicate indefinitely while not being rejected by the immune system. Therapeutic cloning entails the isolation of embryonic stem cells from an embryo created by transplantation of a nucleus from a somatic cell to an enucleated egg. The resulting *in vitro* expanded stem cells are perfectly matched to the patient's immune system. But obtaining, purifying, and expanding stem cell cultures and the control of permanent differentiation processes are issues that still need to be worked out.

17.4 Gene Therapy

The concept of gene therapy has long appealed to biomedical researchers and clinicians because it promised to treat certain diseases at their origins. In the last several years, there have been several trials in which patients have benefited from gene therapy protocols. This progress, however, has revealed important problems, including the problem of insertional oncogenesis.

In the last decade there have been four monogenic diseases for which seminal gene therapy trials have been conducted: ornithine transcarbamylase (OTC) deficiency, hemophilia caused by factor IX deficiency, severe combined immunodeficiency (SCID), and chronic granulomatous disease (CGD). Significant attention has been paid to the serious adverse events that have come from these trials, and in fact these adverse events are used as a basis for formulating future areas of research in gene therapy. It must be highlighted, however, that an increasing number of patients, particularly those with SCID, have benefited from participating in these gene therapy trials. Moreover, for most of these diseases, contemporary therapy is either inadequate or associated with serious adverse events themselves. Evaluation of the gene therapy trials must be done in the context of currently available treatments. Finally, as with any new therapy, for example the development of bone marrow transplantation and solid organ transplantation, there will be growing pains. It would be naively unrealistic to expect that gene therapy will not suffer similar pains. These growing pains have taught us about new obstacles that need to be solved and include the problems with in vivo administration of viral vectors and the problem of insertional oncogenesis.

17.4.1 Problems with In Vivo Administration of Viral Vectors

A first problem is the immune response to *in vivo*-administered viral vectors. Many viral proteins are capable of eliciting an immune response, the magnitude of which may be difficult to predict and varies from patient to patient. In the OTC trial, this variable inflammatory response resulted in the death of one patient; and in the hemophilia trials, the immune response eliminated virally transduced cells thereby abrogating any potential benefit. Transient or long-term immunosuppression is a possible solution to this problem.

A second problem with in vivo administration of viral vectors is that multiple cell types are exposed to the risks of being infected with the administered virus. In the in vivo administration of viral vectors, certain cell types are the desired target because they have the greatest chance of giving clinical benefit. But multiple other cell types, which have a low likelihood of contributing to clinical effectiveness, are exposed to potential infection with the viral vector and the possible deleterious consequences, such as activation of the inflammatory response or insertional oncogenesis if an integrating viral vector is used.

A third problem with the *in vivo* administration of viral vectors is the risk of germ line transduction. While germ line transduction has not been observed so far, if inadvertent germ line transduction did occur, it could create a public outcry against gene therapy in general.

The preclinical assessments of the safety of gene therapy protocols have been inadequate in giving a preclinical assessment of the risk of oncogenesis in human studies. These preclinical safety models have been inadequate for three reasons: (1) analysis of an insufficient number of integration events; (2) observation for an insufficient length of time; and (3) subjective rather than quantitative assessment of risk.

17.5 Functional Genomics

Functional genomics is a field of molecular biology that explores the function and interaction of genes. One approach is to manipulate, activate or deactivate particular gene sequences. The mapping of the human genome, in combination with developments in bioinformatics, has made it possible to identify, swiftly, identical gene sequences in different species. By identifying these gene sequences, experimentation that cannot be executed on humans can be carried out on animals, with useful results that help to determine the function of specific genes or gene sequences. Impressive advances make it feasible to fabricate animals predestined to, for example, heart muscle dysfunction or osteoporosis. Animals are no longer just used as they are; now they are bred and modified to accommodate the needs of science. The 'research animal' is an animal 'produced' to contain gene sequences identical to those suspected to lead to (for instance) cancer in humans. As such, it represents a species modified to accommodate humanness, not just an animal enrolled in research. This branch of research crosses the borders between humans and non-humans, creating animals that can serve as substitutes for human experimentation. Similarly, animals are produced to serve potentially as donors to human beings.

Indeed, despite the lower costs of producing biomolecules in microorganisms, like bacteria and yeast, these organisms do not properly execute several post-translational modifications, such as *N*-linked glycosylations, authentic *O*-linked glycosylations, and correct folding in order to produce a wide range of fully active human proteins. Therefore, many human polypeptides must be produced in mammalian cell systems to be recovered with their full activities. However, principally due to the low productive capacity, the price of human biomolecules produced *in vitro* by mammalian cell culture is extremely high. For this reason many biotechnology companies have been focusing on the production of biopharmaceuticals, at high concentrations, in transgenic livestock bioreactors. So far, transgenic animals are used as productors of pharmaceuticals and biomolecules, alternative source of organ donors, or as a model for human diseases.

Undoubtedly, such experimentation has inflicted suffering on animals. The legitimacy of such affliction on non-humans, however, rests on an anthropocentric ethic, which is rarely disputed in public policy: the notion of an indisputable difference between humans and animals.

17.6 The Concept of Genetic Enhancement

Until now we have focused on potential medical applications of biotechnology. But will surgeons of the future only use biotech techniques to restore and maintain *normal* function? Or will they produce *suprahuman* capabilities (so-called transhumans)? It may be possible through biotechnology at the nanoscale to provide 'suprahuman' capabilities, such as the ability to see in the infrared or ultraviolet portion of the spectrum…or see in the dark using implanted ultrasound sensors. Many writers have sought to draw a sharp line between gene therapy and enhancement in order to protect therapeutic procedures from the moral taint of genetic enhancement, which is often associated with eugenics, 'playing God', creating perfect people, and so on. Maintaining this distinction allows one to acknowledge the negative connotations of genetic modifications while endorsing the positive aspects. But the distinction between therapy and enhancement is blurred. One could argue that the goal of therapy is to treat an existing disease, while the goal of enhancement is to exceed the boundaries of normalcy and health. But many ordinary medical interventions that are designed to prevent disease actually enhance normal human functioning. For example, vaccinations enhance the immune system by causing it to produce cells and antibodies thereby increasing its ability to fight diseases. Furthermore, many socially acceptable medical interventions, such as cholesterol lowering drugs, cardio-pulmonary resuscitation, and hormone replacement therapy, are designed to prevent, forestall, or counteract the normal aging process. One reason why it is so difficult to define 'genetic enhancement' is that the concept of 'enhancement' is based on some understanding of what constitutes a normal healthy human being.

Genetic enhancement would cause the birth of children whose genetic makeup would have been intentionally designed by other human beings, and this would substantially alter the preconditions of 'natural' reproduction, by eliminating the contingency or 'chance' aspect of one's coming into existence to such a degree that the freedom of the future human being would be violated (Habermas 2003). This might deeply alter the moral self-understanding of the human species and influence future generations. But in others' opinion (Ebbesen and Jensen 2006) the main concern is why human genetic makeup or genetic integrity ought to be protected. What is the special moral status of DNA? In their opinion it is hard to claim that DNA has a fundamental intrinsic value in itself without further justification. Habermas emphasises that the crucial thing about interfering with germline DNA is the irreversibility of the procedure. The child would be in a position where he/she cannot say 'yes' or 'no', that is, give informed consent to the procedure. So according to Habermas genetic enhancement on germline cells among other things contradicts the principle of respect for autonomy. Furthermore Habermas operates with a so-called 'right to chance', which is violated if parents intentionally alter the genetic makeup of a future child. But future parents also intentionally choose their partners for reproduction. So, again, the main concern is what constitutes the special status of DNA.

17.7 The Precautionary Principle

As a final comment, the so-called precautionary principle raises a cluster of questions about how prudently to engage in risk-taking. For instance, if technological innovation is necessary in order rapidly to adapt humans to meet environmental threats that would otherwise be catastrophic on a large scale, then the development of biomedical technologies in many forms (including human germ-line genetic engineering) may be required by the precautionary principle, given the prospect of the obliteration of humans in the absence of such enhanced biotechnology.

References

Ebbesen, M., & Jensen, T. G. (2006). Nanomedicine: Techniques, potentials, and ethical implications. *Journal of Biomedicine and Biotechnology, 5*, 515–516.

Habermas, J. (2003). *The future of human nature*. Cambridge: Polity.

Chapter 18
Biodiversity of Romania Stressed by Transgenic Cultures?

Katalin Bartók(✉)

Abstract: Romania, located in the central part of Europe, lies at a conjunction of several biogeographic regions, as well as a number of ecological elements. The plains, hillsides and mountain regions are equally distributed, each accounting for approximately one-third of the whole. Its biomes consist primarily of forests (27%), steppe grasslands (16%), aquatic ecosystems and wetlands (5.8%), as well as alpine and subalpine ecosystems (1.2%). This situation is reflected in its rich and diversified natural world: 3,450 flowering plants, 600 algae, 1,200 lichens and 950 bryophyte species exist in Romania. Some of the most threatened large carnivores are widely distributed as well; one finds here 60% of Europe's brown bears and 40% of its wolf population. Besides these notable examples, over 33,802 other animal species are found here. The author discusses the potential risks posed by experimental and commercial transgenic cultures to the environment and its biodiversity.

Keywords: Biodiversity, GMOs, Romania, transgenic cultures

18.1 Geographic Settings and Climate

Romania is located in Central Europe at an equal distance from the North Pole, the Atlantic Ocean and the Ural Mountains (see Fig. 18.1). The total area of Romania is 23 million 839,100 ha and is a point of conjunction not only of a number of biogeographic regions (arctic, alpine, west and central European, Pannonian, Pontic, Balkan, sub-Mediterranean, and even eastern Colchic, Caucasian, and Turano-Iranian) but of numerous ecological elements from different bioregions, as well. The elevation of the country varies significantly – between sea level (the

Katalin Bartók
Department of Taxonomy and Ecology, Faculty of Biology and Geology, Babeş-Bolyai University, 400015 Cluj-Napoca, Str. Republicii 42, Romania
bartok_katalin@yahoo.com

F. Molfino, F. Zucco (eds.) *Women in Biotechnology,*

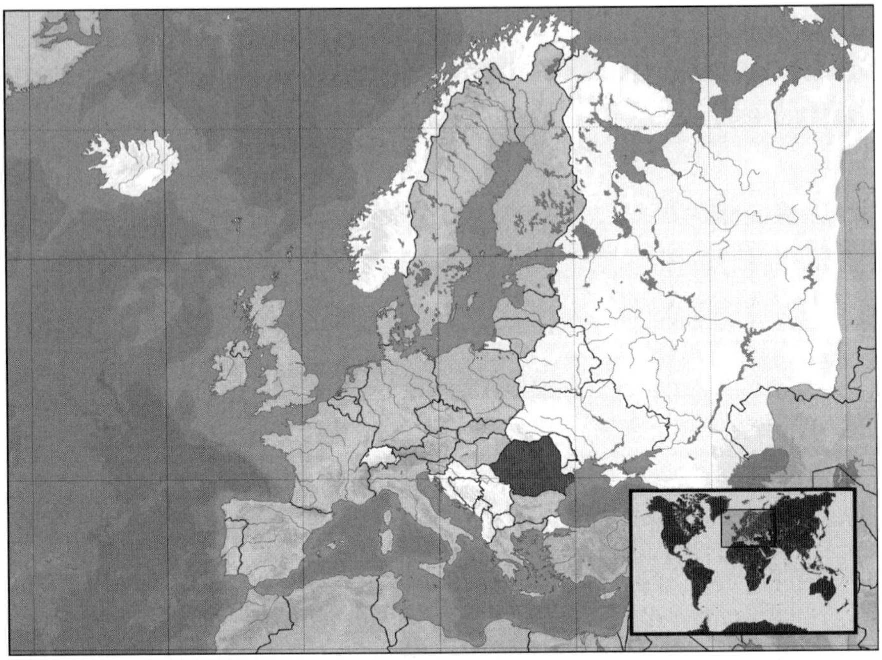

Fig. 18.1 Maps of Europe and Romania (Retrieved and modified October 20, 2007, from http://en.wikipedia.org/wiki/Romania)

Black Sea and Danube Delta) and the highest peaks of the Carpathian Mountains, 2,500 m. In Romania the plains, hillsides and mountain regions are equally distributed in one-third proportions (see Fig. 18.2).

Romania is characterized by a temperate climate, but with significant zonal aspects. The average annual temperature is 8–10°C, with frosty winters (–4°C) and warm summers (22°C); the average annual precipitation is between 400–600 mm.

The hydrological network is rich in Romania; it contains a large portion of the Black Sea Coast (228 km) and over 1,000 km of the Danube River, and numerous tributaries flow through it. Where the river empties into the Black Sea the 580,000 ha of the Danube Delta (113,000 ha permanently covered by water) has been formed.

Most soil types that occur in Europe also occur in Romania, and varying levels of relief have bonded the underlying volcanic, sedimentary and metamorphic rocks.

The main biomes primarily consist of forests (27%), steppe grasslands (16%), aquatic ecosystems and wetlands (5.8%) and alpine and subalpine ecosystems (1.2%).

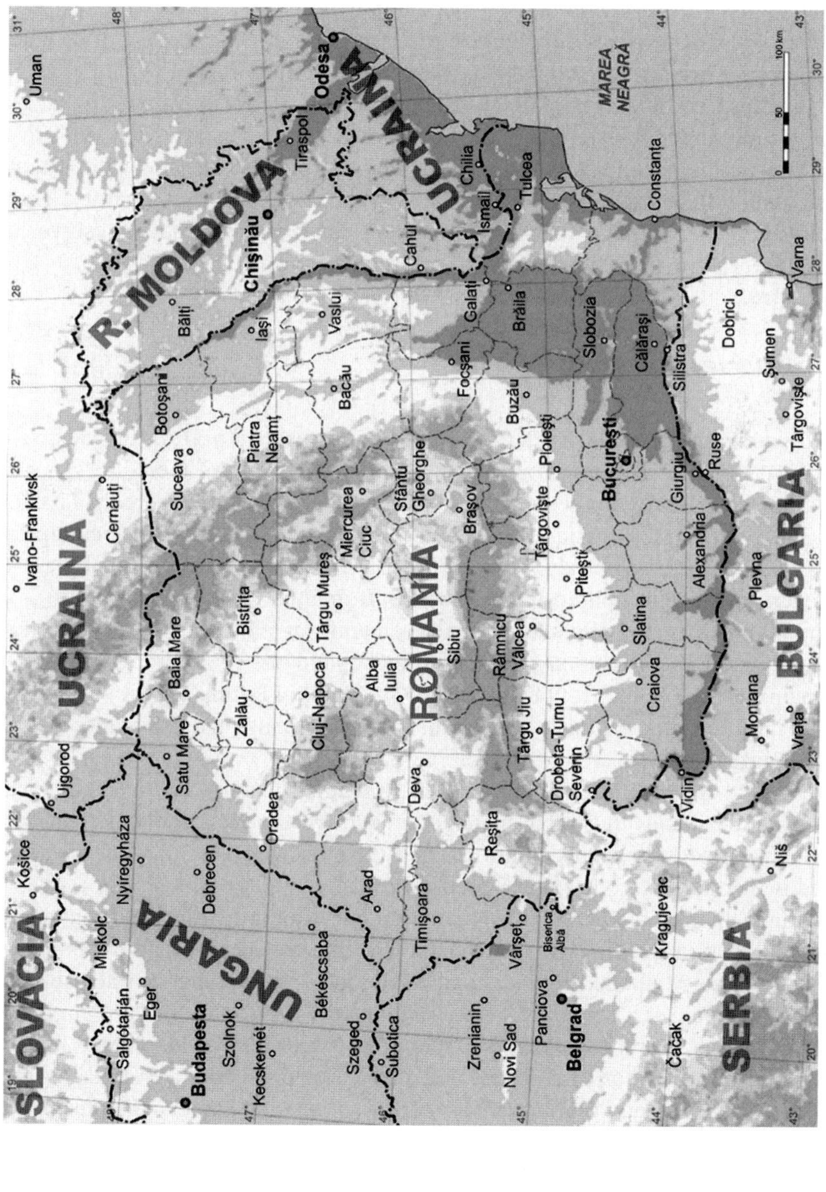

Fig. 18.2 Maps of Romania (Retrieved and modified October 20, 2007, from http://en.wikipedia org/wiki/Romania)

18.2 Biodiversity of Romania

The factors described in the previous section are reflected in this rich and diversified natural world; 47% of the ecosystems are untouched, in their natural or close to natural status; 27% of the country is occupied by forests, 13% of which is protected area. Thus, Romania has the largest forest regions in Europe. One finds here 60% of Europe's brown bears and 40% of the wolf population.

Romania has two biodiversity hotspots; one is the Carpathian Mountains range (Eastern, Southern and Western Carpathians) which are in our times are still covered by forests; the second is the Danube Delta, the biggest and most important protected area in the country, occupying 2.6% of the country's total territory. This is the only part of Romania that is triple protected: it is a biosphere reserve, a Ramsar conservation area and a significant part of the world's patrimony.

18.2.1 Species Diversity

Species diversity in Romania is relatively high. Europe has 12,500 species of flowering plants, 3,450 of them are found only in Romania. Among them, 23 species are declared as natural monuments, 74 species are extinct, 39 endangered, 171 vulnerable and 1,256 rare. Grassland species include 37% of the total species represented. About 600 species of algae, over 700 species of marine and coastal plants, 1,200 lichen and 950 bryophyte species exist in Romania.

A very high percentage of plant species are endemic at 5% (172 species) (Cristea et al. 1996), and 75% of them are found in the Carpathians.

The main endemic centres for plants are the Mountains of Rodna, Bistriț – Ceahlău, Bucegi – Piatra Craiului, Retezat – Godeanu, part of these mountains being declared as national parks.

A great climatic separation line exists in Romanian history, the Ice Age. We identify as relict species those that managed to survive the Pleistocene glaciations in Romanian biogeographical conditions. In Romania there are 75 relict species, classified in Tertiary, Xerotherm and Ice Age relicts.

Romania is an important centre of population density for a variety of threatened and endangered animals. The wolves, bears, lynx and other large carnivore populations were decimated about 150 years ago in Europe, but they are still distributed widely in Romania.

Of the 40% of the European wolf population (*Canis lupus*) 3,000 individuals live in Romania; smaller populations are located only in the northern Iberian Peninsula (2,000 individuals), Apennines and Maritime Alps (400) and Dinarids (1,500 individuals).

The only healthy lynx population in Europe is in the Carpathians (almost 1,500 individuals), which plays an important role in preservation of this species in Europe.

The brown bear (*Ursus arctos*) has also its biggest population in the Romanian Carpathians (6,000 individuals), which means 60% of the European brown bear population; they also play a very important role in species survival in Europe (Bartók 2006a, b).

Other animal species with hunting importance are: 142,250 deer, 33,170 stag, 5,955 red deer, 7,885 chamois, 36,820 wild boar, 1 million 69,600 rabbit, 307,400 pheasant, 7,865 moorhen, 7,865 wildcat, 1,990 lynx (http://enrin.grida.no/htmls/romania/soe2000/rom/cap5/specii.htm).

Romania has, besides its large mammals, over 33,802 other species, subspecies and varieties of animals, of which 33,085 are invertebrates and 707 vertebrates. The number of mammals is 100, bird species 400, and reptiles 23 In addition, 19 amphibians are known in Romania and there are 103 freshwater fish species (Bartók 2006a, b). Of our vertebrates, 24 have been designated as natural monuments.

In the Danube Delta the red-breasted goose (*Branta ruficollis*) hibernates and the Dalmatian pelican (*Pelicanus crispus*) nests; 227 specifically adapted insect species exist in caves, 97% of these are native.

Out of the subendemic species we could mention the red deer, *Cervus elaphus carpathicus*, *Sus scropha attila*, *Lepus europaeus transsylvanicus*, *Rupicapra rupicapra carpatica* and the *Parus montanus transsylvanicus*.

The fauna of Romania contains 2,000 endemic species (Cristea et al. 1996; Bartók 2006a, b).

18.2.2 Genetic Diversity

Because of the varying habitat conditions, Romania has a very high level of genetic diversity. There is a large number of genotypes of pine, Norway spruce, beech and oak, each with different growth and resistance to diseases and pests. The coniferous species (*Picea alba*, *Larix decidua*, *Pinus nigra*) are represented by Carpathian races; *Quercus robur*, *Picea abies* have distinct climatic types; *Quercus robur*, *Quercus petraea* and *Fraxinus excelsior* have edaphic types in Romania.

18.2.3 Ecosystem Diversity

Ecological diversity means the number of species in a certain community of organisms (Bartók 2006a, b).

In Romania, 17 major terrestrial ecosystem formations exist, including all the major ecosystem types existing in Europe. Sixty percent of the Carpathian Mountains are in Romania, which affects the ecological conditions, so it is of major importance. There is also a great diversity of aquatic ecosystems including river floodplains, glacial lakes, coastal wetlands, bogs and mountain rivers. Twenty-two eco-regions have been identified in Romania. The Carpathian Mountains include four vegetation regions: collin with broad-leaved forests, boreal with coniferous forests, subalpine and alpine, which contains grass and small bushes and three vegetation zones: steppe, silvosteppe and nemoral forests.

At lower altitudes (up to 300 m) in more humid regions, broad-leaved forests are dominant, while in the less humid regions are the steppe grasslands, between them is a zone of silvosteppe, containing a mix of forests and grasslands.

The forest and grassland ecosystems in Romania are similar in general structure with its neighbouring countries; they differ in abundance and composition of elements of flora and fauna. Romania is a meeting place of each of these ecosystems and a corridor for the spread of biodiversity (Bartók 2006a, b).

Romania contains five primary habitat types:

- Coastal, marine and sand dunes (with 15 subtypes)
- Freshwater (with 11 subtypes)
- Grasslands and bushes (with 23 subtypes)
- Peat bogs and wetlands (with 7 subtypes)
- Forests (with 28 subtypes)

Thus, Romania has 97 habitat types from the total of 198 European habitats.

18.2.4 The Agriculture – Environment – Biodiversity Relationship in Romanian Agro-Ecosystems

Romania's total territory is constituted by 39.2% of agricultural land, 27% of forests. A few centuries ago 77% was covered by forests, 16% by grasslands, and 5.8% were wetlands and water ecosystems (Bartók 2006a, b).

The agriculture of the country caused some slight ecological damage before the 1960s, but after 1970 until 1989 the intensifying use of the land and industrialized husbandry types became an ecological burden and cause of a higher ecological damage reflected by a biodiversity reduction.

After the political changes in 1989, the whole agricultural landowning system changed, reducing the intensity of farming and the use of pesticides and fertilizers, at the same time causing the growth of unused and waste agricultural areas. Because of the extra proliferation of weeds and pests on these lands, the evolution of different plant associations moved in an unfavourable way. These aspects are very harmful from a nature conservation viewpoint, too.

The medium and long-term strategy of agriculture, food industry and forest development (2001–2005, 2005–2010), elaborated by the Ministry of Agriculture in 2001, includes the development of ecological agriculture. The measures and action directions of the Ministry of Agriculture's strategy are the following:

- From the legislative point of view: application of community legislation on rules, principles of ecologic production and ecologic agro-food production
- From the technical point of view: promotion of an ecologic agriculture concept by specialized associations, universities, agricultural R&D Institutes and stations all over the country; experimental modules of 'ecologic microfarms'
- From the institutional point of view: the setting up of an Office for Ecologic Products as a national authority

In 2000, the number of ecologic agricultural exploitations was 610, from which 160 were zootechnic exploitations, including the breeding of milk cows; in 2004 this number grew about 20%.

The control organisms from EU countries confirmed ecologic production of 13,502 t and 50,000 hl of milk. The total area in the ecologic system will grow from 17.44 thousand hectares (2002) to 198.6 thousand hectares in 2010. In the same time frame, the expected evolution of the areas of breeding milk cows will be from 2.5 (2002) to 15 thousand hectares in 2010; and finally the most spectacular change will occur in the milk sheep area: from 7.6 thousand hectares to 90 thousand hectares in 2010 (Bartók 2006a, b).

The 60 species of autochthonous trees and 30 species of shrubs have an economical importance, producing bark, leaves, flowers, resin and wood with medicinal character or sources of honey.

The spruce trees reach heights of 60 m, the beech 45 m, the pedunculate oak 40 m respectively. It was determined that the main tree species (spruce, fir, oak, common ash, maple, poplar, willow) present an important genetic variation that is valuable in ensuring resistance to disease and pests.

Among the 1,300 grassland plant species, 175 have nutritional value, 70 species medicinal and 180 meliferous value.

Of the forest and grassland animals, 12 species of mammals, 29 freshwater fish and 7 bird species have economical value. The local population in their nutrition and economic health utilizes these biological resources (Bartók 2006a, b).

18.2.5 Forest Biodiversity in Romania

Of the total land area in Romania 27% is covered with forest vegetation (6,245 million hectares). The Romanian forests are populated by very important biodiversity resources: 3,100 native species of plants, 60 native tree species, 10 groups of natural forest formations and 150 types of forest ecosystems.

The forest composition is diversified: conifers make up 31% (23% spruce, 5% fir tree, other coniferous 3%), beech 31%, oaks 18%, other hard broad-leaves 15% and soft broad-leaves 5%. The average growing stock is 215 m^3/ha, and the average growth 5.6 m^3/ha/year. The harvested wood volume was 24–27 million cubic meters between 1951–1976, 22 million cubic meters in 1987 and 14.8 million cubic meters in 1997.

Plantations of new forests were 1,100 ha in 1996 and 900 ha in 1997. The realization of a natural forest composition model is the main goal of the present-day forest management plans (Bartók 2006a, b).

Unlike the Central European countries, where the forest landscape was radically modified, Romania is favoured by the fact that in its mountain and foothill zones, natural forests still exist.

The 85.7% of the examined forest trees in 2000 were healthy, only 13% suffered damages. For that, Romania was rated as moderately damaged forest country (Bartók 2006a, b).

With the guidance of the Genf Conference held in 1995, based on the level of intervention in Romania, we could find three types of forest:

(a) Virgin forests (untouched forests). In Romania, the term 'quasi-virgin forest' is also used, in order to delimit former virgin forests in which sporadic extraction was practiced, but the typical uneven-aged structure was not affected.

In Romania, the official list of protected areas includes 15 virgin and quasi-virgin forests, they make 5% of the total forests, with total area of 3,866 ha.

Among them are some forests that are unique in Europe, for example: beech virgin forest in the Nera forest district (National Park Cheile Nerei – Beuşniţa); spruce, fir, beech virgin forest in Slătioara; virgin oak forests (*Quercus robur*) in Ravna (Maramureş;) and in the National Park Cozia at 1,800 m altitude. The most extensive and largest occurrence can be found of *Pinus cembra* in Retezat National Park; Letea forest in the Danube Delta with oak–ash and *Alnus glutinosa*; spruce, fir and beech forests with *Pinus sylvestris* and *Juniperus sabina* as relicts in a limestone area with frequent karst phenomena. Other national parks include ancient remnants of virgin forests, too.

There are 12 virgin forests: Letea, Bolintin, Runcu, Nera spring, Iauna Craiovei, Retezat, Târnovu, Piatra Craiului, Caraiman, Munţii Călimani, Slătioara and Giumalău. The results were published in the 'Les forêts vierges de Roumanie' professional gazette (Bartók 2006a).

(b) Natural forests. Recent studies recorded 37,000 ha of natural forests, mainly beech (~15,000 ha), also spruce (~11,000 ha), fir forest (~9,000 ha) and sessile oak forests (~800 ha) in Banat.

Spruce natural forests show a large age variation for small diameter categories. The characteristics of this type of forest in Romania are 236–1,565 trees/ha, diameter 16–100 cm, volume 244–700 m^3/ha and height 25–30 m. Their stability is high.

Mixed beech – coniferous natural forests. These mixed forests show no correlation between minimum–maximum age and diameter. The wood volume is high (600–1,100 m^3), maximum height: 50–60 m for fir and spruce and 35–36 m for beech. Their stability is very high.

Beech natural forests. The characteristics of this type of forests in Romania are: 160–588 trees/ha, diameter 39–62 cm, volume 322–1,195 m^3/ha and height 22–36 m. No correlation age–diameter is significant and positive, but a positive correlation was found between height and diameter. The fructification periodicity is 3–4 years to 6–8 years, strongly correlated with altitude. Stability is extremely high.

Oak natural forests and mixed natural forests with oak. This type of forests was very disturbed by anthropogenic activities. The spatial structure consists in three levels: low-layer with hornbeam; intermediate layer with lime, cherry tree, elms, sycamore maple; superior layer with oak species, ash or sometimes elm or ash. Stability is not high, their regeneration is possible because of their longevity, two to three times higher than other tree species.

(c) Man-made or artificial forests. There were no systematic studies on the category. Different studies mentioned, by comparison of qualitative and quantitative aspects, the superiority of natural forest to that of man-made forests.

The negative impact of climatic changes in the case of artificial forest is obvious mainly in coniferous introduced monocultures.

Artificial forests are mainly formed by extraneous species (wattle, pine, scotch and black fir, red oak, etc.) but sometimes include natural species too. Fortunately, these foreign species could approach the natural conditions due to migration of neighbouring species of shrubs and herbaceous plants. Exceptions are the pine and red oak, which have a low capacity of association and, because of the unfavourable bearing surface conditions, they do not naturalize.

In Romania the theoretical concepts of forest conservation have considerably developed in forestry legislation (Romania 2002, 2006). The tool used to promote forest conservation and sustainability was 'functional zoning'. The protected forest areas recorded a current increase of 65%.

One of the most important goals of protecting and managing forests is protection of its biodiversity.

Due to forest management (husbandry), both flora and fauna suffer permanent qualitative and quantitative changes. The quantitative changes appear to be different among animals. In case of birds, their living conditions are directly linked to the forests; therefore, they react immediately to changes of size, exploitation or new forest settlings, as do other small and less mobile animals. The soil fauna instead keeps its characteristics for several years further. The survival of large vertebrates – which exist in large numbers in Romanian forests and have a great hunting importance – depends on the configuration changes of the major areas. Proper forest farming contributes to the conservation and maintenance of its qualitative and quantitative characteristics.

Romania works together with similar institutes at a European level in forest preservation research. Therefore, the PIN MATRA program is included in the Romanian virgin forest inventory, and the COST E4 program participates in the European forest research programs coordination, respectively in the European forests database creation. Besides, the Forest Management Research Institute also coordinates the LIFE, Natura 2000, Phare, Euforgen, FSP, and BioPlatform programs.

18.3 Protected Areas in Romania

Seven percent of Romanian territory has become protected. Protected areas are:

1. Nationally nominated protected areas

   (a) 55 scientific preserves
   (b) 13 national parks
   (c) 13 natural parks
   (d) 234 natural monuments
   (e) 621 nature preserves

2. Internationally nominated protected areas
   (a) Biosphere preserves: 3
   (b) Ramsar sites: 5
   (c) World heritage (natural and cultural); sites: 1 + 672
   (d) Geopark: 1
   (e) Special protected areas (SPA): 119
   (f) Special bird protection areas (SBPA): 28

18.4 Genetically Modified Organisms in Romania

18.4.1 General Information

A genetically modified organism (GMO) is an organism whose genetic material has been altered using genetic engineering techniques generally known as recombinant DNA technology. So DNA molecules from different sources are combined *in vitro* to one molecule and transferred to an organism. This allows the expression of certain traits, altering the phenotype of the organism or proteins it produces.[1]

The transferred gene(s) from the donor organism then functions in specific ways in the host organism, altering both its genetic makeup and its biological behaviour. There is usually more than one donor organism, as DNA sequences from bacteria or viruses are needed to facilitate the transfer of genetic material (as vectors), as control organisms (promoter genes) and as markers to demonstrate that genetic modification has been successful (e.g., antibiotic and herbicide resistance).[2]

The GMOs are self-replicating, cannot be turned off, and no method can take the gene out (Bardocz and Pusztai 2006).

The creation of the first recombinant bacteria took place in 1973, i.e., *Escherichia coli,* expressing a frog gene (Cohen et al. 1973). Herbert Boyer founded the first company to use recombinant DNA technology 'Genetech', and in 1978 they announced the creation of *E. coli* strain producing the human protein insulin.

The first transgenic plant was created in 1983, while the first transgenic agricultural crop in 1996.

Since GM plants grow on open fields, they are often associated with environmental risks. The surfaces occupied by GMOs have been growing every year:

– In 2000 they occupied 40 million hectares, and are expected to grow to 350 million hectares in 2020, the great majority of them in North and South America.

Currently, there is a limited international consensus regarding the acceptability and effective role of modified organisms such as plants or animals. Some groups

[1] Retrieved June 29, 2007, from http://en.wikipedia.org/wiki/GMO

[2] Retrieved June 29, 2007, from http://reports.eea.europa.eu/92-9157-202-0/en/3.9.pdf

advocate the complete prohibition of GMOs, others consider them as a salvation to feed the poor, because they gave unexpectedly high yields, were immune to the usual plant diseases, and needed little care in general. Therefore, most countries require biosafety studies prior to the approval of a new GM plant event, usually followed by a monitoring programme to detect environmental impacts.

The practice of genetic modification as a scientific technique is not restricted in the USA, but in Europe, the coexistence of GM plants with conventional and organic crops has raised many concerns, so measures were required to separate them, and their derived food and feed. European research programmes such as Co-EXTRA, Transcontainer and SIGMEA began to investigate appropriate tools and rules. On the field level these are: Biological containment, isolation distances and pollen barriers.

Several EU member states, including Germany and Austria, ban the cultivation and import of GM seeds outright. The EU legislation does not ban GM products altogether, but insists on strict rules concerning the release of GM seeds into environment and the traceability and labelling of GMOs in food and animal feed. Only seeds approved by European Food Safety Authority may be traded within the EU[3]). European scientists worry that GMOs may yet have unforeseeable and unpredictable consequences on the environment and on health.

18.4.2 The Main Applications of GMOs

The main applications of GMOs[4] are:

1. Food crops
   (a) Herbicide tolerance
   (b) Insect resistance
   (c) Male sterility systems
   (d) Disease resistance
   (e) Delayed softening in fruits
   (f) Altered oil characteristics
   (g) Nitrogen fixation
2. Non-food crops
   (a) Flowers with modified colour
   (b) Trees with altered characteristics
   (c) Plants to produce plastics and pharmaceuticals
   (d) Plants to assist in bioremediation of polluted sites

[3] Retrieved June 29, 2007, from http://www.ens-newswire.com/ens/oct2006/2006-10-30-04.asp
[4] Retrieved April 30, 2007, from.http://reports.eea.europa.eu/92-9157-202-0/en/3.9.pdf

3. Animals
   (a) Increased growth rates
   (b) Therapeutic substances in milk
4. Microorganisms
   (a) Production of enzymes or drugs
   (b) Degradation of pollutants – to clean up contaminated sites

18.4.3 The Main Potential Benefits and Costs Attributed to GMOs[5]

The main potential benefits are:

1. Promoting efficiency in farming
2. Increased yields
3. Providing altered product characteristics to aid in food processing
4. Controlling fertility
5. Reducing fertiliser inputs through nitrogen fixation
6. Reduced pesticide use

The main potential costs are:

1. Direct environmental effects
   (a) If there is gene transfer from the GMO to native flora or fauna, leading to new pests as a result of hybridisation
   (b) Unexpected behaviour of the GMO in the environment if it escapes its intended use and becomes a pest
   (c) Disruption of natural communities, through competition or interference
   (d) Food web effects through harm to non-target species
   (e) Harmful effects on ecosystem processes, if products of GMO interfere with natural biochemical cycles
   (f) Squandering natural biological resources
2. Indirect environmental effects
   (a) Continuation of intensive agricultural systems
   (b) Impacts on biodiversity as a consequence of changes in agricultural practice
   (c) Cumulative environmental impacts from multiple releases and interactions
   (d) Alterations in agricultural practices
3. Health
   (a) New allergens being formed through the inclusion of novel proteins which trigger allergic reactions at some stage

---

[5] Retrieved April 30, 2007.from http://reports.eea.europa.eu/92-9157-202-0/en/3.9.pdf

(b) Antibiotic resistance genes used as 'markers' in the GM food being transferred to gut microorganisms and intensifying problems with antibiotic-resistant pathogens
(c) The creation of new toxins through unexpected interactions between the product of the GM and other constituents

Cross-pollination can result in the crops in neighbouring non-GM farms producing GM seeds. So, how far pollen can travel is important for assessing the risk of genetic pollution by GM crops of non-GM crops, such as organic produce, with attendant safety and economic implications. According to Emberlin (1999) 200 m is a standard separation distance, after Denaeyer and Cristea (2005) 800 m, while Timmons et al. (1995), studying pollen movement between different fields showed that pollen can be dispersed over 1 km.

18.4.4 GMOs in Romania

By 1989 Romania was the largest producer of soya in Europe. At the same time, Romania was the only country in Europe that permitted the cultivation of GM plants, between 1998 and 2006. American biotechnology companies Monsanto Co and Pioneer sold genetically modified seeds to Romania 10 years ago. The environmental organisation Greenpeace in 2004 ranked Romania 11th in a table of the world's biggest producers of GM crops. The increasing use of GM crops hinders organic agriculture, an area in which Romania has the potential to be competitive in the EU market.

In 1998 GM soya, potato, corn and sugar beet were announced for the first time on the Official List of Cultivated Plants in Romania. Soya Roundup Ready was cultivated on 15,000 ha; in 2000 this increased up to 30,000 ha, together with 1,000 ha of potato BtMg, and corn MON863; finally in 2005 the GM cultivated crops covered 130,000–140,000 ha, of which 90% was soya RR.[6]

Roundup Ready is a very powerful pesticide, used to destroy weeds and parasites attacking soya crops, but also destroys every other plant nearby, damaging the environment. At the same time it has a toxic effect on human placenta and embryos. Soya RR is not approved for growing in the EU.

Until 2005 the population was not informed about the existence and cultivation of GM plants, there were no testing laboratories, and no monitoring of its effects. Food containing more than 0.9% GM plants were not labelled.

Who is responsible for it? The Romanian government. In Romania mainly the non- governmental organisations (NGO) are preoccupied by the fate and consequences of GMOs: these are Greenpeace, Ecosens, BioTerra and the Romanian Ecological Agricultural Federation[7]).

---

[6] Retrieved April 30, 2007, from http://omg.ngo.ro/raprom.shtml?AA_SL_Session = 5774cee38e1d5615fc2ae321098458f8

[7] Retrieved April 30, 2007, from http://omg.ngo.ro/raprom.shtml?AA_SL_Session = c90ca301b774130e70fae0a74e12b68

The Romanian legislation was incomplete and did not conform to requirements of the EU. Law 214/2002 adopted the first measures to regulate GM products, when it told manufacturers of GM products to declare this information on packages and labels if the products contain more than 0.9% GM plants. The order 462/2003 of the Minister of Agriculture, Water, Forests and Environment requires all farmers using seeds for biotech crops to report the area planted with such seeds and the yield obtained. According to the Ratification Law 59/2003, Romania is signatory to the Cartagena Protocol on Biosafety, which requires reporting of each commercially approved trans-boundary event (Cionga 2005, USDA Foreign Agricultural Service).

Worldwide, over 100 incidents of illegal or unlabelled GM contamination have been documented in 27 countries on five continents.

During a research tour in Romania (2006), Greenpeace discovered illegal growing of GM soya in 10 out of the 42 Romanian counties. And two of them are in protected areas:

- Near Grindu and Rachelu localities (Tulcea County), at 9 km from the border of the Danube-Delta Biosphere Reservation (one of the largest wetlands on the planet, a unique 'natural museum' of biodiversity) soya RR was cultivated on 25,000 ha.
- Soya RR was also found in the Natural Park of Comana, certified by the Austrian GMOs-test laboratory.

A genetically modified potato field, which can contaminate traditional potato varieties, was found on 3,340 m^2 in Târgu Secuiesc (Harghita County); and also genetically modified plum trees, which contain a gene that is resistant to antibiotics,[8] was found in Bistrița (Bistrița-Năsăud County).

What is the opinion of the Romanian population about GMOs?

The results of the public opinion test there are:

- Seventy-eight percent do not want to buy genetically modified foods.
- Ninety-eight percent think that clear labelling of foods containing GM plants is necessary.
- Forty-seven percent think that GM plants are unhealthy for the population.

The opinion of Greenpeace is that control over GM plants has been lost in Romania, so Romania will become a dustbin of GMOs (http://omg.ngo.ro/raprom.shtml?AA_SL_Session = c90ca301b774130e70fae0a74e12b68).

18.4.5 Some Optimistic News from Romania

The Romanian Ministry of Agriculture announced a ban on the cultivation of GM soya from 1 January 2007. The government is trying to bring food production standards into closer harmony with EU environmental rules. It ordered cuts and

[8]Retrieved June 29, 2007, from http://agbios.com/main.php?action = Show NewsItem§id = 8090)

burning of the production of GM herbicide resistant soya, of which the EU does not approve, and introduced a monitoring and control system for GM crops.[9] But the decontamination process is likely to take years. Romania may also become a test case to see whether GM crop-plant decontamination is possible at all.[10]

In April 2007 were created the National Informational Register of GM plants and the Nationally Authorised Laboratory for testing GMOs, included at the same time on the list of European Reference Laboratory for Testing GMOs.[11]

In Romania two GMOs Free Regions have been declared: the first (2006) 26 localities from Bistrița-Năsăud County, than in 2007 the second in Cluj County, when 14 localities (Huedin microregion) declared themselves as GMOs Free Zones. The GMOs free zones declaration represents the official commitment of local authorities, to the extent of their legal and organisational powers, to cultivate no GMO on their land.

18.5 Conclusions

GMOs can spread through nature and interbreed with natural organisms, contaminating non-GM environments. GMOs have reduced biodiversity on all its levels: produce gene erosion, species and ecosystem reduction and land infertility. GMOs, once released into the environment, cannot be recalled. The biodiversity of Romania is unquestionably stressed by GMOs. The biodiversity of all countries – incl. Romania – must be protected and respected as the global heritage of humankind.

The results of 'GMOs pollution' principally appear in agricultural production. Agriculture plays an important role in the economy of Romania, therefore the principle of 'precaution' should be applied, refusing to allow gene manipulation techniques, and maintaining the use of traditional, sustainable and ecologically friendly agricultural production systems, which preserve human as well as animal health.

References

Bardocz, S., & Pusztai, A. (2006). Nutrition, Health and Future. Manuscript.

Bartók, K. (2006a). Biodiversity and nature conservation in Romania. In G. J. Halasi-Kun (Ed.), *Proceeding seminar on pollution and water resources XXXVI* (pp. 85–103). New York: Columbia University.

Bartók, K. (2006b). *Az élö természet védelme, a biodiverzitás védelme Romániában*. Kolozsvár: Ábel Kiadó.

Cionga, C. (2005). *Romania biotechnology annual 2005*. USDA Foreign Agricultural Service, GAIN Report, RO5008.

[9] Retrieved June 29, 2007, from http://www.ipsnews.net/news.asp?idnews = 35903

[10] Retrieved June 29, 2007 from http://www.ens-newswire.com/ens/oct2006/2006-10-30-04.asp

[11] Retrieved April 30, 2007, from http://omg.ngo.ro

Cohen, S., Chang, A., Boyer, H., & Helling, R. (1973). Construction of biologically functional bacterial plasmids in vitro. *Proceedings of the National Academy of Sciences, 70*, 3240–3244.

Cristea, V., Denaeyer, S., Herremans, J. P., & Goia, I. (1996). *Ocrotrea naturii si protecţia mediului în România*. Cluj-Napoca: Cluj University Press.

Denaeyer, S., & Cristea, V. (2005). Culturile transgenice: o nouă posibilă sursă de poluare? *Environment & Progress, 4*, 159–164.

Emberlin, J. (1999). *The dispersal of maize pollen*. Worcester: Natural Pollen Research Unit.

Romania (2002). Legea 26/1996: Codul Pădurilor. *Monitorul Oficial, 1996*.

Romania (2006). Legea 137/2002. Protecţia Mediului. *Monitorul Oficial, 2002*.

Timmons, A. M., O'Brien, E. T., Charters, Y. M., Dubbels, S. J., & Wilkinson, M. J. (1995). Assessing the risks of wind pollination from fields of genetically modified Brassica napus ssp. oleifera. *Euphytica, 85*, 417–423.

Chapter 19
The Social and Economic Impact of GM Crops: The Case of the Herbicide Tolerance Trait

Suman Sahai(✉)

Abstract: To have any relevance to development of a country's agriculture and to the needs of small farmers, it is necessary for managers of Genetically Engineered (GE) crops to provide real life solutions to the problems inherent in these methods of producing food and fodder. Currently available GM crops are those that were developed for industrialized agriculture in developed countries, complete with their positive and negative traits. At least one of them, the Herbicide Tolerance (HT) trait, is not just irrelevant to our country's needs, it is downright harmful to the agricultural and food systems that strengthen local food security. Using HT crops will reduce the demand for labor in a labor surplus country where agricultural operations such as weeding provide wages, especially for women; it will reduce food and nutrition availability for the poor by eliminating natural leafy greens, it will strike at human and veterinary health by destroying locally available medicinal plants, and it will increase risks of crop failure by making it impossible to engage in mixed cropping, which distributes risk and provides more food and nutrition.

Keywords: GE crops, herbicide tolerance, developing country, agriculture, loss of nutrition, loss of medicinal plants, loss of women's wages

19.1 Socio Economic Considerations Under the Cartagena Protocol

There is broad acknowledgment of the fact that genetically modified crops and foods could have an impact on the environment and human health. It is thus recognized that their impact on these two social concerns needs to be measured.

Suman Sahai
Gene Campaign, J-235/A, Lane W-15C, Sainik Farms, Khanpur,
New Delhi – 110062, India
genecamp@vsnl.com

F. Molfino, F. Zucco (eds.) *Women in Biotechnology,*

What is less recognized is the fact that such Genetically Engineered (GE) crops are anticipated to have a social and economic impact on farm families, particularly in developing countries. Here the particularly vulnerable groups will be the small and marginal farmers; especially those in rain-fed areas[1] and women-led farm families, a result of men migrating to urban areas for work.

The Cartagena Protocol on Biosafety (Mackenzie et al. 2003) requires that risk assessment and monitoring of Genetically Modified (GM) crops must be done where there is uncertainty about their environmental impact.

The risk assessment must be carried out in a sound manner using available scientific information and must include collection of detailed data on the location, climate and ecological characteristics of the region where GM crops are introduced.

On the subject of assessing the socio-economic impact of GM crops, the Protocol is less clear than in the case of the environment. Article 15 read together with Annex III[2] gives the impression that risk assessment is confined only to environmental and health effects. However, the Protocol recognizes that risks involved with respect to GM crops and Living Modified Organisms (LMOs in the Protocol) have a broader scope. This is acknowledged in Article 26[3] which says that countries may take into account socio-economic concerns arising from LMOs when taking a decision either on domestic use or on import of GM products. Socio-economic concerns can arise from the impact of GM crops and foods on the conservation and sustainable use of biological diversity and especially with regard to the value of biological diversity to indigenous and local communities (Convention on Biological Diversity 1992).

---

[1] About seventy percent of the annual rice production in India comes from rain-fed, non-irrigated, areas.

[2] Article 15. Risk Assessment

1. Risk assessments undertaken pursuant to this Protocol shall be carried out in a scientifically sound manner, in accordance with Annex III and taking into account recognized risk assessment techniques. Such risk assessments shall be based, at a minimum, on information provided in accordance with Article 8 and other available scientific evidence in order to identify and evaluate the possible adverse effects of living modified organisms on the conservation and sustainable use of biological diversity, taking also into account risks to human health.
2. The Party of import shall ensure that risk assessments are carried out for decisions taken under Article 10. It may require the exporter to carry out the risk assessment.
3. The cost of risk assessment shall be borne by the notifier if the Party of import so requires.

Annex III. Risk Assessment

Objective

1. The objective of risk assessment, under this Protocol, is to identify and evaluate the potential adverse effects of living modified organisms on the conservation and sustainable use of biological diversity in the likely potential receiving environment, taking also into account risks to human health.

[3] Article 26. Socio-Economic Considerations

Article 26[4] identifies the types of socio-economic considerations that parties may take into account in reaching decisions on imports or under its domestic measures implementing the protocol. It requires that such considerations be taken into account consistent with a Party's other international obligations.

The range of socio-economic considerations indicated in Article 26 is restricted to those arising from the impact of LMOs on the conservation and sustainable use of biological diversity, especially with regard to the value of biological diversity to indigenous and local communities. The language of the Protocol indicates that not all socio-economic considerations may be taken into account, only those arising from the impact of LMOs on biodiversity. Thus, where the introduction of LMOs affects biological diversity in such a way that social or economic conditions are affected, a party can use Article 26 to justify taking such impacts into account for purposes of making decisions either for domestic application or imports.

Socio-economic considerations with respect to the value of biological diversity to indigenous and local communities may also refer to the impact of introduction of LMOs on the ability of those communities to make use of the biological diversity upon which their survival and traditional livelihood depends.

The freedom to use socio-economic considerations while taking a decision on LMOs is limited by the condition that such an action must be consistent with the country's other international obligations. This is particularly relevant in the case of a Party's international obligation regarding international trade. For example, if a Party is a member of the World Trade Organization (WTO), then it has to ensure that its obligations under the WTO Agreements are not violated as a result of any application of socio-economic considerations in making import decisions on LMOs.

Socio-economic considerations can also be taken into account in taking decisions regarding measures for implementation of the Protocol. However, the Protocol

---

1. The Parties, in reaching a decision on import under this Protocol or under its domestic measures implementing the Protocol, may take into account, consistent with their international obligations, socio-economic considerations arising from the impact of living modified organisms on the conservation and sustainable use of biological diversity, especially with regard to the value of biological diversity to indigenous and local communities.
2. The Parties are encouraged to cooperate on research and information exchange on any socio-economic impacts of living modified organisms, especially on indigenous and local communities.

[4] Article 26. Socio-Economic Considerations

1. The Parties, in reaching a decision on import under this Protocol or under its domestic measures implementing the Protocol, may take into account, consistent with their international obligations, socio-economic considerations arising from the impact of living modified organisms on the conservation and sustainable use of biological diversity, especially with regard to the value of biological diversity to indigenous and local communities.
2. The Parties are encouraged to cooperate on research and information exchange on any socio-economic impacts of living modified organisms, especially on indigenous and local communities.

does not give any guidance on exactly how these considerations can be taken into account (Mackenzie et al. 2003). In the case of indigenous and local communities, some participants think that possible ways of taking socio-economic considerations into account could include procedures for assessing and addressing socio-economic impacts in risk assessment and management and prior public consultation processes with respect to decisions on import, especially with respect to communities that will be directly affected by the import.

Though the protocol admits consideration of socio-economic concerns, the scope is greatly restricted and is limited to effects on biodiversity. Even then, it can be further curtailed by a Party's international obligations, chiefly with respect to the WTO. Unlike Annex III that provides details of scientific Risk Assessment, there is no such provision for socio-economic risk evaluation. The scope of the protocol as compared to other international agreements is not yet clear and in the absence of a clear guideline on how to use Article 26 in taking import decisions conflicting situations can arise. It is for developing countries to be proactive in developing the contours of more liberal socio-economic risk assessment parameters and to negotiate for their inclusion in the Protocol. A number of socio-economic aspects likely to emerge from the introduction of GM crops in developing countries are discussed below.

19.2 Genetically Engineered Crops Competing with Traditional Crops

GM technology is providing alternative ways of producing commodities that have traditionally been supplied by developing countries. This has the potential of taking away the economic base of farmers who produce such products. Many crops of developing countries have a global value because of the special chemical compounds they contain like aromatic or other special oils. If GM technology will modify common, commodity crops to produce these higher value chemicals, it will lead to displacement of the developing country products from such niche markets from where they can earn incomes, leading in turn to economic deprivation.

It is not that biotechnology is the sole factor responsible for eroding the economic base of poor farmers, but it contributes its share. The production of High Fructose Corn Syrup (HFCS) from corn, a major GM crop bred for industrial applications like this, has undermined sugar prices and distorted sugar markets for cane sugar producers of developing countries like India, Pakistan, Cuba, Brazil, etc. The massive subsidies of the EU and US for agriculture is a source of grave distortion in international agricultural trade but the creation of HFCS is also a major contributing factor in displacing cane sugar producers from the global sugar market.

Biotechnology research taking place in coconut for instance will have highly negative economic implications for farmers in coconut producing regions. Coconut provides high value oil used for edible and for industrial purposes. The main advantage of this oil is the high lauric acid that it contains. The US alone imports upwards of US$350 million worth of coconut and palm oil annually (Bennet 1999). Now agricultural biotechnology is creating GM canola, a form of rapeseed, to produce the same special high lauric acid oil as coconut (Nichterlein 1997).

Research by Integrated Coffee Systems Incorporated is directed to developing a variety of coffee in which maturation of the bean can be controlled artificially. This is to facilitate mechanical harvesting. Coffee is one cash crop that employs a large number of farm labourers for harvesting because the ripening of the coffee beans is uneven and picking has to be done over an extended period. This activity supports over seven million farm families in Africa and Latin America. The uniform ripening, mechanically harvested coffee variety would certainly reduce livelihood options in these regions. Developing country policy makers need to be especially vigilant about the potential for such devastating economic impacts when adopting biotechnologies.

The development of herbicide tolerant (HT) crops has to be seen in the same light. This technology, based on proprietary herbicides, claims its goal is to 'reduce drudgery' on the farm, especially for rural women. This claim has little to do with rural reality in most parts of the developing world. Weeding is an assured avenue for earning casual wages in rural areas, especially for women. Sometimes it is the only source of wages. Farm operations like sowing, weeding, harvesting and winnowing are the key sources of rural employment. Agricultural labour constitutes the largest section of the labour force in developing countries; in India and other South Asian countries, the agricultural labour force is growing at the rate of six to seven percent per annum. The herbicide tolerance trait is essentially a labour saving and hence a labour displacing trait and would have economic implications were it to be introduced in labour surplus developing countries.

Then there is the question of the social and economic definition of 'weeds'. Weeds are considered a nuisance in the monoculture agricultural systems of industrial nations where several thousand hectares of wheat or corn would be planted solely for the cereal. The fields are not supposed to yield anything else. In the case of developing countries, the fields are supposed to and do yield a myriad other goods apart from the chief crop that has been planted. The local flora, the so-called weeds, has several useful functions critical to the well being of rural communities.

What those doing the weeding in an agricultural field collective, fulfils two important nutritional roles. The plants that constitute weeds are largely nutritious leafy greens, which are a valued source of nutrition in the family's diet. A typical wheat field in India or Bangladesh would yield at least twenty types of leafy greens over the cropping season. These greens provide nutrition in a fresh and easily available form, at no cost. This has to be seen in the context of rural poverty where many farm families would not be able to buy many vegetables from the market but they are able to access them for free from fields and field boundaries. This access to free nutrition is one of the reasons why nutritional status is better among the rural poor than among the urban poor who have to buy all their food.

Plants collected during weeding that are not consumed by the family serve as fodder for the livestock that rural families maintain as additional income sources. India has a very large livestock population; it has the largest cattle population in the world. India is also a fodder-starved country and increasing fodder availability is one of the key concerns of the agricultural research system. For rural families the livestock they can keep is critical for increasing incomes either through milk or the sale of the animals for meat. The fodder that is collected during weeding is fodder that is obtained for free. If rural families had to buy all the fodder that was needed to maintain their cows, goats or pigs, many would not be able to afford keeping animals and would have to forego the extra income.

Apart from this, using herbicide tolerant crops would make it impossible to plant crops on the field bunds, as is done in many parts of Asia both for additional food and for increasing farm incomes. Typically, farmers will grow crops like yams, ginger or vegetables on the bunds surrounding rice fields. Thus two or three kinds of produce are available from the field in the same season. This advantage would be lost if the package of herbicide tolerant crop varieties and herbicide use would be implemented. In addition, the practice of intercropping and mixed farming would suffer a setback. Traditionally farmers plant more than one crop in the field. Sugar cane for instance is interspersed with lentils or mustard and it is not uncommon to find farmers planting mustard along with wheat, to be harvested one after the other or linseed together with lentils. Mixed cropping is widely practiced, with differing combinations of crops depending on the region.

So-called weeds are also the medicinal plants that rural families depend on for the health and veterinary care needs of themselves and their animals. The introduction of herbicide tolerant crops with accompanying herbicide use would kill the surrounding vegetation and deprive rural communities of the medicinal plants that form the basis of indigenous healing traditions.

A GM trait like herbicide tolerance has serious social and economic impacts on rural families and has the potential to reduce household nutrition and incomes. Loss of weeding opportunities means the family income will suffer due to loss of wage labour from weeding, loss of income from products derived from additional livestock, the man-days lost in collecting fodder from elsewhere or expenditure on buying fodder. This loss of income or additional financial burden will have an impact on other aspects of a family's life; less money may mean pulling out a girl child from the school, less money for school books or fewer clothes.

Household nutrition will suffer due to reduced intake of nutrients like vitamins and minerals, resulting from the loss of green vegetables from the diet, especially in the case of women who in any case receive less nutrition than men and children when food is scarce in poor families. This reduced family income, arising from the loss of supplementary crops like yams and vegetables that are grown on field bunds surrounding the principal crop and which can be sold in the market, impacts women most since they are placed at the bottom of the social hierarchy.

A major impact of HT crops is likely to be felt at the level of health and veterinary care for families and their livestock, due to the loss of locally available medicinal plants.

This will result in increased expenditure that the rural family will have to incur on procuring treatment from the commercial sector and the loss of man-days in travelling to the nearest formal health facility to seek medical and veterinary help. Given that government medical facilities are scarce in rural areas, the destruction of medicinal plants will compel the rural population to access expensive, even unreliable medicines from private sources that are often inhabited by unqualified people if not outright quacks. The loss of medicinal plants will deny the community its ability to be self reliant in procuring health and veterinary care and will place financial burdens on those needing to acquire these services elsewhere.

One of the serious outcomes of introducing the herbicide tolerant trait is the development of new weeds because the herbicide is known to shift easily into other crops. In the case of canola, all the kinds of HT genes that have been used are found to have migrated into other non-GM canola (Biotechnology Australia 2003). The crop of most concern in developing countries though is rice. Several studies (Gressel 2002; Baorong et al. 2003) have shown that the HT trait shifts quickly to rice relatives, specifically, from Oryza sativa, the cultivated rice, to Oryza rufipogon, a wild relative also called red rice, which is a commonly found weed in the areas where rice is cultivated. The shift of the HT trait into a rice weed like O. rufipogon will have economic implications because red rice is a strong competitor of cultivated rice and tends to take over rice fields. Socio-economic indicators should be developed to assess the economic impact on the farmer of poor rice yields and the impact on food reserves of shortfalls in rice production in countries that continue to have food security concerns. In these countries, including India, a shortfall in rice production will impact directly and seriously on the country's ability to provide adequate food to the poor, as well as its ability to hold buffer stocks of food grains to meet emergencies. A related cost will have to be factored in as well, that of having to buy food grains from the international market, in order to maintain the grain buffer stocks which are central to planning for food security.

In the case of the *Bacillus thuringensis* (Bt) crops,[5] the cost of introducing this technology under developing countries goes clearly against the small farmer, more so against women farmers who are subsistence level farmers with negligible farm budgets. Unlike the large cotton plantations of the US for which this technology was developed in the first place, resource-poor farmers with small land holdings, usually under rain-fed conditions, cultivate cotton in developing countries. These farmers by and large do not implement the insect refuge of twenty percent that is required to maintain the Bt pest resistant strategy (Sahai and Rahman 2003). The refuge is not maintained because it is uneconomical for a farmer to divert twenty percent of his or her total land area, especially when a holding is less than 1 ha. The Bt technology is therefore being implemented in a situation where from the point of view of local agriculture practices, it should not.

The socio-economic impact of quick resistance development on farmers who have changed their farming practice to accommodate Bt cotton will need to be assessed. In addition, there is likely to be a negative fallout of the indiscriminate manner in which the Bt gene is being deployed in a large number of crops (Sharma et al. 2003) which will be planted in all the major crop seasons. With a Bt crop in the field all the

[5] The Bt crops are transgenic crops containing a gene from *Bacillus thuringiensis* that are coded for the protein crystals in Bt which are toxic to some insect larvae.

time, the pests will not only develop resistance to the toxin but also start moving to other crops as fresh hosts. This is likely to result in lower agricultural productivity, loss of farm incomes and nutritional stress for rural families, particularly women.

The introduction of GM crops will impact negatively on efforts underway to diversify the agriculture base and move away from standard crop rotations like the rice-wheat rotation. Diversification is being promoted to enhance the crop basket with greater variety, find markets for non-classical products to increase incomes and minimise the risks faced by farmers. In addition to this, farmers' ability to adopt organic agriculture will be hampered by the presence of GM crops. The competitive advantage that small resource-poor farmers have is the ability to go organic more easily (because of little or no investment in agrochemicals) than those farmers engaged in heavy chemical intensive agriculture. Small farmers, especially women farmers, are in a good position to take advantage of the growing market for organic foods, provided they can remain GM free. Loss of organic markets particularly by small farmers must be measured and the impact assessed on their incomes and livelihoods.

Finally there are the issues of the impact of GM crops on traditional farming practices and the indigenous knowledge that is associated with it. These practices are natural to rural communities. Women-led households are not at a great disadvantage with traditional practices and are able to cope when they have to lead farm operations in the absence of men who have migrated. GM agriculture is alien, expensive and brings with it a package of negative impacts on the household. It will also erode, in time, the wealth of indigenous knowledge that the rural community, especially women, rely on to create better food and livelihood options for their families and their community.

References

Baorong, L., Zhiping, S., & Jiakuan, C. (2003). Can transgenic rice cause ecological risks through transgene escape? *Progress in Natural Science, 13*(1), 17.

Bennet, D. J. (1999). *Ethical aspects of agricultural biotechnology* (pp. 42–54). The Hague: Cambridge Biomedical Consultants.

Biotechnology Australia (2003). *Agricultural biotechnology, herbicide tolerant crops in Australia.* Canberra: Bureau of Rural Sciences.

Convention on Biological Diversity (1992). Retrieved October 31, 2007, from http://www.cbd.int

Gressel, J. (2002). Preventing, delaying and mitigating gene flow from crops - Rice as an example. In *The 7th international symposium on the biosafety of genetically modified organisms, Beijing* (pp. 59–77). Retrieved October 31, 2007, from http://www.isbr.info/symposia/docs/isbgmo.pdf

Mackenzie, R., Burhenne-Guilmin, F., La Vina, A. G. M., Werksman, J. D., Ascencio, A., Kinderlerer, J., et al. (2003). *An explanatory guide to Cartagena protocol on biosafety.* Cambridge: IUCN.

Nichterlein, K. (1997). Biotechnology advances in oil crops affecting coconut production. *BINAS News, 3,* 3–4.

Sahai, S., & Rahman, S. (2003). Mahyco-Monsanto's Bt cotton fails to perform. *Current Science, 85*(4), 426–427.

Sharma, M., Charak, K. S., & Ramanaiah, T. V. (2003). Agricultural biotechnology research in India: Status and policies. *Current Science, 84*(3), 297–302.

Afterword

Making Our Own Paths, Setting Our Own Agendas and Putting Them into Action

Gillian Youngs(✉)

Abstract: This short contribution offers some concluding thoughts on the WONBIT conference indicating how the proceedings can help lead to deeper engagement of women and feminist perspectives in scientific processes, including where women are currently least represented. It argues that the conference offers a number of ways forward with direct policy implications and practical possibilities.

Keywords: Biotechnology, science, male/female, science/nature, competitiveness, risk, feminist perspectives, women

The WONBIT conference, 'Women in Biotechnology, Scientific and Feminist Approaches', was distinctive in its reach both in subject matter and policy implications. One of its greatest contributions is the showcasing of a wide range of women scientists and social scientists, who are working on biotechnology and associated areas across the world. As an event it provided important evidence of the complex roles of women in the production of science and diverse critiques of it, including from women's and feminist perspectives. It therefore demonstrates that the need to increase the number of women working in and influencing the scientific domain is a move founded on well established and extensive achievements of women in this sphere. Part of the problem of attracting more women into the sphere is perhaps insufficient public recognition of such achievements and their importance to science writ large.

The purpose of this short contribution is to highlight a number of aspects the conference as a whole identifies. The first relates to the comprehensiveness of women's engagements with science as innovators, producers and consumers. From listening to the papers presented it became clear to me that we need to think about

Gillian Youngs
Department of Media and Communication, University of Leicester, University Road, LE1 7RH, UK gy4@le.ac.uk

F. Molfino, F. Zucco (eds.) *Women in Biotechnology,*

women across these three areas in relation to knowledge, goods and policy. So when considering women and feminist perspectives it is essential to look across scientific processes in a holistic way, from their conception (including in pure and basic research), their applications through diverse technologies, and the consumption of the goods that result from them. The debates at the WONBIT conference clarified for me the complexity of the women and science issue along such lines.

My assessment is that this is an area of particular significance for policy makers at all levels, national, EU and global, where the imperatives of increasing women's involvement in scientific and technological fields have been major preoccupations of long standing. The content of the WONBIT conference provided some helpful signposts on the path to different ways in which we can think about women *in* science and women *and* science. The more complete the trajectory of such focus the better, from where scientific processes begin to where they end. Certainly key are education, discovery, invention, innovation, application, production, use and consumption. Women are involved throughout these processes although much more heavily in consumption rather than discovery and innovation.

Despite such imbalances in women's roles at this point, it was evident from the conference that they have knowledge to bring to bear on the full spectrum of scientific processes, and the possibilities and risks involved in them, and that this stands as one of the most challenging realms for contemporary public policy. Even where women are least represented, policymakers need to find ways to engage them in the issues at stake, and to take account of their perspectives and insights. We cannot wait for women to have an equal place and equal influence in science for their vital knowledge to be part of the policy considerations relating to it. Science is part of, and impacts on, society as a whole and assessments of those impacts needs to extend well beyond the hierarchies of expertise associated with them, especially when these are highly gendered.

The following developments came to mind for me as possible future ways forward along these lines:

- Expansion of global networks of women scientists, social scientists, and professionals working in the biotechnology sector to share knowledge, build strategies and harness opportunities. Lobbies associated with such networks could help to bring women's and feminist assessments and viewpoints into general science and policy realms where they are minimally represented or missing at this stage. These networks could harness new virtual technologies of the world wide web to expand women's knowledge about science in general and women's roles and potential within it, to engage them more actively in debates and policy processes about it, and to support efforts to attract more women and girls to science and technology educational and career paths.
- Women's significant roles and presence in biotechnology and environmental activism could be harnessed and connected to encourage more innovation towards a sustainable world, and a greater leading role by women in the scientific community, policy and political world to transform science and technology from a production and consumption to a sustainability framework.

- Redefinitions of 'stakeholders' to recognize the inequalities impacting on women's inputs into scientific processes across the board, leading to greater public policy emphasis on women and science at every stage from invention, through production and consumption. Recognition that if women are not present or minimally present in areas of science, their perspectives and concerns still need to be integrated into them. Their role of stakeholder as heavily located at the consumer end (where least control and expert knowledge is available) needs to be transformed into more holistic contributions throughout scientific processes even those that are still male-dominated.
- More attention to women's and feminist perspectives on science especially on the question of 'value'. This needs to be understood from social and cultural standpoints as well as from material and economic ones. Debate surrounding fields such as biotechnology should be expanded along the former two areas, as the latter two tend to dominate. Women and feminists, inside science as well as in society more generally, have much to contribute to a debate about 'value' in this context.
- Greater critical awareness of assumptions surrounding the rigid male/female binary of thought leading to essentialist notions (biological or cultural) surrounding the roles and identities of men and women, the characteristics and proclivities of masculinity and femininity. Challenging of mutually associated binaries of male/female and science/nature, rationality/emotion, which contribute to assumptions about men's roles and identities as 'naturally' more scientific and rational, and women's as more related to nature and emotion. More challenging of heteronormative assumptions in policy, scientific and market orientations.
- Greater attention to the 'individualization' of risk from scientific and industrial processes (including through consumption and the socio-economic and cultural factors influencing it, and leading to unequal effects across different social or geographical groups).
- Focus on the biopolitics of contemporary change and the digitization of life, the increasing role of data in framing identities and social and market roles, patterns of surveillance and self-surveillance or disciplining that result or could be possible in the future.
- Awareness that the continuing growth of scientific processes in shaping societies and lived realities contributes to the further embedding of disembodied enlightenment rationalities into life. As the human is increasingly captured and understood as data, including through new virtual communications technologies as much as systems like genetic profiling, the rationales of science and technology are ever more intricately interwoven into life and being.
- Recognition that there are many ways in which science is a new ideology driving not only the established wealthy economies but the major new global growth economies such as China and India. There is a high price to pay for being anti-science or resistant to science as economies, cultures or individuals. Competitiveness (national, regional, global) is linked heavily to science and technology.

These are just a few aspects I would highlight from the WONBIT conference and they hardly do justice to the content of the conference as a whole. However, I do hope they sufficiently show that new thinking and approaches are being generated from women's and feminist roles and perspectives, with direct and immediate policy implications and practical possibilities.

Note: I discuss many issues related to the above points in my latest book *Global Political Economy in the Information Age: Power and Inequality*. London: Routledge, 2007.

Some Remarks on Women and Science

Heidi Diggelmann(✉)

Abstract: The WONBIT conference has realized a first step towards its goal, namely to bring together representatives from different fields of research to discuss possible consequences of new scientific developments for humanity and the environment. It became evident that patience, mutual respect and tolerance are needed to enter a fruitful dialogue between partners with so different backgrounds. Particular attention has been drawn to the necessity of developing a simple common language that can be understood by all partners. In interdisciplinary projects the early and continuous participation of all partners is necessary to establish mutual confidence. Furthermore, the need for an adequate representation of women in decision making processes was considered important to improve the dialogue and it became clear that the procedures presently used in the selection of leaders have to be modified to reach this goal.

Keywords: Common language and science, criteria for women nominations, adequate women representation

The goal of the WONBIT conference as formulated in its initial statement was a very ambitious one – it proposed to create a platform for a public debate on recent developments in biology that are of general relevance for our societies. In contrast to other meetings with similar general goals, the WONBIT program was based on a series of specific assumptions and convictions that rendered it unique. It was clearly recognized that all of the available human resources will be needed to tackle the large number of complex problems we are confronted with and that women have so far not been given an appropriate place in public debate and in decision-making processes. It also expressed a strong conviction that new ideas and creative

Heidi Diggelmann
Avenue de Cour 51, 1007 Lausanne, Switzerland
Heidi.Diggelmann@chuv.ch

solutions to complex challenges will most likely arise at the intersection of the traditional disciplines using interdisciplinary approaches between humanities, social sciences, natural sciences, biology and medicine. The organisers hoped that putting together women with different experiences, interests and backgrounds would create a unique exciting atmosphere with lively, potentially controversial discussions leading to novel ideas and approaches for the future.

It seems to me that the meeting has succeeded in creating a positive atmosphere of curiosity and interest for the different aspects that were presented by the speakers and discussants; this openness resulted in a certain permeability between the world of humanities (philosophy, ethics) and the more technically oriented so called 'hard' sciences. Communication between disciplines which all have their specific, highly coded languages, cultures and mentalities is very difficult. The process of trying to find a common language takes a lot of time and effort; misunderstandings and misinterpretations are frequent and it is important to develop a tolerance for these problems. The issue is even more complex than using the right words and understanding their meaning.

To understand each other we have to open up to the cultural differences that are hidden behind the words. As an example I would like to mention just one source of conflicts. The word 'science' in the Anglo-Saxon context generally describes activities in natural and biomedical sciences but does not include humanities (in contrast, the German word 'Wissenschaft' is used in a global sense and includes humanities and social sciences). The exclusion of humanities in the discussion of 'science' and 'scientific' issues has serious consequences on the perception of these fields and leads to an underestimation of their role and to a lack of respect for their point of view. It is therefore not so astonishing that many representatives from the humanities are very sceptical when finally asked to collaborate in projects put forward, e.g., by biologists or medical doctors. They fear that their contribution (ethical-egal-social-competence) is not taken seriously, but will only be used to obtain acceptance of the project from politicians or the general public. This fear is justified by the fact that representatives of humanities are often brought into the picture in the late phase of the elaboration of a project or only upon request of an evaluation body. For true interdisciplinarity to operate, all partners of a project should therefore be involved from the moment of its conception. Development of a language and a concept understood by all is a prerequisite for the formulation of a common project and will finally allow all partners to identify themselves with its content and goal. This whole process takes time, patience and energy, but it is necessary to create an atmosphere of confidence and respect between the different partners. The organizers of the WONBIT conference did just that. It was a first successful step. Many new questions arose and ideas of new contents and formulations have been put forward. Here just a few ideas and questions that have come up on different occasions during the last two days.

- How can we deal with the fact that technological developments are very fast and can hardly be stopped, while reflection about possible consequences and elaboration of a public consensus need time?

- Is the slow democratic process capable of creating transparency and a just decision process concerning the use of new technologies?
- How can we develop mechanisms to try to anticipate future developments and initiate public discussion at an early time point?

I would like to address a few more specific issues on women and biotechnology.

The deplorable loss of human resources as represented by the leaky career pathway of women in science is acknowledged by everyone. Despite the often equal and sometimes very high representation of women amongst successful students in our universities, these well-trained scientists leave the system in an almost continuous fashion over the years and become progressively under-represented in the upper segments of the career ladder. Women rarely occupy more than 10–20% of the top positions in academic institutions and almost everywhere else for that matter. The causes for this situation have been analysed in a large number of studies. I think that the times of studies are now over: based on their results, urgent measures have to be decided to try to prevent these losses in the future and to encourage the reinsertion of the competent women who have left the system for whatever reason. Many creative ideas have been formulated and shown to be successful on a small scale. The promotion of women in science is officially recognized as a basic task of our institutions, but too often, the implementation of these principles depends on a few particularly motivated individuals and slows down as soon as the pressure exerted by them is released.

To improve the present situation *it is of utmost importance that more women become part of the decision-making processes* in all compartments of society, e.g., academia, the private and public sector of economy, administration and politics. Mentoring and the establishment of networks allow women to acquire the necessary professional and social competence to become eligible for positions with high responsibilities. But everything has to be done at the right time. First women (and men) need to establish themselves in their respective fields, because moving up the ladder and being accepted as a leader depends (or should depend) mostly on an excellent reputation in the given field. A premature step-up might expose the candidate, particularly women, to the criticism of incompetence. Women in science are generally critical towards quota regulations to increase their representation in decision-making bodies. They want to be nominated for their competence and not because they are female. My experience has shown that there are enough competent women available, but persons who are now in charge of nominations (mainly males) tend to think of their (mostly male) colleagues when it comes to propose candidates. It is at this early stage that women have to be informed about opportunities and encouraged to apply. Only if competent women are presented in sufficient numbers to a decision-making group can we hope to improve female representation at the next higher level.

Very much remains to be done about the *criteria generally applied for promotions and nominations*. They clearly reflect and favour traditional linear career pathways within a specialized field. For interdisciplinary research however a wider spectrum of competences is needed, but performance criteria do not take into account these specific requirements, e.g., that it takes time to establish oneself in a new field and to publish first-rate papers. In addition to the lower productivity as measured by classical criteria, scientists who change fields are often considered as lacking focus or being superficial. It is not unexpected that this harsh judgement is often applied to women. Some women might be specifically interested in interdisciplinary research in the first place, but many of them enter this pathway accidentally when they move with their partners to places that do not offer adequate opportunities in their original field and look for adjacent domains that could enrich or complete their experience and allow them to remain professionally active. They only realize much later that such breaks or reorientations might have serious negative consequences on future promotions or on the general outcome of their career.

For nominations to leading positions the usual criteria of scientific excellence generally applied by the selection committees in academia *are insufficient.* A number of additional qualities, e.g., social competence, generosity, management skills, capacity to deal with potential conflicts, willingness to communicate and interact with the public, etc, are of utmost importance. These special qualities, often found in women, are, however, not seriously taken into account in the regular peer review system which values mainly individual performance. I propose that the criteria applied in the evaluation process for nominations and promotions to leading positions be modified to take into account these new elements. I am convinced that women would profit from such a change. Also the current nomination procedures should be more open to accommodate unusual career pathways in order to avoid the automatic exclusion of a large number of talented and highly motivated candidates.

The last aspect I want to raise very briefly might be considered shocking in the context of this conference: is there really a gender specific approach to the problems of our society?

It was great to meet so many women to discuss important issues. We have expressed our concern for the present and the next generation of inhabitants of this planet, for their wellbeing, their health, their needs, their aspirations. We are worried about the environment, its deterioration. But we will not be able to progress alone. Many men share our concerns and are responsive to our questioning and challenging the present status of our society. If we want to succeed we have to identify and motivate sensitive men to join us in the 'interdisciplinary project' for a better world.

Index